Monografie Matematyczne

Instytut Matematyczny Polskiej Akademii Nauk (IMPAN)

Volume 64

Founded in 1932 by
S. Banach, B. Knaster, K. Kuratowski,
S. Mazurkiewicz, W. Sierpiński, H. Steinhaus

Managing Editor:
Przemysław Wojtaszczyk, IMPAN and Warsaw University

Monografie Matematyczne
Instytut Matematyczny Polskiej Akademii Nauk (IMPAN)

Volume 64
(New Series)

Founded in 1932 by
S. Banach, B. Knaster, K. Kuratowski,
S. Mazurkiewicz, W. Sierpiński, H. Steinhaus

Volumes 31–62 of the series
Monografie Matematyczne were published by
PWN – Polish Scientific Publishers, Warsaw

Paweł Walcak

Dynamics of Foliations, Groups and Pseudogroups

Springer Basel AG

Author's address:

Paweł Walczak
Faculty of Mathematics
University of Łódź
ul. Banacha 22
90-238 Łódź
Poland
e-mail: pawelwal@math.uni. lodz.pl

2000 Mathematics Subject Classification 57R30, 54H20, 20F05

A CIP catalogue record for this book is available from the Library of Congress, Washington D.C., USA

Bibliographic information published by Die Deutsche Bibliothek
Die Deutsche Bibliothek lists this publication in the Deutsche Nationalbibliografie;
detailed bibliographic data is available in the Internet at <http://dnb.ddb.de>.

ISBN 978-3-0348-9611-5 ISBN 978-3-0348-7887-6 (eBook)
DOI 10.1007/978-3-0348-7887-6

Originally published by Birkhäuser Verlag, Basel, Switzerland in 2004

Cover design: Micha Lotrovsky, CH-4106 Therwil, Switzerland
Printed on acid-free paper produced of chlorine-free pulp. TCF∞

9 8 7 6 5 4 3 2 1

www.birkhäuser-science.com

To Zofia

Contents

Preface

Recent years brought increasing interest in the dynamics of general systems such as groups, pseudogroups, relations and foliations. Starting from different points and working in different areas, many authors realized that several methods of the classical theory of dynamical systems and ergodic theory can be adopted to study of such general systems. In fact, this approach is very natural. The theory of classical (smooth) dynamical systems can be described as a geometric theory of ordinary differential equations and the theory of foliations as a geometric theory of partial differential equations. Several properties of foliations, also those of dynamical nature, can be reflected by their holonomy pseudogroups. Also, any group G of transformations (in particular, a classical dynamical system generated by an invertible transformation) can be extended to a pseudogroup $\mathcal{G}$ of local transformations. In several situations, $\mathcal{G}$ carries more information about the dynamics of the system than the group G itself. Moreover, in the compact world, several objects (fundamental groups, holonomy pseudogroups, etc.) are generated by finite sets. Therefore, our groups and pseudogroups will be always finitely generated.

"Dynamics" is a very extensive term. The number of methods used in dynamical systems and the ergodic theory is very large. In this book, we focus on some of them only. Roughly speaking, we concentrate on notions and results related to different ways of measuring growth of the systems under consideration. More precisely, Chapter 1 contains a brief review of pseudogroups, their morphisms, foliations and their holonomy. Chapter 2 is devoted to the study of growth types of different objects: groups, pseudogroups, Riemannian manifolds and foliations. In particular, we focus on the expansion growth of groups and pseudogroups, the notion related to the entropy which becomes a central topic in Chapter 3. After a brief review of the classical theory of topological and measure theoretic entropies of transformations (Section 3.1) we focus on the geometric entropy of foliations. A good part of Chapter 3 (as well as Section 4.6 of Chapter 4) is occupied by the proof of the following result:

The entropy $h(\mathcal{F})$ of a codimension-one foliation $\mathcal{F}$ of a compact manifold is positive if and only if $\mathcal{F}$ has some resilient leaves. Consequently, if $h(\mathcal{F}) = 0$, then the Godbillon–Vey class $\mathrm{gv}(\mathcal{F})$ of $\mathcal{F}$ vanishes.

This result shows that there is a close link between the topology and dynamics of foliations. Its second part is based upon an unpublished result ([**Du1**], [**Du2**]) on vanishing Godbillon–Vey invariants of codimension-one foliations.

Geometric entropy generalizes topological entropy of maps. To generalize measure theoretic entropy is not so obvious: Several groups and pseudogroups of transformations admit no non-trivial invariant measures. In Chapter 4, we collect some results about the existence of invariant and quasi-invariant measures. In particular, we focus on harmonic (in the sense of [**Gar1**]) measures on foliated Riemannian manifolds and Patterson–Sullivan quasi-invariant measures for Kleinian groups. Chapter 5 is devoted to Hausdorff dimension and related invariants while Chapter 6 contains some applications. For instance, using geometric entropy of foliations we show after Attie, Hurder [**AH**] and Zeghib [**Ze**] how to construct a complete Riemannian manifold of bounded geometry which cannot be quasi-isometric to a leaf of any compact foliated manifold $(M, \mathcal{F})$ and we describe after Ghys [**Gh3**] the generic (with respect to any harmonic measure) topological type of a leaf of such an $(M, \mathcal{F})$.

The book is designed for mathematicians working with groups, foliations, dynamical systems and related fields as well as for graduate and postgraduate students interested in these topics. The reader is assumed to know enough general and algebraic topology, measure theory and manifolds. All the information needed to read the book is rather standard, should be contained in graduate courses and can be found in textbooks (like [**En**] or [**Ke**] for general topology, [**GHa**] for algebraic topology, [**Do**] for measure theory and [**Co**] for manifolds). Some knowledge of Riemannian geometry (geodesics and geodesic flow, in particular), classical dynamical systems and ergodic theory would be helpful. Our favourite sources of information in these areas are [**Kl**] and [**GKM**] for Riemannian geometry, [**Pa**] for smooth dynamical systems, [**Ma**] and [**Wa**] for ergodic theory. Of course, the number of good books covering these areas is incredibly large and the reader can choose his favourites as well. Since our review of foliation theory in Chapter 1 is rather brief, the reader who is not familiar with foliations should consult other books: [**CL**], [**HH**] or [**Go2**], for example.

Certainly, the bibliography here is not complete. It contains only the items (books and papers) which are mentioned in the body of this book. The complete (at the moment of publication) bibliography in the area of foliations is contained in [**To**] and consists of more than 2400 items. It seems that a complete bibliography in the area of dynamical systems would be even larger. This shows that making our bibliography complete is rather hopeless and — in the presence of existence of several databases available easily from the Internet — needless.

Finally, I would like to express my thanks to everyone who had an influence on this text. In particular, my best thanks are due to Etienne Ghys, Rémi Langevin and Takashi Inaba who collaborated with me in the study of entropies and dimensions, to Lawrence Conlon and Takashi Tsuboi for long term invitations to the Washington University in St. Louis and to the University of Tokyo, to Steven Hurder, Shigenori Matsumoto, Paul Schweitzer, Fabiano Brito and many others for several helpful discussions, to colleagues from the Department of Geometry of the Faculty of Mathematics at the University of Łódź who participated in my seminar and listened to lectures based on some parts of this text. Special thanks are due to the referee for all the time he spent with the book, his careful reading and all the remarks which allowed this book to be more readable. I am also grateful to my wife Zofia and our children Maria, Błażej and Szymon for the everyday support and accepting the style of life of a mathematician's family. Szymon deserves also special thanks for his help in computer drawing. Last but not least, I would like to acknowledge my gratitude to the University of Łódź and to the Polish Committee of Scientific Research (KBN, grant # 1037/P03/96/10) for financial support during preparation of the manuscript.

Chapter 1

Dynamical systems

This is an introductory chapter providing a review of several dynamical systems considered throughout the subsequent chapters. These systems include general pseudogroups (Section 1.1), foliations and their holonomy pseudogroups (Section 1.3), pseudogroups of some special sort (Section 1.4) as well as, due to the richness of dynamics provided by different aspects of hyperbolicity, hyperbolic spaces and groups in Gromov's sense (Section 1.5).

1.1 Pseudogroups

The notion of a pseudogroup generalizes that of a group of transformations. Given a space X, any group of transformations of X consists of maps defined globally on X, mapping X bijectively onto itself and such that the composition of any two of them as well as the inverse of any of them belongs to the group. The same holds for a pseudogroup with this difference, that the maps are not defined globally but on open subsets, so the domain of the composition is usually smaller than those of the maps being composed.

To make the above precise, let us take a topological space X and denote by $\mathrm{Homeo}\,(X)$ the family of all homeomorphisms between open subsets of X. If $g \in \mathrm{Homeo}\,(X)$, then D_g is its domain and $R_g = g(D_g)$.

Definition 1.1.1. A subfamily $\mathcal{G}$ of $\mathrm{Homeo}\,(X)$ is said to be a *pseudogroup* if it is closed under composition, inversion, restriction to open subdomains and unions. More precisely, $\mathcal{G}$ should satisfy the following conditions:

(i) $g \circ h \in \mathcal{G}$ whenever g and $h \in \mathcal{G}$,

(ii) $g^{-1} \in \mathcal{G}$ whenever $g \in \mathcal{G}$,

(iii) $g|U \in \mathcal{G}$ whenever $g \in \mathcal{G}$ and $U \subset D_g$ is open,

(iv) if $g \in \mathrm{Homeo}\,(X)$, $\mathcal{U}$ is an open cover of D_g and $g|U \in \mathcal{G}$ for any $U \in \mathcal{U}$, then $g \in \mathcal{G}$.

Moreover, we shall always assume that

(v) $\mathrm{id}_X \in \mathcal{G}$ (or, equivalently, $\cup\{D_g; g \in \mathcal{G}\} = X$).

In particular, if f_1 and f_2 are homeomorphisms between open subsets of X and $f_1 = f_2$ on the intersection of the domains, then $f = f_1 \cup f_2$, the map defined on the union of the domains by $f = f_i$ on the domain of f_i $(i = 1, 2)$, belongs to $\mathcal{G}$ whenever it maps its domain homeomorphically onto its image.

Hereafter, the members of pseudogroups are usually defined on the largest possible domains. Therefore, $D_{g\circ h} = D_h \cap h^{-1}(D_g)$, $D_{g^{-1}} = R_g$ and so on.

Definition 1.1.2. If two pseudogroups $\mathcal{G}$ and $\mathcal{G}' \subset \mathrm{Homeo}\,(X)$ satisfy the condition $\mathcal{G} \subset \mathcal{G}'$, then $\mathcal{G}$ is a *subpseudogroup* of $\mathcal{G}'$.

For any set $\Gamma \subset \mathrm{Homeo}\,(X)$ satisfying

$$\bigcup\{D_g \cup R_g; g \in \Gamma\} = X \tag{1.1.1}$$

there exists a unique smallest pseudogroup $\mathcal{G}(\Gamma)$ containing Γ: $g \in \mathcal{G}(\Gamma)$ if and only if $g \in \mathrm{Homeo}\,(X)$ and for any $x \in D_g$ there exist maps $g_1, \ldots, g_n \in \Gamma$, exponents $e_1, \ldots, e_n \in \{\pm 1\}$ and an open neighbourhood U of x, such that $U \subset D_g$ and

$$g|U = g_1^{e_1} \circ \cdots \circ g_n^{e_n}|U. \tag{1.1.2}$$

Definition 1.1.3. $\mathcal{G}(\Gamma)$ is said to be *generated* by Γ.

Remark 1.1.4. The condition (1.1.1) implies (v). It could be omitted and then to generate a pseudogroup we should begin with replacing Γ by $\Gamma' = \Gamma \cup \{\mathrm{id}_X\}$.

In this book, we shall deal first and foremost with finitely generated pseudogroups.

Definition 1.1.5. A pseudogroup $\mathcal{G}$ is *finitely generated* if $\mathcal{G} = \mathcal{G}(\Gamma)$ for a finite set Γ. If $\Gamma = \{f_1, \ldots, f_n\}$, then we write $\mathcal{G}(f_1, \ldots, f_n)$ instead of $\mathcal{G}(\Gamma)$.

A generating set Γ is *symmetric* when $\mathrm{id}_X \in \Gamma$ and $\Gamma^{-1} = \{g^{-1}; g \in \Gamma\} \subset \Gamma$. For any Γ, $\Gamma' = \Gamma \cup \Gamma^{-1} \cup \{\mathrm{id}_X\}$ is symmetric and $\mathcal{G}(\Gamma') = \mathcal{G}(\Gamma)$. Therefore, any finitely generated pseudogroup admits a finite symmetric generating set.

Remark 1.1.6. As far as we know, the notion of a pseudogroup appeared for the first time in the paper by Veblen and Whitehead [**VW**] in the context of geometric objects studied in differential geometry. The first formal definition was given by Gołąb [**Gb**] who defined it as a family of local homeomorphisms closed under composition and restriction to open subdomains and such that for each member of this family and for any point of its domain the inverse of its restriction to some neighbourhood of this point is in the pseudogroup. As was observed by Waliszewski [**Wls**], under this definition generating could be a problem: there exist sets of local homeomorphisms which are not contained in a smallest pseudogroup in this sense.

The definition given here appeared later and now seems to be quite natural and reasonable. Let us mention also that pseudogroups have been brought into the theory of foliations by Haefliger [**Hae1**] (see also [**Hae3**]) whose role in developing of this theory cannot be overestimated.

In an analogy with homomorphisms of groups we shall define now morphisms of arbitrary pseudogroups. Pseudogroups with these morphisms form a category.

Let $\mathcal{G} \subset \text{Homeo}(X)$ and $\mathcal{H} \subset \text{Homeo}\ (Y)$ be pseudogroups.

Definition 1.1.7. A family $\Phi \subset \text{Homeo}(X,Y)$, $\text{Homeo}(X,Y)$ being the set of all homeomorphisms $\phi : D_\phi \to R_\phi$ between open sets $D_\phi \subset X$ and $R_\phi \subset Y$, is a *morphism* of $\mathcal{G}$ into $\mathcal{H}$, $\Phi : \mathcal{G} \to \mathcal{H}$, if the domains D_ϕ, $\phi \in \Phi$, cover X and

$$\phi \circ g \circ \psi^{-1} \in \mathcal{H} \text{ for all } g \in \mathcal{G},\ \phi \text{ and } \psi \in \Phi.$$

Φ is an *isomorphism* if $\Phi^{-1} := \{\phi^{-1}; \phi \in \Phi\}$ is a morphism of $\mathcal{H}$ into $\mathcal{G}$. In this case,

$$\bigcup\{D_\phi; \phi \in \Phi\} = X,\ \bigcup\{R_\phi; \phi \in \Phi\} = Y$$

and

$$\phi \circ g \circ \psi^{-1} \in \mathcal{H}\ \ (\phi \in \Phi, \psi \in \Psi)\ \Leftrightarrow\ g \in \mathcal{G}.$$

Note that if spaces X and Y are compact and pseudogroups $\mathcal{G} \subset \text{Homeo}(X)$ and $\mathcal{H} \subset \text{Homeo}(Y)$ are isomorphic, then there exists a finite family Φ_0 establishing an isomorphism between $\mathcal{G}$ and $\mathcal{H}$. In fact, if $\Phi : \mathcal{G} \to \mathcal{H}$ is any isomorphism, $\phi_1, \dots, \phi_{m+n} \in \Phi$, $\bigcup_{i=1}^m D_{\phi_i} = X$ and $\bigcup_{j=m+1}^{m+n} R_{\phi_j} = Y$, then one could take $\Phi_0 = \{\phi_1, \dots, \phi_{m+n}\}$. Also, any isomorphism $\Phi : \mathcal{G} \to \mathcal{H}$ can be enlarged to the family $\tilde{\Phi} = \{h \circ \phi \circ g;\ g \in \mathcal{G},\ h \in \mathcal{H} \text{ and } \phi \in \Phi\}$ which maps $\mathcal{G}$ isomorphically onto $\mathcal{H}$ again. $\tilde{\Phi}$ satisfies the condition

$$g \in \mathcal{G},\ h \in \mathcal{H} \text{ and } \phi \in \Phi \Rightarrow h \circ \phi \circ g \in \tilde{\Phi},$$

so it becomes an isomorphism of pseudogroups in the sense of [**Hae2**].

1.2 First examples

$\text{Homeo}(X)$ and $\mathcal{G}(\text{id}_X)$ are, respectively, the largest and the smallest pseudogroup on a space X.

If X is a C^r-differentiable manifold ($r = 1, 2, \dots, \infty, \omega$) and $s \le r$, then $\text{Diff}^s(X)$, the family of all C^s-diffeomorphisms between open subsets of X, is a pseudogroup. If X is orientable, then the family $\text{Diff}^s_+(X)$ of all orientation preserving diffeomorphisms is a subpseudogroup of $\text{Diff}^s(X)$.

If X is equipped with a G-structure ($G \subset \text{Gl}(n, \mathbb{R})$, $n = \dim X$), then the family $\text{Diff}^s_G(X)$ of C^s-diffeomorphisms preserving this G-structure is a pseudogroup on X. In particular, if (X, g) is a Riemannian manifold ($G = \text{O}(n)$), then one can consider the pseudogroup Iso (X) of isometries between open subsets of X: $f \in \mathrm{Iso}(X)$ iff $f \in \mathrm{Diff}^1(X)$ and $f^*g = g$.

Also, if (X, d) is an arbitrary metric space, then all the local isometries of X generate a pseudogroup Iso (X) on X. Note that (because of condition (iii) in Definition 1.1.1) an element f of Iso (X) does not need to map D_f onto R_f isometrically: $f \in \mathrm{Iso}(X)$ iff any point $x \in D_f$ admits a neighbourhood U such that $f|U : U \to f(U)$ is an isometry. Moreover, all the locally Lipschitz maps $f \in$ Homeo (X) form a pseudogroup Lip (X): $f \in \mathrm{Lip}(X)$ iff for any point x of D_f there exists a neighbourhood U of x and a constant $c \geq 1$ such that

$$\frac{1}{c} \cdot d(f(y), f(z)) \leq d(y, z) \leq c \cdot d(f(y), f(z)) \tag{1.2.1}$$

for all y and $z \in U$. As before, the maps of Lip (X) do not need to be globally Lipschitz: inequality (1.2.1) does not need to hold for all y and z of D_f. Clearly, Iso (X) is a subpseudogroup of Lip (X).

If ∇ is a connection on a manifold X, then Aff (X), the family of all the diffeomorphisms $f \in \mathrm{Diff}^1(X)$ preserving ∇ is a pseudogroup. If Ω is a volume form on X, then all the diffeomorphisms $f \in \mathrm{Diff}^s(X)$ $(s \geq 1)$ preserving Ω, $f^*\Omega = \Omega$, form a subpseudogroup of $\mathrm{Diff}^s(X)$. Also, if μ is a Borel measure on an arbitrary topological space X, then all the maps $f \in$ Homeo (X) preserving μ (i.e., such that $\mu(f(Z)) = \mu(Z)$ for any Borel set $Z \subset D_f$) form a pseudogroup.

In general, the pseudogroups mentioned above are not finitely generated. Obviously, any group Γ acting continuously on X generates a pseudogroup $\mathcal{G}(\Gamma)$. If Γ is finitely generated, then $\mathcal{G}(\Gamma)$ is finitely generated as well. Conversely, if $\mathcal{G}(\Gamma)$ is finitely generated and the action of Γ on X is locally free in the sense that any two elements γ_1 and γ_2 of Γ coincide whenever $\gamma_1|U = \gamma_2|U$ for some non-empty open subset U of X, then Γ itself is finitely generated.

Also, arbitrary (global) homeomorphisms $f_1, \dots, f_n$ of a topological space X generate a pseudogroup. Moreover, if $f_i : X \to X$ are local homeomorphisms (in the sense that for any $i \leq n$ and any $x \in X$ there is a neighbourhood U of x for which $f|U \in$ Homeo (X)), then all the restrictions $f_i|U$ which belong to Homeo (X) generate a pseudogroup which will be denoted by $\mathcal{G}(f_1, \dots, f_n)$. If X is compact, then this pseudogroup is finitely generated.

Also, if $\mathcal{G}$ is a pseudogroup on X and $Y \subset X$ is *$\mathcal{G}$-invariant* (i.e., $g(x) \in Y$ for all $g \in \mathcal{G}$ and $x \in D_g \cap Y$), then the maps $g|Y \cap D_g$, $g \in \mathcal{G}$, generate a pseudogroup denoted by $\mathcal{G}|Y$. Clearly, if $\mathcal{G} = \mathcal{G}(g_1, \dots, g_n)$ is finitely generated, then $\mathcal{G}|Y = \mathcal{G}(g_1|Y, \dots, g_n|Y)$ is also finitely generated. This construction could be done for any (possibly non-invariant) subspace Y of X: $\mathcal{G}|Y$ would become generated by all the maps of the form $g|Y \cap g^{-1}(Y)$, $g \in \mathcal{G}$.

Note that the closure of an $\mathcal{G}$-invariant set is also $\mathcal{G}$-invariant. In particular, the closure of a *$\mathcal{G}$-orbit* $\mathcal{G}(x)$,

$$\mathcal{G}(x) = \{g(x);\ g \in \mathcal{G} \text{ and } x \in D_g\}, \tag{1.2.2}$$

is $\mathcal{G}$-invariant. Closed invariant sets are partially ordered by inclusion. If X is compact, then ordered families of non-empty closed invariant sets admit minimal elements: if $\mathcal{C}$ is such a family, then $\cap\mathcal{C}$ is closed, non-empty, $\mathcal{G}$-invariant and contained in all the members of $\mathcal{C}$. Therefore, by Zorn's Lemma, any non-empty closed invariant set contains minimal (with respect to the inclusion) non-empty closed invariant sets which are called just *minimal* here. If $\mathcal{G}$ has all the orbits dense in X, X is the only closed $\mathcal{G}$-invariant set and $\mathcal{G}$ is also called *minimal.* Obviously, any closed orbit is minimal.

Definition 1.2.1. Minimal sets different from X and single orbits are called *exceptional.*

The problem of existence of exceptional minimal sets for dynamical systems has been attracting interest for a long time. Probably, the first example of such a set is contained in [**De**]. We will describe it briefly below but first we shall explain why such sets are "rare" if the systems are "nice".

For this purpose, let us recall that given any homeomorphism f of the circle S^1 the rotation number $\rho(f) \in \mathbb{R}/\mathbb{Z}$ is defined by the formula

$$\rho(f) = \lim_{n\to\infty} \frac{1}{n}|\tilde{f}^n(t) - t| + \mathbb{Z} \tag{1.2.3}$$

where $t \in \mathbb{R}$ and $\tilde{f} : \mathbb{R} \to \mathbb{R}$ is a lift of f. It is well defined since if $\hat{f}$ is another lift of f, then the limits in (1.2.3) obtained from $\tilde{f}$ and $\hat{f}$ differ by an integer. Therefore, one can speak without confusion about circle homeomorphisms with rational (or, irrational) rotation number. It is not hard to see that homeomorphisms with $\rho(f) \in \mathbb{Q}$ have periodic orbits. If f is a C^2-diffeomorphism of S^1, then — by the classical Denjoy Theorem ([**De**], see also [**CaC1**], [**Go1**], [**Tam**] etc.) — f is topologically conjugate to the rotation $R_{2\pi\rho(f)}$. Therefore pseudogroups generated by C^2-diffeomorphisms $f : S^1 \to S^1$ (i.e., pseudogroups of the form $\mathcal{G}(f)$ with f as above) admit no exceptional minimal sets. The situation is quite different when f is only of the class C^1.

Example 1.2.2. Let $I = [0, 1]$ and α be an irrational number. Choose any decreasing sequence $(l_n, n \in \mathbb{N})$ of positive numbers l_n such that $l_{n+1}/l_n \to 1$ when $n \to \infty$ and $l = \sum_{n\in\mathbb{N}} l_n < \infty$. For each $n \in \mathbb{Z}$ let $a_n = \{n\alpha\}$ be the fractional part of $n\alpha$, i.e., $0 < a_n < 1$ and $n\alpha - a_n \in \mathbb{Z}$. For any $m \in \mathbb{N}$ let J_m be the segment obtained from I by cutting it at the points a_n and gluing in there segments I_n of length $l_{|n|}$, $|n| \leq m$. Obviously, the length of J_m equals $1 + 2l_1 + \cdots + 2l_m$, the segments J_m converge with $m \to \infty$ to a segment J of length $1 + 2l$ and I can be considered as a compact subset of J. Next, choose differentiable functions f_n mapping I_n onto I_{n+1} in such a way that $f_n'(t) = 1$ in some neighbourhoods of the end points of I_n and $f_n'(t)$ lies between $\frac{l_{n+1}}{l_n} - \frac{1}{n}$ and $\frac{l_{n+1}}{l_n} + \frac{1}{n}$ (Figure 1.1), and define f by the formulae

$$f(t) = \begin{cases} \{t + \alpha\} \text{ when } t \in I, \\ f_n(t) \text{ when } t \in J_n. \end{cases} \tag{1.2.4}$$

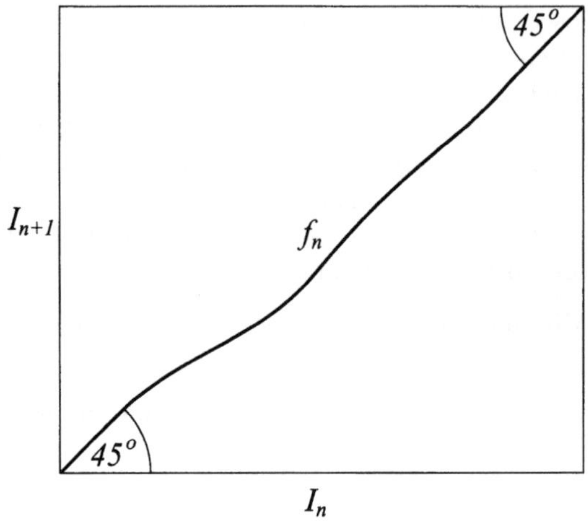

Figure 1.1: Graph of f_n.

It is easy to see that f generates a homeomorphism, denoted by f again, of the circle obtained by identification of the end points of J. f is differentiable, f' is positive and continuous. Therefore, f is a C^1-diffeomorphism of the circle. The set I is f-invariant, closed and minimal. In fact, the orbit $\{f^n(x), n \in \mathbb{Z}\}$ of any point $x \in J$ accumulates on I. Of course, $I \neq J$ and I consists of infinitely many orbits. Hence, I is exceptional. Moreover, f admits no periodic or dense orbits. (The reader not convinced by this rough description of f and its properties is referred to [**CL**], pp. 55–59, where all the details are given.)

1.3 Foliations, laminations and holonomy

The simplest example of a foliation is provided by a single submersion $F : M \to N$, M and N being manifolds. Such a foliation consists of the connected components of the fibres $f^{-1}(x)$, $x \in N$, which are $(n-q)$-dimensional ($n = \dim M$, $q = \dim N$) submanifolds of M.

General foliations are defined by submersions locally. Their fibres should fit nicely when they intersect to produce differentiable submanifolds, leaves of a foliation. This property can be expressed by the existence of local homeomorphisms of the base space which should connect one of such submersions to another. That is, if the domains of two such submersions f and g overlap, then any point of the intersection of these domains admits a neighbourhood on which $f = \phi \circ g$ for some local homeomorphism ϕ of the base space. All the maps ϕ like that form a pseudogroup called the holonomy pseudogroup of the foliation. In the case of a foliation defined by a single submersion its holonomy is trivial, generated just by the identity.

The property discussed above can be also expressed in terms of good (called distinguished) charts on the foliated manifold. To make the above rough description precise we shall formulate a series of definitions.

So, let M be an n-dimensional manifold, possibly with non-empty boundary ∂M. (All the manifolds here are supposed to be Hausdorff, paracompact and smooth, i.e., C^∞-differentiable.)

Definition 1.3.1. A p-dimensional C^r-*foliation* $\mathcal{F}$ ($r = 0, 1, \ldots, \infty$) on M is a decomposition of M into connected submanifolds (called *leaves*) such that for any $x \in M$ there exists a C^r-differentiable chart $\phi = (\phi', \phi'') : U \to \mathbb{R}^n = \mathbb{R}^p \times \mathbb{R}^q$ defined on a neighbourhood U of x and satisfying the condition

(i) for any L of $\mathcal{F}$ the connected components (called *plaques*) of $L \cap U$ are given by the equation $\phi'' = \text{const}$.

Charts satisfying (i) are said to be *distinguished* by $\mathcal{F}$. An atlas built of distinguished charts is said to be *foliated*. The number $q = \operatorname{codim} \mathcal{F} = n - p$ is the *codimension* of $\mathcal{F}$.

The manifold topology of leaves is in general stronger than these induced from M. Leaves for which these topologies coincide are said to be *proper*. Certainly, compact leaves are proper. Simple examples show that non-compact proper leaves may exist. In terms of distinguished charts U this property can be expressed as follows: A leaf L is proper if and only if any plaque $P \subset U \cap L$ has an open neighbourhood $V \subset U$ such that $V \cap L = P$.

If the boundary ∂M of M is non-empty, then any point $x \in \partial M$ should be contained in the domain of a distinguished chart $\phi : U \to \mathbb{R}^n$ which maps (1) U either into $\mathbb{R}^p \times \mathbb{R}^q_+$ or into $\mathbb{R}^p_+ \times \mathbb{R}^q$ (where $\mathbb{R}^k_+ = \{u_1, \ldots, u_k); u_k \geq 0\}$) and (2) $U \cap \partial M$ either into $\mathbb{R}^p \times \mathbb{R}^q_0$ or into $\mathbb{R}^p_0 \times \mathbb{R}^q$ (where $\mathbb{R}^k_0 = \{(u_1, \ldots, u_k); u_k = 0\}$). The plaques containing x coincide either with $\phi^{-1}(\mathbb{R}^p \times \{u_0\})$ or with $\phi^{-1}(\mathbb{R}^p_+ \times \{v\})$, where $u_0 \in \mathbb{R}^q_0$ and $v \in \mathbb{R}^q$. Therefore, the boundary ∂M decomposes as $\partial^\top M \cup \partial^\pitchfork M$, where $\partial^\top M$ is a union of some leaves while $\mathcal{F}$ is transverse to ∂M at any point $x \in \partial^\pitchfork M$. $\partial^\top M$ is called the *tangential boundary* while $\partial^\pitchfork M$ the *transverse boundary* of $(M, \mathcal{F})$ (Figure 1.2). Clearly, both $\partial^\top M$ and $\partial^\pitchfork M$ are unions of connected components of ∂M. (The situation can be different if we allow M to have some "corners" (see [**CaC1**].))

If $\mathcal{A}$ is an atlas on M such that

(ii) for any ϕ and $\psi \in \mathcal{A}$ the map $\phi \circ \psi^{-1} : V \to \mathbb{R}^p \times \mathbb{R}^q$ ($V \subset \mathbb{R}^p \times \mathbb{R}^q$) is of the form

$$\phi \circ \psi^{-1}(x, y) = (\alpha(x, y), \gamma(y)), \tag{1.3.1}$$

where $x \in \mathbb{R}^p$ and $y \in \mathbb{R}^q$,

then there exists a unique foliation $\mathcal{F}$ on M such that all the charts of $\mathcal{A}$ are distinguished by $\mathcal{F}$. Any foliated atlas is contained in a maximal (with respect to the relation "$\subset$") one. Such a maximal foliated atlas $\mathcal{A}$ determines a unique

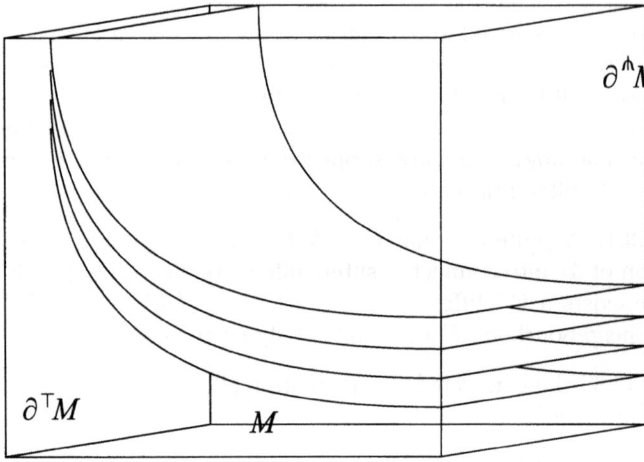

Figure 1.2: Tangential and transverse boundaries.

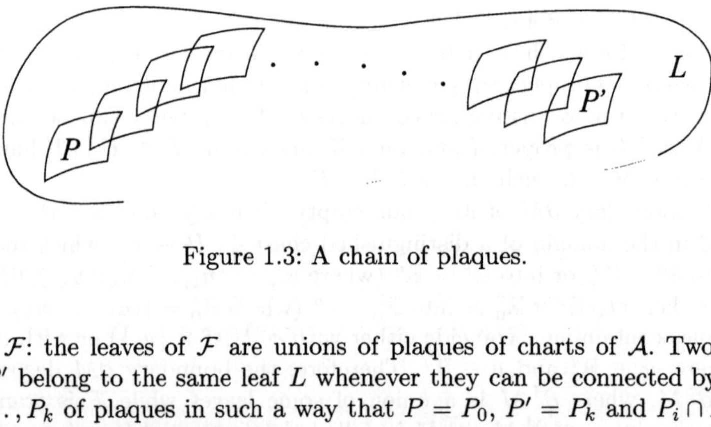

Figure 1.3: A chain of plaques.

foliation $\mathcal{F}$: the leaves of $\mathcal{F}$ are unions of plaques of charts of $\mathcal{A}$. Two plaques P and P' belong to the same leaf L whenever they can be connected by a chain $P_0, P_1, \ldots, P_k$ of plaques in such a way that $P = P_0$, $P' = P_k$ and $P_i \cap P_{i+1} \neq \emptyset$ for $i = 0, 1, \ldots, k-1$ (Figure 1.3).

If $\mathcal{F}$ is a C^r-foliation on M and $r \geq 1$, then the set $E = T\mathcal{F}$ of all the vectors $v \in TM$ tangent to the leaves of $\mathcal{F}$ becomes a C^{r-1}-subbundle of TM. This subbundle is *involutive* in the following sense: If $\mathcal{X}(E)$ denotes the family of all the E-valued vector fields on M, then

$$X, Y \in \mathcal{X}(E) \Rightarrow [X, Y] \in \mathcal{X}(E), \tag{1.3.2}$$

where $[\cdot, \cdot]$ is the Lie bracket on M. The classical Frobenius Theorem (see, for instance, [**St**], Section III.5) says that for any involutive C^r-subbundle E of a manifold M there exists a C^r-foliation $\mathcal{F}$ for which $E = T\mathcal{F}$. Since any 1-dimensional bundle is involutive, it determines a foliation. In particular, any nowhere vanishing vector field X on M determines a foliation with leaves being orbits of its flow.

Similarly, if $\Omega(M) = \oplus_{k=0}^{n}\Omega^k(M)$ is the graded algebra of differential forms, $d : \Omega(M) \to \Omega(M)$ is the exterior differentiation and $\mathcal{J} \subset \Omega(M)$ is an ideal such that

$$d\mathcal{J} \subset \mathcal{J} \tag{1.3.3}$$

and

$$\ker \mathcal{J}(x) = \{v \in T_xM; \omega(v) = 0 \text{ for all } \omega \in \mathcal{J}^1\} \tag{1.3.4}$$

has the same dimension, say q, for all $x \in M$, then the subbundle $\ker \mathcal{J}$ of TM is involutive and determines a foliation $\mathcal{F}$ of codimension q. In particular, the equation

$$\omega = 0, \tag{1.3.5}$$

where ω is a nowhere-vanishing 1-form satisfying

$$d\omega = \eta \wedge \omega \tag{1.3.6}$$

for some 1-form η, defines a foliation of codimension 1. This happens, for instance, when ω is closed ($d\omega = 0$).

Definition 1.3.2. A foliated atlas $\mathcal{A}$ is said to be *nice* (also, *nice* is the covering of M by the domains D_ϕ of the charts $\phi \in \mathcal{A}$) if

(iii) the covering $\{D_\phi; \phi \in \mathcal{A}\}$ is locally finite,

(iv) for any $\phi \in \mathcal{A}$, $R_\phi = \phi(D_\phi) \subset \mathbb{R}^n$ is an open cube,

(v) if ϕ and $\psi \in \mathcal{A}$, and $D_\phi \cap D_\psi \neq \emptyset$, then there exists a chart χ distinguished by $\mathcal{F}$ and such that R_χ is an open cube, D_χ contains the closure of $D_\phi \cup D_\psi$ and $\phi = \chi | D_\phi$.

Since manifolds are supposed to be paracompact here, they are separable and hence nice coverings are denumerable. Nice coverings on compact manifolds are finite.

Lemma 1.3.3. *For any foliation $\mathcal{F}$ on any manifold M, nice coverings of M exist.*

Proof. Let $\mathcal{A}$ be a foliated atlas on M such that R_ϕ is a cube for any $\phi \in \mathcal{A}$. Denote by $\mathcal{U}$ the open covering of M by the domains of all the charts of $\mathcal{A}$. Since M is paracompact, we can find a locally finite open covering $\mathcal{V}$ of M subordinated to $\mathcal{U}$ and such that for any $V \in \mathcal{V}$ there exists $U \in \mathcal{U}$ for which

$$\mathrm{St}\,(V, \mathcal{V}) = \bigcup \{W \in \mathcal{V};\ W \cap V \neq \emptyset\} \subset U. \tag{1.3.7}$$

For any $V \in \mathcal{V}$, choose $U \in \mathcal{U}$ satisfying (1.3.7) and let $\phi' = \phi | V$, where ϕ is a chart of $\mathcal{A}$ defined on U. Denote by $\mathcal{A}'$ the foliated atlas built of all the charts ϕ' obtained in this way. Now, for any $\phi' \in \mathcal{A}'$ find a locally finite covering of $R_{\phi'}$ by open cubes Q contained in $R_{\phi'}$ together with the closures, and construct a new atlas $\mathcal{A}''$ consisting of all the restrictions of the charts $\phi' \in \mathcal{A}'$ to the sets of the form ${\phi'}^{-1}(Q)$. $\mathcal{A}''$ and the covering $\mathcal{U}'' = \{D_{\phi''};\ \phi'' \in \mathcal{A}''\}$ are nice. □

Now, fix a nice covering $\mathcal{U}$ of a foliated manifold $(M, \mathcal{F})$. For any $U \in \mathcal{U}$, let T_U be the space of the plaques of $\mathcal{F}$ contained in U. Equip $T_U = U/(\mathcal{F}|U)$ with the quotient topology: two points of U are equivalent iff they belong to the same plaque. T_U is homeomorphic (C^r-diffeomorphic when $\mathcal{F}$ is C^r-differentiable and $r \geq 1$) to an open cube $Q \subset \mathbb{R}^q$ ($q = \operatorname{codim} \mathcal{F}$) *via* the map ϕ'', where $\phi = (\phi', \phi'') : U \to \mathbb{R}^p \times \mathbb{R}^q$ is a distinguished chart on U.

Definition 1.3.4. The disjoint union

$$T = \bigsqcup \{T_U; U \in \mathcal{U}\} \tag{1.3.8}$$

is called a *complete transversal* for $\mathcal{F}$.

Transversality refers to the fact that, if $\mathcal{F}$ is differentiable, each of the spaces T_U can be mapped homeomorphically onto a C^r-submanifold $T'_U \subset U$ transverse to U: if $x \in T'_U$ and L is the leaf of $\mathcal{F}$ passing through x, then

$$T_x M = T_x T'_U \oplus T_x L. \tag{1.3.9}$$

Completeness of T means that every leaf of $\mathcal{F}$ intersects at least one of the submanifolds T'_U.

Definition 1.3.5. Given two sets U and $V \in \mathcal{U}$ such that $U \cap V \neq \emptyset$ the *holonomy map* $h_{VU} : D_{VU} \to T_V$, D_{VU} being the open subset of T_U which consists of all the plaques P of U for which $P \cap V \neq \emptyset$, is defined in the following way:

$$h_{VU}(P) = P' \Leftrightarrow \textit{the plaques } P \subset U \textit{ and } P' \subset V \textit{ intersect.} \tag{1.3.10}$$

From (v), it follows that the map h_{VU} is well defined on an open subset D_{VU} of T_U. In fact, any plaque of U intersects at most one plaque of V. From (1.3.1), it follows that h_{VU} maps D_{VU} homeomorphically (C^r-diffeomorphically when $\mathcal{F}$ is C^r-differentiable and $r \geq 1$) onto an open subset D_{UV} of T_V. Clearly,

$$h_{UV} = h_{VU}^{-1} \text{ and } h_{WU} = h_{WV} \circ h_{VU} \tag{1.3.11}$$

whenever all the holonomy maps involved are defined. All the maps h_{UV} ($U, V \in \mathcal{U}$) generate a pseudogroup $\mathcal{H}$ on T. $\mathcal{H}$ is called the *holonomy pseudogroup* of $\mathcal{F}$. The following result is essentially due to Haefliger (see [**Hae2**] or [**Hae3**], for instance).

Proposition 1.3.6. *The holonomy pseudogroups $\mathcal{H}$ and $\mathcal{H}'$ corresponding to two nice coverings $\mathcal{U}$ and $\mathcal{U}'$ of a foliated manifold $(M, \mathcal{F})$ are isomorphic.*

Proof. First, assume that $\mathcal{U}'$ is subordinated to $\mathcal{U}$. Then, the covering $\mathcal{U}'' = \mathcal{U} \cup \mathcal{U}'$ is nice. Let Φ be the family of all the maps of the form

$$h \circ h''_{UU'} \circ h', \tag{1.3.12}$$

where $h \in \mathcal{H}$, $h' \in \mathcal{H}'$, $U \in \mathcal{U}$, $U' \in \mathcal{U}'$, $U' \subset U$ and $h''_{UU'} \in \mathcal{H}''$, the holonomy pseudogroup $\mathcal{H}''$ corresponding to $\mathcal{U}''$. Since each set U' of $\mathcal{U}'$ is contained in some U of $\mathcal{U}$, $T = \sqcup T_U$ meets all the leaves and for any two points x and $y \in T$ of the same leaf L there exists a holonomy map $h \in \mathcal{H}$ mapping x to y, the domains (resp., ranges) of all the maps of Φ cover $T' = \bigsqcup T'_{U'}$ (resp., T) and, therefore, Φ becomes a pseudogroup isomorphism: $\Phi : \mathcal{H}' \to \mathcal{H}$.

Now, for any $\mathcal{U}$ and $\mathcal{U}'$ one can find a nice covering $\mathcal{U}''$ subordinated to both coverings. The holonomy pseudogroup $\mathcal{H}''$ corresponding to $\mathcal{U}''$ is isomorphic to both holonomy pseudogroups, $\mathcal{H}$ and $\mathcal{H}'$. □

Remark 1.3.7. In a similar way, the holonomy of $\mathcal{F}$ can be defined on any submanifold T of M ($\dim T = \operatorname{codim} \mathcal{F}$) transverse to $\mathcal{F}$ and intersecting all the leaves. T like that is also said to be a complete transversal for $\mathcal{F}$. The holonomy pseudogroup on T is again isomorphic to those arising from nice coverings.

For any leaf L of $\mathcal{F}$ and any point $x \in L \cap U$, (U, ϕ) being a distinguished chart of a nice covering $\mathcal{U}$, one can consider the germs $[h]_x$ at $\phi(x)$ of holonomy maps h of the form

$$h = h_{U_0U_1} \circ \cdots \circ h_{U_{k-1}U_k},$$

where $U_0 = U_k = U$. If $\gamma : [0,1] \to L$ is a leaf curve covered by plaques $P_0, P_1, \ldots, P_k$, $P_i \subset U_i$, in such a way that $\gamma([t_i, t_{i+1}]) \subset P_i$ for some t_i's such that $0 = t_0 < t_1 < \cdots < t_k = 1$, then the map h defined above is called a *holonomy along* γ and is denoted sometimes by h_γ. This notation can be accepted because of the following.

If two chains $U_0, U_1, \ldots, U_k$ and $U'_0, U'_1, \ldots, U'_l$ cover homotopic leaf curves $\gamma, \gamma' : [0,1] \to L$, $\gamma(0) = \gamma'(0) = x$ and $\gamma(1) = \gamma'(1) = y$, (in particular, the same curve γ), then the corresponding germs $[h]_x$ and $[h']_x$ are equal. Therefore, we have a properly defined holonomy homomorphism

$$\Phi_L : \pi_1(L, x) \ni [\gamma] \mapsto [h_\gamma]_x = [h]_x,$$

where h is any holonomy map corresponding to a chain of charts covering γ. Then, $H_L = \operatorname{im} \Phi_L$ is called the *holonomy group* of L.

Holonomy groups of L corresponding to different points x, different charts U and different nice coverings $\mathcal{U}$ are isomorphic. They describe the behaviour of $\mathcal{F}$ in a neighbourhood of L while the whole holonomy pseudogroup $\mathcal{H}$ reflects some global properties of the foliation.

To explain better the notion of holonomy it seems to be good to begin with the construction of the *suspension* of a single map.

Let f be a diffeomorphism of a compact manifold T. Consider $\tilde{M} = T \times \mathbb{R}$ and identify two points (x, t) and (x', t') of $\tilde{M}$ whenever there exists $n \in \mathbb{Z}$ such that $t' - t = n$ and $x' = f^n(x)$. This identification is an equivalence relation and one can consider the quotient space M. M is a manifold which has a natural structure of a fibre bundle over S^1 (with the typical fibre T). The vector field d/dt

projects to M. Trajectories of this projection constitute a 1-dimensional foliation transverse to the fibres of M. Any fibre T of this bundle can serve as a complete transversal of our foliation. The holonomy pseudogroup on T is isomorphic to $\mathcal{G}(f)$, the pseudogroup generated by the map f. The foliation obtained in this way is called the *suspension* of f (or, of the group $\mathbb{Z}$ generated by f and acting on $\tilde{M}$).

The following example shows how to generalize the suspension construction to arbitrary (finitely presented) groups.

Example 1.3.8. Let Γ be a finitely generated group isomorphic to the fundamental group $\pi_1(B)$ of a compact manifold B. (Recall that any finitely presented group is isomorphic to the fundamental group of a certain compact manifold.) Take another manifold F and a homomorphism h of Γ into the group of C^r-diffeomorphisms of F. Γ acts on the universal covering $\tilde{B}$ *via* the covering maps and on $\tilde{B} \times F$ in the following way:

$$\gamma(x, z) = (\gamma(x), h(\gamma)(z)) \quad (x \in \tilde{B}, z \in M). \tag{1.3.13}$$

The foliation $\tilde{\mathcal{F}} = \{\tilde{B} \times \{z\};\ z \in F\}$ projects (*via* the canonical projection) onto a C^r-foliation $\mathcal{F}$ of $M = (\tilde{B} \times F)/\Gamma$. The holonomy pseudogroup $\mathcal{H}$ of $\mathcal{F}$ is isomorphic to $\mathcal{G}(\Gamma_0)$, where $\Gamma_0 = \mathrm{im}\,(h)$. Therefore, any pseudogroup of the form $\mathcal{G}(\Gamma)$, Γ being a finitely presented group of diffeomorphisms of some manifold, is isomorphic to the holonomy pseudogroup of a foliation.

This foliation $\mathcal{F}$ is called the *suspension* of the homomorphism h. It is easy to see that the manifold M is a fibre bundle over B and F coincides with its fibre. Finally, $\mathcal{F}$ is transverse to the fibres of this bundle and $\dim \mathcal{F} = \dim B$. This is why $(M, \mathcal{F})$ is said to be a *foliated bundle* . Note that any compact foliated bundle (i.e., a fibre bundle with a foliation transverse to the fibres and of dimension equal to the dimension of the base) can be obtained as a suspension of a homomorphism h as above.

Now, let $f : N \to M$ be a differentiable map transverse to a C^1-foliation $\mathcal{F}$ of M. This means that

$$T_{f(x)}M = f_*(T_xN) + T_{f(x)}\mathcal{F} \quad (x \in N) \tag{1.3.14}$$

where $T_y\mathcal{F}$ is the fibre of $T\mathcal{F}$ over y (i.e., $T_y\mathcal{F} = T_yL_y$, L_y being the leaf of $\mathcal{F}$ through y). For any $z \in f(N)$, $f^{-1}(L_z)$ is a submanifold of N. The connected components of all such submanifolds provide a foliation $f^*\mathcal{F}$ of N, $\operatorname{codim} f^*\mathcal{F} = \operatorname{codim} \mathcal{F}$. The foliation $f^*\mathcal{F}$ is called the *pullback* of $\mathcal{F}$ *via* f. In particular, if $(M, \mathcal{F})$ has non-empty transverse boundary, then the inclusion map $\partial^{\pitchfork} M \hookrightarrow M$ is transverse to $\mathcal{F}$ and $\mathcal{F}$ induces the foliation $\mathcal{F}^{\pitchfork}$ of $\partial^{\pitchfork} M$.

Proposition 1.3.9. *If $f(N)$ intersects all the leaves, then the holonomy pseudogroup $\mathcal{H}'$ of $\mathcal{F}' = f^*\mathcal{F}$ is isomorphic to a subpseudogroup of $\mathcal{H}$, the holonomy pseudogroup of $\mathcal{F}$.*

Proof. Let $T' \subset N$ be a complete transversal for $\mathcal{F}'$, $T = f(T')$. T is a complete transversal for $\mathcal{F}$ and $\tilde{f} = f|T' : T' \to T$ is a local diffeomorphism. In fact, if

$x \in N$, then

$$T_xN = T_x\mathcal{F}' \oplus T_xT' \quad \text{and} \quad f_*(T_x\mathcal{F}') \subset T_{f(x)}\mathcal{F}. \tag{1.3.15}$$

From (1.3.14) and (1.3.15) we get

$$T_{f(x)}M = \tilde{f}_*(T_xT') + T_{f(x)}\mathcal{F}. \tag{1.3.16}$$

Therefore, $\dim \tilde{f}_*(T_xT') \geq \mathrm{codim}\mathcal{F}$. On the other hand, $\dim \tilde{f}_*(T_xT') \leq \dim T_xT'$ $= \mathrm{codim}\mathcal{F}$. Consequently, $\dim \tilde{f}_*(T_xT') = \dim T_xT'$ and the differential $\tilde{f}_*$ turns out to be an isomorphism of tangent spaces.

Let Φ be the family of all the restrictions $\tilde{f}|W$, where $W \subset T'$ is any open set which is mapped by $\tilde{f}$ diffeomorphically onto $\tilde{f}(W)$. Φ generates an isomorphism of $\mathcal{H}'$, the holonomy pseudogroup of $\mathcal{F}'$ acting on T', onto the subpseudogroup of $\mathcal{H}$, the holonomy pseudogroup of $\mathcal{F}$ acting on T, generated by all the maps of the form $(\tilde{f}|W') \circ h' \circ (\tilde{f}|W)^{-1}$, where $h' \in \mathcal{H}'$, W and W' are sufficiently small open subsets of T'. □

Holonomy pseudogroups can be defined in the same way for laminations (called foliated spaces in [**CaC1**]). Laminations are "generalized foliations". The simplest example is that of a "bunch of leaves" $\mathcal{L}$ (of a regular foliation $\mathcal{F}$) with the compact union $X = \cup\mathcal{L}$. The holonomy pseudogroup of such a lamination is isomorphic to the restriction of the holonomy pseudogroup of $\mathcal{F}$ to the intersection of a complete transversal T (for $\mathcal{F}$) with X. The generalization consists in replacing a transverse manifold T by a "transverse topological space". In other words, foliations are built of plaques parametrized by points of a manifold, while laminations consist of plaques parametrized by points of a topological space. More precisely, we have the following.

Definition 1.3.10. A *p-dimensional lamination* $\mathcal{L}$ is a separable locally compact space X equipped with an open covering $\mathcal{U}$ and homeomorphisms (*charts distinguished* by $\mathcal{L}$) $\phi : U \to D_U \times T_U$ ($U \in \mathcal{U}$), where D_U is an open subset of $\mathbb{R}^p$ and T_U is a topological space (a *transversal*). For any other distinguished chart $\psi : V \to D_V \times T_V$, the transition map $\phi \circ \psi^{-1}$ is of the form (1.3.1) with $x \in D_V$ and $y \in T_V$. As for foliations, the sets $\phi^{-1}(D_u \times \{y\})$, $y \in T_U$, are called *plaques* and the plaques glue together to form maximal connected sets (*leaves* of $\mathcal{L}$) $L \subset X$. Usually, some smoothness conditions are required. We assume that the maps α in (1.3.1) are C^r-differentiable in the first variable and that all their partial derivatives with respect to the first variable are continuous in all the variables.

As for foliations, if $\mathcal{U}$ is a nice covering of X (i.e. a covering by distinguished charts such that any plaque of one chart intersects at most one plaque of another), we can define the holonomy maps h_{VU} by (1.3.10) and generate a pseudogroup $\mathcal{H}$ (the *holonomy pseudogroup* of $\mathcal{L}$) acting on the space $T = \bigsqcup T_U$.

Example 1.3.11. If $\mathcal{F}$ is a foliation of a manifold M and X is a closed *saturated* subset of M (i.e., X is the union of a family of leaves of $\mathcal{F}$), then the leaves contained in X form a lamination of X. If $\mathcal{A}$ is a foliated atlas on $(M, \mathcal{F})$, then the maps $\phi|X \cap D_\phi$, $\phi \in \mathcal{A}$, become charts distinguished by $\mathcal{L}$. If T is a complete transversal for $\mathcal{F}$, then $T \cap X$ becomes a complete transversal for $\mathcal{L}$ and the holonomy pseudogroup of $\mathcal{L}$ coincides with $\mathcal{H}|T \cap X$, $\mathcal{H}$ being the holonomy pseudogroup of $\mathcal{F}$ on T. Other examples of laminations can be found in [**Ca1**] and [**Su6**].

We shall complete this section with several definitions.

Definition 1.3.12. A subset A of M is said to be *$\mathcal{F}$-saturated* when any leaf of a foliation $\mathcal{F}$ which intersects A is entirely contained in A. The union $\tilde{A}$ of all the leaves intersecting an arbitrary set $A \subset M$ is obviously saturated. $\tilde{A}$ is called the *saturation* of A.

Therefore A is saturated if and only if $\tilde{A} = A$.

Definition 1.3.13. A non-empty closed saturated set is said to be *minimal* whenever it contains no proper subsets that are non-empty, closed and saturated.

Obviously, single closed leaves are always minimal. If all the leaves of $\mathcal{F}$ are dense, then M itself is minimal and we call the foliation $\mathcal{F}$ *minimal* too. Similarly to Definition 1.2.1 we have

Definition 1.3.14. Minimal subsets of $(M, \mathcal{F})$ different from M and single leaves are called *exceptional.*

As we shall see later (Chapters 3 and 4), existence of exceptional minimal sets implies complexity of geometry, topology and dynamics of foliations.

It is clear that the terminology above is parallel to that for pseudogroups. For instance, a set A is saturated whenever its intersection with T, a complete transversal, is invariant under the action of the holonomy pseudogroup; A is minimal whenever $A \cap T$ is closed, holonomy invariant and contains no proper subsets enjoying these properties; A is exceptional whenever $A \cap T$ is exceptional (in the sense of Definition 1.2.1) for $\mathcal{H}$, the holonomy pseudogroup $\mathcal{F}$ acting on T, etc. The Zorn Lemma implies immediately the following

Proposition 1.3.15. *The closure of any leaf L of a foliation $\mathcal{F}$ on a compact manifold M contains a minimal set.* □

Minimal sets on non-compact foliated manifolds need not exist.

From remarks of Section 1.2 it follows that suspensions of C^2-diffeomorphisms of S^1 admit no exceptional minimal sets. Suspending the Denjoy diffeomorphism of Example 1.2.2 we obtain a C^1-foliation of the 2-dimensional torus T^2 which has an exceptional minimal set. From Kneser [**Kn**] classification of 1-dimensional C^2-foliations of T^2 it follows that such foliations admit no exceptional minimal sets. This result can be generalized to arbitrary surfaces: Arbitrary 1-dimensional C^2-foliation of a surface without boundary admits no exceptional minimal sets ([**Sc**], see also [**Go1**]).

1.4 Markov pseudogroups

Shifts and subshifts play an important role in the classical theory of dynamical systems [**Wa**]. They can carry a pretty rich dynamics while they are relatively easy to consider. Thus they provide a good class of examples.

As was observed by Cantwell and Conlon [**CC5**] there exists a nice class of pseudogroups with the dynamics semiconjugate to subshifts of finite type. They call them Markov. Roughly speaking, Markov pseudogroups are generated by maps which either can be composed "everywhere" or the domain of their composition is empty. To make this clear we shall accept the following definitions.

Definition 1.4.1. A finite subset $\mathcal{S} = \{h_1, \ldots, h_m\}$ of Homeo(X), $h_j : D_j \to R_j$, together with nonempty compact sets $K_j \subset R_j$ is called a *Markov system* if

(i) $R_i \cap R_j = \emptyset$ when $i \neq j$,

(ii) either $K_i \subset Q_j$ or $K_i \cap D_j = \emptyset$,

where $Q_j = h_j^{-1}(K_j)$. If $\mathcal{S}$ is Markov and $\bigcup_i (D_i \cup R_i) = X$, then $\mathcal{G}(\mathcal{S})$ is called a *Markov pseudogroup.*

Definition 1.4.2. For any Markov system $\mathcal{S} = \{h_1, \ldots, h_m\}$ one has its *transition matrix* $P = [p_{ij}; i, j = 1, \ldots, m]$:

(iii) $p_{ij} \in \{0, 1\}$ and $p_{ij} = 1$ iff $K_j \subset Q_i$,

and its *Markov invariant set* Z_0:

(iv) $Z_0 = Z \setminus \mathrm{int}\,(Z)$, where $Z = \bigcap_{n=1}^{\infty} \bigcup \{K_g; g \in \mathcal{S}_n\}$, $\mathcal{S}_n = \{h_{i_1} \circ \cdots \circ h_{i_n}; i_1, \ldots, i_n \leq m\}$ and $K_g = g(Q_{i_n})$ when $g = h_{i_1} \circ \cdots \circ h_{i_n}$.

Conditions (i)–(iii) imply that $K_g \neq \emptyset$ iff $g = h_{i_1} \circ \cdots \circ h_{i_n}$ and

$$p_{i_k i_{k+1}} = 1 \tag{1.4.1}$$

for any $k \in \{1, \ldots, n-1\}$. This is why infinite sequences $(i_1, i_2, \ldots)$ satisfying (1.4.1) for all $k \in \mathbb{N}$ are called *P-admissible.* Hereafter $\mathcal{X}_P \subset \{1, \ldots, m\}^{\mathbb{N}}$ will denote the space of all P-admissible sequences equipped with the topology induced from the product topology on $\{1, \ldots, m\}^{\mathbb{N}}$ (and the discrete topology on $\{1, \ldots, m\}$). The *one-sided shift* $\sigma : \{1, \ldots, m\}^{\mathbb{N}} \to \{1, \ldots, m\}^{\mathbb{N}}$ defined by

$$\sigma((x_j)_{j=1}^{\infty}) = (y_j)_{j=1}^{\infty}, \quad \text{where} \quad y_j = x_{j+1}, \tag{1.4.2}$$

is continuous and leaves the sets $\mathcal{X}_P$ invariant: $\sigma(\mathcal{X}_P) \subset \mathcal{X}_P$ for any transition matrix P.

Definition 1.4.3. $\sigma_P = \sigma | \mathcal{X}_P$ is called a *subshift of finite type* or a *topological Markov chain.*

For any $a > 1$ we can define a distance function d_a on $\mathcal{X}_P$ by

$$d_a(x, y) = \sum_{n=1}^{\infty} \frac{|x_n - y_n|}{a^n} \tag{1.4.3}$$

when $x = (x_1, x_2, \ldots)$ and $y = (y_1, y_2, \ldots)$. With this distance $\mathcal{X}_P$ becomes a compact metric space of diameter $\leq m/(a-1)$ and σ_P becomes Lipschitz:

$$d_a(\sigma(x), \sigma(y)) \leq a d_a(x, y) \tag{1.4.4}$$

with the equality for sequences $x = (x_1, x_2, \ldots)$ and $y = (y_1, y_2, \ldots)$ such that $x_1 = y_1$. In other words, $\mathcal{G}(\sigma_P) \subset \operatorname{Lip}(\mathcal{X}_P)$.

Now, given a Markov system $\mathcal{S}$ with the transition matrix P, one can define the maps $T : \bigcup_i R_i \to \bigcup_i D_i$, $\tau : Z_0 \to Z_0$ and $h : Z_0 \to \mathcal{X}_P$ by the formulae

$$T|R_i = h_i^{-1}, \tag{1.4.5}$$

$$\tau = T|Z_0, \tag{1.4.6}$$

and

$$h(x) = j = (j_1, j_2, \ldots) \text{ iff } x \in C_j, \tag{1.4.7}$$

where

$$C_j = \bigcap_{n=1}^{\infty} K_{g_n}, \; g_1 = h_{j_1} \text{ and } g_n = g_{n-1} \circ h_{j_n} \in \mathcal{S}_n \text{ for } n = 2, 3, \ldots. \tag{1.4.8}$$

These definitions are correct. Clearly, $T(Z) \subset Z$ and T is a local homeomorphism, so $T(Z_0) \subset Z_0$; if $x \in \bigcap_{n=1}^{\infty}(K_{g_n} \cap K_{g'_n})$, $g_n = g_{n-1} \circ h_{j_n}$ and $g'_n = g'_{n-1} \circ h_{j'_n}$, then $x \in R_{j_1} \cap R_{j'_1}$ and — by property (i) — $j_1 = j'_1$, $h_{j_1}(x) \in R_{j_2} \cap R_{j'_2}$ and — by the same reason — $j_2 = j'_2$, etc. Evidently,

$$\sigma \circ h = h \circ \tau, \tag{1.4.9}$$

so the dynamical system (Z_0, τ) is semiconjugate to $(\mathcal{X}_P, \sigma_P)$.

Example 1.4.4. (a) If $P = I$, the identity matrix, then $h_i(K_i) \subset K_i$, $Z = \bigcup_{i=1}^{m} K_i^{\infty}$ with $K_i^{\infty} = \bigcap_{n=1}^{\infty} h_i^n(K_i)$ and $Z_0 = \bigcup_{i=1}^{m} \partial K_i^{\infty}$, where ∂A denotes the set-theoretic boundary of a set $A \subset X$. Moreover, $\mathcal{X}_P = \{(j, j, j, \ldots); j = 1, \ldots m\}$ is a finite set, $\sigma = \mathrm{id}_{\mathcal{X}_P}$ and $h|\partial K_i^{\infty} = \text{const}$. In particular, if $X, K_1, \ldots, K_m$ and $Q_1, \ldots, Q_m$ are closed segments on $\mathbb{R}$, then $K_1^{\infty}, \ldots, K_m^{\infty}$ are either segments or singletons and the Markov invariant set Z_0 is finite.

(b) If $X = [-\frac{1}{3}, \frac{4}{3}]$, $m = 2$, $D_1 = D_2 = X$, $h_1(x) = \frac{1}{3}x$, $h_2(x) = \frac{1}{3}(x+2)$, $K_1 = [0, \frac{1}{3}]$ and $K_2 = [\frac{2}{3}, 1]$, then $p_{ij} = 1$ for all i and $j \in \{1, 2\}$, $\mathcal{X}_P = \{0, 1\}^{\mathbb{N}}$, $Z = Z_0$ is the standard Cantor set of all the reals x of the form

$$x = \sum_{i=1}^{\infty} \frac{x_n}{3^n}, \quad x_n \in \{0, 2\}, \tag{1.4.10}$$

and $h : Z_0 \to \mathcal{X}_P$ sends x given by (1.4.10) to the sequence $j = (j_1, j_2, \dots)$, where $j_n = 1 + \frac{1}{2}x_n$. The map h is a homeomorphism and τ is topologically conjugated to σ, the one-sided shift in $\{1,2\}^{\mathbb{N}}$.

The similar observations can be made in the case of arbitrary m, $m \geq 2$, maps $h_1, \dots, h_m : [0,1] \to K_i$, $K_1, \dots, K_m$ being pairwise disjoint closed subintervals of $[0,1]$ (compare [**Mat1**]). Modifying Example (b) by replacing $X = [0,1]$ with $\tilde{X} = X \times Y$, Y being any topological space, and the maps h_i with $\tilde{h}_i$, $\tilde{h}_i(x,y) = (h_i(x), y)$, we obtain the Markov invariant set $\tilde{Z}_0 = Z_0 \times Y$ and the corresponding maps $\tilde{\tau}$ and $\tilde{h}$ given by $\tilde{\tau}(x,y) = (\tau(x), y)$, $\tilde{h}(x,y) = h(x)$. Again, $\tilde{h} \circ \tilde{\tau} = \sigma \circ \tilde{h}$ but $\tilde{h}$ is not a homeomorphism.

Now, we shall provide some conditions sufficient for Z_0 to contain an exceptional minimal set.

First, recall (see [**Gan**]) that a matrix $A = (a_{ij})$ with positive entries is *irreducible* if for any i and j there exists $n \geq 1$ such that the (i,j)-th entry $a_{ij}^{(n)}$ of A^n is positive:

$$a_{ij}^{(n)} > 0. \tag{1.4.11}$$

The dynamics of topological Markov chains with irreducible transition matrices P is well known ([**Pry**], [**Wa**], etc.). Here, we shall strengthen (1.4.11) replacing it by the following: for any i and j there exists n such that

$$p_{ij}^{(n)} \geq 2. \tag{1.4.12}$$

This condition implies that given two numbers i and $j \in \{1, \dots, m\}$ one can find two different P-admissible sequences $(x_1, \dots, x_n)$ and $(y_1, \dots, y_n)$ with $x_1 = y_1 = i$ and $x_n = y_n = j$. It is obviously satisfied by the pseudogroup of Example 1.4.4(b). As we shall see, it implies the existence of exceptional minimal sets inside of Z_0.

Proposition 1.4.5. *If P satisfies* (1.4.12), *then $\mathcal{X}_P$ is a minimal $\mathcal{G}(\sigma_P)$-invariant Cantor set.*

Proof. Obviously, $\mathcal{X}_P$ is compact and 0-dimensional. If $x = (x_1, x_2, \dots) \in \mathcal{X}_P$, then for any $n \geq 1$ one can find $z_n = (z_1^{(n)}, z_2^{(n)}, \dots) \in \mathcal{X}_P$ such that $z_n \neq x$, $z_1^{(n)} = x_1, \dots, z_n^{(n)} = x_n$. Obviously, $z_n \to x$ as $n \to \infty$, so $\mathcal{X}_P$ admits no isolated points. Also, given arbitrary sequences $x = (x_1, x_2, \dots)$ and $y = (y_1, y_2, \dots)$ in $\mathcal{X}_P$, and any $k \geq 1$, one can find a P-admissible sequence w of the form

$$w = (y_1, \dots, y_k, z_1, \dots, z_l, x_1, x_2, \dots), \quad z_1, \dots, z_l \in \{1, \dots, m\}.$$

Obviously, $w \in \mathcal{G}(\sigma_P)(x)$ and $d_2(w,y) \leq m/2^{k+1}$. Consequently, $y \in \overline{\mathcal{G}(\sigma_P))(x)}$ and $\overline{\mathcal{G}(\sigma_P)(x)} = \mathcal{X}_P$. This shows that $\mathcal{X}_P$ is minimal and completes the proof. □

Proposition 1.4.6. *If $S = \{h_1, \dots, h_m\}$ is a Markov system with the transition matrix P, and $K_1, \dots, K_m$ and X are segments on the real line $\mathbb{R}$, then the following conditions are equivalent.*

(1) *$\mathcal{X}_P$ is a minimal $\mathcal{G}(\sigma_P)$-invariant Cantor set.*

(2) *The Markov invariant set Z_0 of $\mathcal{S}$ contains an exceptional minimal $\mathcal{G}(\mathcal{S})$-invariant set C such that $h(C) = \mathcal{X}_P$.*

Proof. Assume that $\mathcal{X}_P$ is a minimal $\mathcal{G}(\sigma_P)$-invariant Cantor set and let B be the set of all the points $z \in Z_0$ such that $z \notin \overline{\mathcal{G}(\mathcal{S})(x)}$ for some $x \in Z_0$. Obviously, both sets, B and $C = Z_0 \setminus B$, are $\mathcal{G}(\mathcal{S})$-invariant. From the definition of B it follows immediately that C is minimal. Also, $C \neq X$ since $\operatorname{int} C \subset \operatorname{int} Z_0 = \emptyset$.

Next, take any point $z \in B$. Then, $z \in \partial C_j$, C_j being defined by (1.4.8) for some $j = (j_1, j_2, \dots) \in \mathcal{X}_P$, and there exists a point $x \in Z_0$ such that $x \in \partial C_k$ for some $k = (k_1, k_2, \dots) \in \mathcal{X}_P$ and the $\overline{\mathcal{G}(\mathcal{S})\text{ orbit}}$ of x does not cluster on z. On the other hand, since $\mathcal{X}_P$ is minimal, $j \in \overline{\mathcal{G}(\sigma_P)(k)}$. It follows that for any $n \in \mathbb{N}$ there exist l, r and indices $i_1, \dots, i_r$ such that $\alpha_n = (j_1, \dots, j_n, i_1, \dots i_r, k_l, k_{l+1}, \dots)$ lies in $\mathcal{X}_P$. Clearly, $j \notin \mathcal{G}(\sigma_P)(k)$ and, therefore, $\alpha_n \neq j$ for all n. Put

$$g_n = h_{j_1} \circ \dots \circ h_{j_n} \circ h_{i_1} \circ \dots \circ h_{i_r} \circ h_{k_l} \circ (h_{k_1} \circ \dots \circ h_{k_l})^{-1}.$$

Then $C_k \subset D_{g_n}$, $g_n(C_k) \subset h_{j_1} \circ \dots \circ h_{j_n}(D_{j_n})$ and $C_j \cap g_n(C_k) = \emptyset$. It follows that $\partial C_j \cap \overline{\mathcal{G}(\mathcal{S})(x)} \neq \emptyset$ and $\# \partial C_j > 1$, i.e. C_j is a non-degenerate interval: $\operatorname{int} C_j \neq \emptyset$. Since X is separable, B is contained in a countable union of finite sets ∂C_j, so B is at most countable. Since $Z_0 = h^{-1}(\mathcal{X}_P)$ cannot be countable, C cannot reduce to a single orbit.

Finally, $h(\bigcup_j \partial C_j) = \mathcal{X}_P$ and $\partial C_j \cap C \neq \emptyset$ for all $j \in \mathcal{X}_P$, hence $h(C) = \mathcal{X}_P$.

The implication "(2) $\Rightarrow$ (1)" is obvious. □

Corollary 1.4.7. *If P satisfies* (1.4.12), *then the Markov invariant set Z_0 of any Markov system on the real line $\mathbb{R}$ with the transition matrix P contains an exceptional minimal $\mathcal{G}(\mathcal{S})$-invariant set.* □

The importance of Markov pseudogroups consists in the following result.

Theorem 1.4.8. *If $\mathcal{G}$ is a Markov pseudogroup on a circle such that its Markov invariant set Z_0 contains a $\mathcal{G}$-invariant minimal set C, then there exist a closed foliated 3-manifold $(M, \mathcal{F})$, $\dim \mathcal{F} = 2$, an exceptional minimal set $E \subset M$, a complete transversal T and a homeomorphism $h : E \cap T \to C$ which conjugates $\mathcal{G}|C$ to $\mathcal{H}|E \cap T$, $\mathcal{H}$ being the holonomy pseudogroup of $\mathcal{F}$ acting on T.*

The proof given below is due to Cantwell and Conlon (see [**CC7**]). Alternative proofs can be found in [**In1**] and [**Ta**].

Proof. Let $S = \{h_1, \dots, h_m\}$ be a Markov system on S^1 which satisfies the assumptions of the theorem. Keep the notation introduced in Definition 1.4.1 and later on in Section 1.4.

Take first $M_0 = \Sigma \times S^1$, where Σ is S^2 with $2m$ discs $\Delta_1, \dots, \Delta_{2m}$ removed. Thus, M_0 is a compact 3-manifold with boundary ∂M_0 which consists of $2m$ tori $T_1, \dots, T_{2m}, T_k = \Delta_k \times S^1$. Equip M_0 with the product foliation $\mathcal{F}_0 = \{\Sigma \times \{\theta\}; \theta \in S^1$. $\mathcal{F}_0$ is transverse to ∂M_0 and induces a foliation by circles on each of the T_k's.

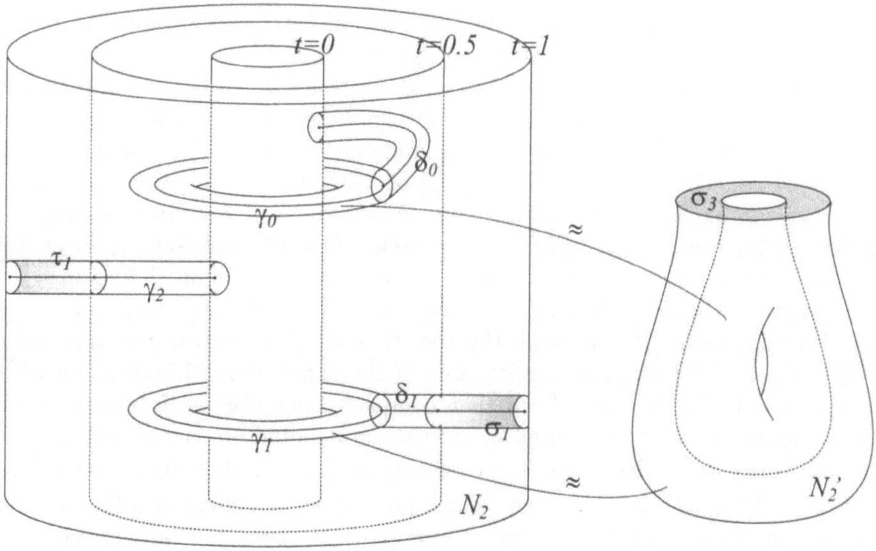

Figure 1.4: A step in a construction of an exceptional minimal set.

(By the foliation induced on T_k we mean the pullback (compare Section 1.3 and Proposition 1.3.9) $\iota_k^*\mathcal{F}_0$, where $\iota_k : T_k \to M_0$ is the canonical inclusion.)

Construct now another foliated 3-manifold M_1 with boundary $\partial M_1 = \tau' \cup \tau''$ being the union of two tori, to past its m copies to M_0.

To this end, consider first the product $N_1 = T^2 \times [0,1]$, equip $N_{1,1} = T^2 \times [\frac{1}{2}, 1]$ with the product foliation $\{T^2 \times \{t\}; \frac{1}{2} \leq t \leq 1\}$ and fill $N_{1,0} = T^2 \times (0, \frac{1}{2})$ with the cylinders spiralling towards $T^2 \times \{0\}$ and $T^2 \times \{\frac{1}{2}\}$. Take $T^2 \times \{0\}$ as a leaf and denote the obtained foliation of N_1 by $\tilde{\mathcal{F}}_1$.

Then, remove from N_1 a small tubular neighbourhood $N(\gamma_2)$ of a segment γ_2 which is transverse to $\tilde{\mathcal{F}}_1$ and has its end points on $T^2 \times \{0\}$ and $T^2 \times \{1\}$ respectively. Remove also transverse to $\tilde{\mathcal{F}}_1$ tubular neighbourhoods $N(\gamma_i)$, $i = 0, 1$, of two disjoint meridians γ_i on $T^2 \times \{t_0\}$, where $0 < t < \frac{1}{2}$. Assume also that $N(\gamma_i)$'s are small enough to be entirely contained in $T^2 \times (0, \frac{1}{2})$. Turbulize $\tilde{\mathcal{F}}_1$ (see Section 3.5.3) along γ_0 and γ_1 to get $\tilde{\mathcal{F}}_1'$, a foliation tangent to $\partial N(\gamma_0) \cup \partial N(\gamma_1)$. (While turbulizing, choose orientations carefully to make the next step manageable!) Remove now transverse to $\tilde{\mathcal{F}}_1'$ tubular neighbourhoods $N(\delta_i)$ of segments δ_i, $i = 0, 1$, which connect some points of $\partial N(\gamma_i)$ to points of $T^2 \times \{i\}$. Let $N_2 = N_1 \setminus (N(\gamma_0) \cup N(\gamma_1) \cup N(\delta_0) \cup N(\delta_1))$ and $\tilde{\mathcal{F}}_2$ be the restriction of $\tilde{\mathcal{F}}_1'$ to N_2. (Figure 1.4 depicts a piece of N_1 where all these modifications take place.)

Next, equip $N_2' = (T^2 \setminus \Delta) \times [0,1]$, $\Delta \subset T^2$ being a disc, with the product foliation $\tilde{\mathcal{F}}_2' = \{(T^2 \setminus \Delta) \times \{t\}; 0 \leq t \leq 1\}$ and paste the foliated manifold $(N_2', \tilde{\mathcal{F}}_2')$ to $(N_2, \tilde{\mathcal{F}}_2)$ in such a way that each component $(T^2 \setminus \Delta) \times \{i\}$, $i = 0, 1$, of $\partial N_2'$

becomes identified with $\partial N(\gamma_i) \cap N_2$ (see Section 3.5.2 and Figure 1.4). Denote the foliated manifold obtained at this stage by $(N_3, \tilde{\mathcal{F}}_3)$.

The boundary ∂N_3 contains two tori, $T^2 \times \{0\}$ and $T^2 \times \{1\}$, each of them with two discs removed. So, we can paste one of them to another in such a way that two ends of $N(\gamma_2)$ as well as one end of $N(\delta_0)$ and one end of $N(\delta_1)$ become identified. N_4, the obtained manifold, is now equipped with $\tilde{\mathcal{F}}_4$, the foliation obtained by transverse gluing (see Section 3.5.2 again) of two copies of $\tilde{\mathcal{F}}_3$. The boundary ∂N_4 consists of two tori τ_1 and τ_2 foliated by circles. One of them, say τ_1, arises from $\partial N(\gamma_2)$ while another one, τ_2 from $\delta_0 \cup \delta_1 \cup \sigma_3$ (compare Figure 1.4 again).

Finally, produce $(M_1, \mathcal{F}_1)$ by taking two copies $(N_4', \mathcal{F}_4')$ and $(N_4'', \mathcal{F}_4'')$ of $(N_4, \mathcal{F}_4)$ and pasting them along the tori τ_1' and τ_1'' in such a way that points of $\tau_1' \cap N_{1,1}$ in N_4' (compare the first step of the construction!) become identified with points of $\tau_1'' \cap N_{1,1}$ in N_4''. M_1 is a compact 3-manifold with boundary ∂M_1 which consists of two tori foliated by circles and denoted now by τ' and τ''. Each of these tori can be decomposed into the union of four domains, $\tau' = \bigcup_{i=1}^4 \sigma_i'$ while $\tau'' = \bigcup_{i=1}^4 \sigma_i''$, in such a way that a part of M_1 is equipped with a product foliation by leaves with two boundary components contained respectively in σ_1' and σ_1'', another part of M_1 — with a product foliation by leaves with boundaries contained in σ_3' and yet another part — with a product foliation by leaves with boundaries contained in σ_3''.

To complete the construction of $(M, \mathcal{F})$, the foliated manifold satisfying the conditions of our Theorem, take m copies of $(M_1, \mathcal{F}_1)$, denote them by $(P_j, \hat{\mathcal{F}}_j)$, $j = 1, \ldots, m$, and paste them to $(M_0, \mathcal{F}_0)$ by identifying each τ_j' with T_j and τ_j'' with T_{m+j}, where obviously τ_j' and τ_j'' are the boundary components of ∂P_j corresponding to τ' and τ''. Perform this identification in such a way that the σ_1'-domains on τ_j''s, $j = 1, \ldots, m$, become identified with $\partial\Delta_j \times Q_j$, the σ_3'-domains on τ_j''s become identified with $\partial\Delta_j \times (S^1 \setminus D_j)$, the σ_1'' domains on τ_j'''s — with $\partial\Delta_{m+j} \times K_j$ and the σ_3''-domains on τ_j'''s — with $\partial\Delta_{m+j} \times (S^1 \setminus R_j)$. Clearly, the identifications $\sigma_1' \cong \partial\Delta_j \times Q_j$ and $\sigma_1'' \cong \partial\Delta_{m+j} \times K_j$ can be arranged in such a way that $h_j : Q_j \to K_j$ become holonomy maps for $\mathcal{F}$. Since $\bigcup\{K_j; p_{ij} = 1\} \subset Q_i$, $\bigcup\{K_j; p_{ij} = 0\} \subset S^1 \setminus D_i$ and $\bigcup\{K_j; j \neq i\} \subset S^1 \setminus R_i$ $(i, j \leq m)$, there are no other holonomy maps in a neighbourhood of the Markov invariant set $Z_0 \subset S^1$ than these generated by $h_1, \ldots, h_m$. Therefore, the $\mathcal{F}$-saturation E of C, an invariant set contained in Z_0, is an exceptional minimal set for $\mathcal{F}$ and has all the required properties. □

1.5 Hyperbolic spaces and groups

Hyperbolicity is one of the most important properties in dynamics and geometry. Hyperbolic dynamical systems have been studied extensively for a long time and provide one of the best known classes of systems with rich dynamics. Compact hyperbolic manifolds carry (in contrast to those of positive curvature) complicated topological structure, in particular, their fundamental groups have complicated algebraic structure as well as rich dynamics. Certainly, these two notions of hyperbolicity (dynamical and geometric) differ substantially. However, there exist several connections between them. For example, the induced action of the fundamental group of a hyperbolic (in the geometric sense) manifold on the ideal boundary of its universal covering is often hyperbolic (in the dynamical sense).

Not long ago, Gromov [**Gro2**] introduced the notion of hyperbolicity for general metric spaces and, in particular, finitely generated groups. His definition is very geometric and applies to Riemannian manifolds of negative curvature as well as to their fundamental groups, so its occurence is very natural. Of course, the hyperbolic plane H^2 (and all the hyperbolic spaces H^n, $n \geq 3$) should become hyperbolic in any reasonable sense. Triangles built of geodesic segments in H^2 are "thin" (in the sense described in Definition 1.5.1). Making them "as thin as possible" we end up with a graph that has the shape of the letter "Y". It consists of three edges and four vertices and, as a limit of hyperbolic triangles, should become hyperbolic too. This graph equipped with the length metric d (the distance between two points equals the sum of the lengths of segments connecting them) has the following property: given arbitrary three points x_1, x_2 and x_3 of Y and a base point x_0 (say, the common origin of the graph's edges), the sums $a_i = d(x_j, x_0) + d(x_k, x_0) - d(x_j, x_k)$, where $i = 1, 2, 3$ and (i, j, k) is an even permutation of the set $(1, 2, 3)$, satisfy — for any such permutation — the inequality

$$a_i \geq \min\{a_j, a_k\}.$$

A δ-approximation ($\delta > 0$) of this condition serves as another definition of Gromov's hyperbolicity (see Proposition 1.5.11 below).

Gromov's hyperbolic spaces and groups became of great interest as soon as they were introduced. In this section, we provide the reader with basic definitions and these properties of hyperbolic spaces and groups, their ideal boundaries and actions which are related to the aspects of dynamics considered in this book. The reader interested in more of this theory is referred to [**GH**].

Let (X, d) be a metric space. Given any curve $\gamma : [0, 1] \to X$ its length $l(\gamma)$ is defined by

$$l(\gamma) = \sup\{\sum_{i=0}^{n} d(\gamma(t_i), \gamma(t_{i+1}); 0 \leq t_0 \leq t_1 \leq \cdots \leq t_n \leq 1, n \in \mathbb{N}\}. \tag{1.5.1}$$

Such a curve γ is said to be a *geodesic segment* when $l(\gamma) = d(\gamma(0), \gamma(1))$. The space (X, d) is said to be *geodesic* when any two points of X can be joined by a (not necessarily unique) geodesic segment. The classical Hopf–Rinow Theorem (see [**GKM**] or [**GLP**], for example) shows that any complete Riemannian manifold equipped with the standard distance function becomes a geodesic metric space. Finally, a *geodesic triangle* in a metric space X is a triple $\Delta = (\gamma_0, \gamma_1, \gamma_2)$ of geodesic segments for which $\gamma_{i+1}(0) = \gamma_i(1)$ for $i = 0, 1, 2$ (with the obvious convention $\gamma_3 = \gamma_0$).

Definition 1.5.1. A geodesic metric space is said to be *hyperbolic* (in Gromov's sense) when there exists a number $\delta \geq 0$ such that for any geodesic triangle $\Delta = (\gamma_0, \gamma_1, \gamma_2)$ and any $t \in [0,1]$ the inequality

$$\min\{d(\gamma_0(t), \gamma_i(s)); 0 \leq s \leq 1, i = 1, 2\} \leq \delta \tag{1.5.2}$$

holds. Condition (1.5.2) is called *Rips condition* and δ is called a *Rips constant* for X. Roughly speaking, (1.5.2) says that geodesic triangles in X are δ-thin.

Remark 1.5.2. The original Gromov definition of hyperbolicity [**Gro2**] differs from the above but we shall see later (Prop. 1.5.11) that they are equivalent.

The following result motivates the term "hyperbolic".

Theorem 1.5.3. *Any complete simply connected Riemannian manifold of sectional curvature bounded from above by a negative constant is hyperbolic.*

Proof. Toponogov's Comparison Theorem (see [**CE**] or [**GKM**], for example) implies that it is enough to prove inequality (1.5.2) for manifolds M of constant negative sectional curvature K, say $K = -1$. Since every geodesic triangle in such a manifold is entirely contained in a 2-dimensional totally geodesic submanifold, we may assume without losing generality that $M = H^2$, the hyperbolic plane.

In H^2, the classical Gauss–Bonnet Theorem shows that the area A bounded by any geodesic triangle $\Delta = (\gamma_0, \gamma_1, \gamma_2)$ is at most π. If Δ does not satisfy (1.5.2), then there exists $t_0 \in [0,1]$ such that $d(\gamma_0(t_0), \gamma_i(s)) > \delta$ for all $s \in [0,1]$ and $i = 1, 2$. Let $x_0 = \gamma(\delta/2)$ be the point inside Δ on the normal geodesic γ originated at $\gamma_0(t_0)$ and perpendicular to γ_0. By the cosine rule in hyperbolic geometry, $d(x_0, \gamma_0(t)) \geq \delta/2$ for all $s \in \mathbb{R}$. Also, $d(x_0, \gamma_i(s)) \geq \delta/2$ for all $s \in [0,1]$ and $i = 1, 2$. Therefore, the disc $B(x_0, \delta/2)$ is entirely contained in the region bounded by Δ and its area A_δ satisfies $A_\delta \leq A \leq \pi$. If $\delta > 0$ is large enough, then $A_\delta > \pi$, a contradiction. Hence, the Rips condition with such a constant δ has to be satisfied everywhere in H^2. □

To get more examples of hyperbolic spaces let us consider a finitely generated group G equipped with a finite symmetric generating set G_1. Set $G_0 = \{e\}$ and denote by G_n, as before, the set of all the elements of G of the form $g_1 \cdot \dots \cdot g_n$ with $g_i \in G_1$. Equip G with the distance function d defined by

$$d(g, g') = \min\{n; g^{-1}g' \in G_n\}. \tag{1.5.3}$$

Next, connect any two elements g and hg ($g \in G$, $h \in G_1$) by a segment of length 1, denote by $\mathcal{C} = \mathcal{C}(G, G_1)$ the graph obtained this way and extend the distance function d to $\mathcal{C}$ in a natural way. $(\mathcal{C}, d)$ becomes a geodesic metric space called a *Cayley graph* of G.

Definition 1.5.4. A group G is called *hyperbolic* when its Cayley graph $\mathcal{C}$ is hyperbolic.

To see that this definition is correct (i.e., that hyperbolicity is independent of the choice of G_1) let us consider two arbitrary metric spaces (X, d) and (X', d') and accept the following.

Definition 1.5.5. A map $f : X \to X'$ is *quasi-isometric* when there exist constants $\lambda \geq 1$ and $c \geq 0$ such that

$$\frac{1}{\lambda}d(x,y) - c \leq d'(f(x), f(y)) \leq \lambda d(x,y) + c \tag{1.5.4}$$

for all x and $y \in X$. Such an f is said to be a *quasi-isometry* whenever there exists $\varepsilon > 0$ such that the image $f(X)$ is ε-dense in X', that is $\operatorname{dist}(x', f(X)) \leq \varepsilon$ for all $x' \in X'$. If $\varepsilon = \lambda = c$, the map f is called an *ε-quasi-isometry.* The spaces X and X' are *quasi-isometric* when there exist quasi-isometries $f : X \to X'$ and $f' : X' \to X$ and a constant $c \geq 0$ such that

$$d(f'(f(x)), x) \leq c \quad \text{and} \quad d'(f(f'(x')), x') \leq c \tag{1.5.5}$$

for all $x \in X$ and $x' \in X'$.

Note that the maps f and f' above need not be continuous.

It is easy to see that the Cayley graphs $\mathcal{C}$ and $\mathcal{C}'$ of a finitely generated group G defined by two different generating sets G_1 and G_1' are quasi-isometric. In fact, one can take $f = f' = \mathrm{id}_G$ and gets (1.5.4) and (1.5.5) with $c = \lambda = k$, where k is any natural number such that $G_1 \subset G_k'$ and $G_1' \subset G_k$.

Proposition 1.5.6. *If two geodesic metric spaces are quasi-isometric and one of them is hyperbolic, then the other one is hyperbolic as well.*

Proof. Assume that (X, d) is hyperbolic, $\delta \geq 0$ is its Rips constant, $f : X \to X'$ and $f' : X' \to X$ are maps satisfying conditions (1.5.4) and (1.5.5). Let $\Delta' = (\gamma_0', \gamma_1', \gamma_2')$ be a geodesic triangle in X' and $\Delta = (\gamma_0, \gamma_1, \gamma_2)$ any geodesic triangle in X such that $\gamma_i(0) = f'(\gamma_i(0))$ for $i = 0, 1, 2$.

Take any $t \in [0,1]$. There exist $s \in [0,1]$ and $j \in \{1,2\}$ such that $d(\gamma_0(t), \gamma_j(s)) \leq \delta$. Then,

$$\begin{aligned} d'(f(\gamma_0(t)), \gamma_0'(t)) &\leq d'(f(\gamma_0(t)), f(\gamma_0(0))) + d'(f(\gamma_0(0)), \gamma_0'(0)) + d(\gamma_0'(0), \gamma_0'(t)) \\ &\leq \lambda t + c + d'(f(\gamma_0(0)), \gamma_0'(0)) + t \leq \lambda + 1 + 2c \end{aligned}$$

and, similarly,

$$d'(f(\gamma_j(s)), \gamma_j'(s)) \leq \lambda + 1 + 2c$$

by (1.5.4) and (1.5.5). Therefore,

$$\begin{aligned} d'(\gamma_0'(t), \gamma_j'(s)) &\leq 2(\lambda + 1 + 2c) + d'(f(\gamma_0(t)), f(\gamma_j(s)) \\ &\leq 2(\lambda + 1 + 2c) + \lambda d(\gamma_0(t), \gamma_1(s)) + c \leq 5c + 2(1 + \lambda) + \lambda\delta, \end{aligned}$$

i.e., X' satisfies the Rips condition with the constant $\delta' = 5c + 2(1 + \lambda) + \lambda\delta$. □

Example 1.5.7. Any finite group and any finite extension of a cyclic group $\mathbb{Z}$ is hyperbolic. (Hyperbolic groups of this type are called elementary.) Also, any non-abelian free group is hyperbolic (with a Rips constant zero).

In Section 2.2, we shall prove that the fundamental group of any compact Riemannian manifold M is quasi-isometric to the universal covering $\tilde{M}$ of M. This implies the following.

Corollary 1.5.8. *The fundamental group of a compact Riemannian manifold of negative sectional curvature is hyperbolic.* □

Our next goal is to define the boundary of a hyperbolic space and to equip it with a reasonable topology and metric. So, let (X, d) be a geodesic hyperbolic space for which all the closed bounded sets are compact. By a *geodesic ray* in X we mean any quasi-isometric map $\gamma : [0, \infty] \to X$. Two rays γ_0 and γ_1 are said to be *asymptotic* ($\gamma_0 \sim \gamma_1$) when

$$\sup_{0 \le t < \infty} d(\gamma_0(t), \gamma_1(t)) < \infty. \tag{1.5.6}$$

The triangle inequality implies immediately that the relation "$\sim$" is an equivalence. Moreover, it is not too hard to prove (and we leave this to the reader) that two asymptotic geodesic rays γ_0 and γ_1 with the same origin satisfy the condition

$$\sup_{t \ge 0} d(\gamma_0(t), \gamma_1(t)) \le 8\delta, \tag{1.5.7}$$

δ being a Rips constant for X. Similarly, for any two asymptotic distance minimizing geodesics γ_0 and γ_1 there exists $s \in \mathbb{R}$ such that

$$\sup_{t \in \mathbb{R}} d(\gamma_0(t), \gamma_1(t - s)) \le 16\delta. \tag{1.5.8}$$

The property essential for the proofs of (1.5.7) and (1.5.8) is that any hyperbolic space X with a Rips constant δ is 8δ-convex in the sense that the inequality

$$d(\sigma_0(t), \sigma_1(t)) \le (1 - t)d(\sigma_0(0), \sigma_1(0)) + td(\sigma_0(1), \sigma_1(1)) + 8\delta \tag{1.5.9}$$

holds for all geodesic segments $\sigma_i : [0, a_i] \to X$, $i = 0, 1$, parametrized by arc length, i.e., in such a way that $d(\sigma_i(t), \sigma_i(0)) = td(\sigma_i(1), \sigma_i(0))$ for any t.

Definition 1.5.9. The *ideal boundary* (called also the *boundary at infinity* or just the *boundary*, for short) ∂X of X is defined as $R(X)/\sim$, the set of all the equivalence classes of $\sim$, $R(X)$ being the family of all geodesic rays on X.

This definition coincides with the standard one, used in classical hyperbolic or Riemannian geometry for H^n or arbitrary Hadamard manifolds (i.e., complete and simply connected Riemannian manifolds of non-positive curvature). In fact, in the case of an Hadamard manifold (M, g) any such geodesic ray γ is asymptotic to a genuine geodesic c. Such c can be obtained as an accumulation point of the sequence (c_n) of geodesics defined uniquely by $c_n(0) = \gamma(0)$ and $c_n(n) = \gamma(n)$ for $n = 1, 2, \ldots$. That is, c satisfies $c(0) = \gamma(0)$ and $\dot{c}(0) = \lim_{k\to\infty} \dot{c}_{n_k}(0)$ for some sequence $n_k \to \infty$. Therefore, in the standard case of the Poincaré model of H^n, ∂H^n coincides with the Euclidean sphere which bounds the unit ball representing H^n. Also, ∂M is homeomorphic to the unit sphere whenever M is a so-called *visibility manifold* (see [**Eb**]), for example, an Hadamard manifold of sectional curvature bounded by negative constants.

Since any geodesic ray γ_0 is asymptotic to a ray γ_1 with prescribed origin, say $\gamma_1(0) = x_0$, ∂X can be represented as the quotient $R(X, x_0)/\sim$, where $R(X, x_0)$ is the set of all geodesic rays originated at x_0. Also, as we shall see later, ∂X can be described in terms of converging to infinity sequences (x_n) of points of X. This approach is useful when one wants to equip ∂X with a suitable topology (and metric). To this end we shall need

Definition 1.5.10. The (based at a point x_0) *Gromov product* on X is a map $(\cdot|\cdot) : X \times X \to \mathbb{R}$ given by

$$(x|y) = \frac{1}{2}\left(d(x, x_0) + d(y, x_0) - d(x, y)\right). \tag{1.5.10}$$

The Gromov product provides a convenient condition equivalent to hyperbolicity.

Proposition 1.5.11. *A geodesic metric space (X, d) satisfies the Rips condition with a constant $\delta \geq 0$ if and only if the inequality*

$$(x|z) \geq \min\{(x|y), (y|z)\} - \eta \tag{1.5.11}$$

holds for some $\eta \geq 0$, any base point x_0 and all x, y and z in X.

Comparing (1.5.10) and (1.5.11) one can express hyperbolicity of X by saying that for any quadrangle with vertices x_0, x, y and z in X the largest sum of two sides or diagonals with disjoint end points does not exceed the smallest among such sums enlarged by 2δ:

$$\begin{aligned}\max\{&d(x_0, x) + d(y, z), d(x_0, y) + d(x, z), d(x_0, z) + d(x, y)\}\\ &\leq \min\{d(x_0, x) + d(y, z), d(x_0, y) + d(x, z), d(x_0, z) + d(x, y)\} + 2\delta.\end{aligned}$$

Proof. Let Δ be a geodesic triangle with vertices x, y and z. There exists a tripod T_Δ consisting of three line segments S_x, S_y and S_z with common vertex at, say, the origin of the Euclidean plane and an isometry $f_\Delta : \Delta \to T_\Delta$ mapping the points x, y and z onto the end points of the corresponding segments of T_Δ. The length of S_x is equal to the (based at x) Gromov product $(y|z)$. The lengths of S_y and S_z can be expressed similarly.

Rips condition (1.5.2) can be expressed in the form

$$d(u, w) \leq d_0(f_\Delta(u), f_\Delta(w)) + \delta, \tag{1.5.12}$$

d_0 being the natural distance in T_Δ. More precisely, (1.5.12) implies (1.5.2) while (1.5.2) implies (1.5.12) with δ replaced by 4δ. Moreover, it is easy to see that (1.5.12) is equivalent to the following:

$$d(u,w) \le \delta \quad \text{whenever} \quad u, w \in \Delta \quad \text{and} \quad f_\Delta(u) = f_\Delta(w). \tag{1.5.13}$$

(i) First, assume that (1.5.11) holds with some $\eta \ge 0$. Take two points u and w on geodesic segments joining x to y and x to z, respectively. Assume that $f_\Delta(u) = f_\Delta(w)$ and put $t = d(x,u) = d(x,w)$. Then

$$d_0(f_\Delta(x), f_\Delta(u)) = d_0(f_\Delta(x), f_\Delta(w)) = t \le (y|z)$$

and

$$(u|y) = (w|z) = t.$$

From (1.5.11) it follows that

$$(u|w) \ge \min\{(u|y), (y|z), (z|w)\} - 2\eta = t - 2\eta.$$

Also, $d(u,w) = 2t - 2(u|w)$. Consequently, $d(u,w) \le 4\eta$ and Δ satisfies the Rips condition with the constant $\delta = 4\eta$.

(ii) Now, assume that Δ satisfies (1.5.2) with some $\delta \ge 0$ and choose any base point x_0 for the Gromov product $(\cdot,\cdot)$. Let $t = \min\{(x|z), (z|y)\}$. Find points x', y' and z' on geodesic segments joining x_0 to x, y and z and such that $d(x_0, x') = d(x_0, y') = d(x_0, z') = t$. Consider the geodesic triangle Δ_x with vertices x_0, x and z, and the corresponding isometry $f_x = f_{\Delta_x}$. Since $d(x', x_0) = d(z', x_0) \le (x|z)$, $f_x(x') = f_x(z')$ and $d(x', z') \le \delta$ by (1.5.13). Similarly, $d(y', z') \le \delta$ and so $d(x', y') \le 2\delta$.

If $t > (x|y)$, then there exist points z_x and z_y on the geodesic segment joining x to y and such that $f_0(z_x) = f_0(x')$ and $f_0(z_y) = f_0(y')$, where $f_0 = f_{\Delta_0}$, Δ_0 being the geodesic triangle with vertices x_0, x and y. Again, (1.5.13) implies that $d(x', z_x) \le \delta$ and $d(z_y, y') \le \delta$. Consequently,

$$\begin{aligned} 4\delta &\ge d(x', y') + 2\delta \ge d(z_x, z_y) = d(x,y) - d(x, z_x) - d(y, z_y) \\ &= d(x,y) - d(x,x') - d(y,y') \\ &= d(x,y) - d(x_0,x) + d(x_0,x') - d(x_0,y) + d(x_0,y') = 2t - 2(x|y) \end{aligned}$$

and

$$(x|y) \ge t - 2\delta. \tag{1.5.14}$$

If $t \le (x|y)$, then (1.5.14) is satisfied trivially. This yields (1.5.11) with $\eta = 2\delta$. □

Coming back to the boundary ∂X of a hyperbolic space X let us say that a sequence (x_n) of points of X *diverges to infinity* when $(x_n|x_m) \to \infty$ as $m, n \to \infty$. Two such sequences (x_n) and (y_n) are equivalent,$(x_n) \sim (y_n)$, when

$$\lim_{n,m\to\infty} (x_n|y_m) = \infty. \tag{1.5.15}$$

From (1.5.10) and (1.5.11) one can see easily that the relation "$\sim$" is reflexive, symmetric and transitive. Also, it is not too hard to show that there is a canonical bijection of $R(X)/\sim$ onto $S_\infty(X)/\sim$, where $S_\infty(X)$ is the collection of all sequences divergent to infinity. It could be defined by

$$[\gamma]_\sim \mapsto [(\gamma(n))]_\sim.$$

So, ∂X can be identified with the quotient $S_\infty(X)/\sim$. We write $x_\infty = \lim_{n\to\infty} x_n$ whenever the sequence (x_n) represents a point $x_\infty \in \partial X$.

We can extend the Gromov product to ∂X by the formula

$$(x_\infty|y_\infty) = \sup \liminf_{m,n\to\infty} (x_n|y_m), \tag{1.5.16}$$

where the supremum is taken over the set of all the sequences (x_n) and (y_m) such that $x_n \to x_\infty$ and $y_m \to y_\infty$. Clearly,

$$(x_\infty|y_\infty) = \infty \quad \text{iff} \quad x_\infty = y_\infty \quad \text{in} \quad \partial X, \tag{1.5.17}$$

$$(x_\infty|y_\infty) = (y_\infty|x_\infty) \tag{1.5.18}$$

and, by (1.5.11),

$$(x_\infty|z_\infty) \geq \min\{(x_\infty|y_\infty), (y_\infty|z_\infty)\} - \delta \tag{1.5.19}$$

for all x_∞, y_∞ and z_∞ of ∂X.

For any $\varepsilon > 0$, x_∞ and y_∞ of ∂X put

$$\rho_\varepsilon(x_\infty, y_\infty) = \exp(-\varepsilon(x_\infty|y_\infty)) \tag{1.5.20}$$

and

$$\begin{aligned} &d_\varepsilon(x_\infty, y_\infty) \\ &= \inf\left\{\sum_{i=0}^{n-1} \rho_\varepsilon(x_\infty^i, x_\infty^{i+1}); x_\infty^i \in \partial X, x_\infty^0 = x_\infty, x_\infty^n = y_\infty, n \in \mathrm{N}\right\}. \end{aligned} \tag{1.5.21}$$

Certainly, $d_\varepsilon(x_\infty, y_\infty) \leq \rho_\varepsilon(x_\infty, y_\infty)$ for all $x_\infty, y_\infty \in \partial X$. From (1.5.17)–(1.5.19) it follows that if ε is small enough ($\varepsilon < 1/\delta$) then d_ε is a distance function on ∂X. The only difficulty is to show the implication "$d_\varepsilon(x_\infty, y_\infty) = 0 \Rightarrow x_\infty = y_\infty$" which can be achieved as an immediate consequence of the inequality

$$d_\varepsilon(x_\infty, y_\infty) \geq c \cdot \rho_\varepsilon(x_\infty, y_\infty), \tag{1.5.22}$$

where $c = 3 - 2\exp(\varepsilon\delta)$. (Hint: (1.5.22) follows from the inequality

$$\sum_{i=0}^{n-1} \rho_\varepsilon(x_\infty^i, x_\infty^{i+1}) \geq c\rho_\varepsilon(x_\infty, y_\infty)$$

which can be proved by induction on n while using the inequality

$$\rho_\varepsilon(x_\infty, y_\infty) \leq \exp(\varepsilon\delta) \cdot \max\{\rho_\varepsilon(x_\infty, z_\infty), \rho_\varepsilon(z_\infty, y_\infty)\}$$

which is evidently equivalent to (1.5.19).) Finally, it is not too hard to show that the metric space $(\partial X, d_\varepsilon)$ is compact. To this end just take a sequence (x_∞^n) in ∂X and a family of sequences $(x_i^n)_{i=1}^\infty$ such that $x_i^n \to x_\infty^n$ for $n = 1, 2, \dots$, choose a subsequence (n_k) of integers for which all the sequences $(x_i^{n_k})$, $i = 1, 2, \dots$, have limits x_i in X, show that $(x_i) \in S_\infty(X)$, set $x_\infty = \lim_{i\to\infty} x_i$ and show that $\rho_\varepsilon(x_\infty^{n_k}, x_\infty) \to 0$ when $k \to \infty$.

Let us discuss some examples.

Example 1.5.12. (i) If M is a simply-connected n-dimensional Riemannian manifold of bounded negative curvature, then — as we mentioned before — ∂M is homeomorphic to the sphere S^{n-1}. To get a suitable homeomorphism fix $x_0 \in M$, for any unit vector $v \in T_{x_0}M$ take a unique geodesic γ with $\gamma(0) = x_0$ and $\dot\gamma(0) = v$ and map v to the point $\gamma(\infty) = \lim_{n\to\infty} \gamma(n)$.

(ii) If G is a hyperbolic group and $X = \mathcal{C}(G, G_1)$ is its Cayley graph, then $\partial X = \emptyset$ when G is finite, $\#\partial X = 2$ if and only if G contains an infinite cyclic subgroup of finite index and $\#\partial X \geq 3$ (in fact, as we shall see later, ∂X is infinite and uncountable) when G is non-elementary. In fact, if G is infinite, one can choose elements g_n of G such that $d(e, g_n) = 2n$, find the midpoints g_n' of geodesics joining e to g_n, put $h_n = (g_n')^{-1}$ and $k_n = h_n g_n$. Then, $(h_n|h_n) = (k_n|k_n) = n$ while $(h_n|k_n) = 0$, $(\cdot|\cdot)$ being the Gromov product based at e. Therefore, suitable subsequences of (h_n) and (k_n) represent two different points of ∂X, i.e., $\#\partial X \geq 2$. Now, if ∂X consist exactly of two points, say x_∞ and y_∞, then one can choose a minimizing geodesic $\gamma_0 : \mathbb{R} \to X$ such that $\gamma_0|(-\infty, 0]$ represents x_∞ while $\gamma_0|[0, \infty)$ represents y_∞. Any other geodesic $\gamma_1 = g \circ \gamma_0$ for some $g \in G$ is asymptotic to γ_0, so it satisfies (1.5.8) with some s. Mapping g to $\gamma_0(s)$ we obtain a quasi-isometry. So, G is quasi-isometric to $\mathbb{N}$ and another exercise left to the reader is to show that, consequently, G contains a cyclic subgroup of finite index.

Now, take any isometry f of X and extend it to ∂X by the formula

$$f(x_\infty) = \lim_{n\to\infty} f(x_n) \quad \text{whenever} \quad x_\infty = \lim_{n\to\infty} x_n. \tag{1.5.23}$$

Since f is an isometry, the Gromov product based at a point x_0 satisfies

$$|(f(x)|f(y)) - (x|y)| \leq d(x_0, f(x_0)) \tag{1.5.24}$$

and thus the definition (1.5.23) is correct: If $(x_n) \sim (y_m)$, then $(x_n|y_m) \to \infty$ with $m, n \to \infty$, $(f(x_n)|f(y_m)) \to \infty$ and $(f(x_n)) \sim (f(y_m))$. Moreover, (1.5.24) implies that

$$e^{-\varepsilon d(x_0, f(x_0))} \cdot \rho_\varepsilon(x_\infty, y_\infty) \leq \rho_\varepsilon(f(x_\infty), f(y_\infty)) \leq e^{\varepsilon d(x_0, f(x_0))} \cdot \rho_\varepsilon(x_\infty, y_\infty) \tag{1.5.25}$$

and this yields the similar condition for the distance d_ε. In this way we have proved

Proposition 1.5.13. *Any isometry of a hyperbolic space extends to a Lipschitz homeomorphism of its boundary.* □

As in classical hyperbolic geometry, isometries of X can be classified according to the behaviour of their extensions on ∂X.

Definition 1.5.14. An isometry f of a hyperbolic space X is

(i) *elliptic* when the orbits $\{f^n(x)\}$ of some (equivalently, any) points $x \in X$ are bounded,

(ii) *hyperbolic* when ∂X contains a point x_∞ fixed by f and such that the *force* $\Phi(f, x_\infty)$ of f at x_∞ defined by

$$\Phi(f, x_\infty) = \lim_{n\to\infty} \left(\frac{d(f^n(y_\infty), f^n(z_\infty))}{d(y_\infty, z_\infty)} \right)^{\frac{1}{n}}, \quad y_\infty, z_\infty \in \partial X \setminus \{x_\infty\}, \quad (1.5.26)$$

differs from 1.

(iii) *parabolic* if neither elliptic nor hyperbolic.

Here, $d = d'_\varepsilon$ is a metric on $\partial X \setminus \{x_\infty\}$ given by the formulae analogous to (1.5.20) and (1.5.21), where the standard Gromov product is replaced by the relative one $(\cdot|\cdot)_0$,

$$(x|y)_0 = \frac{1}{2}\left(b_\infty(x, x_0) + b_\infty(y, x_0) - d(x, y)\right),$$

where $b_\infty : X \times X \to \mathbb{R}$ is the *Busemann function* at x_∞; b_∞ is given by

$$b_\infty(x, \gamma) = \limsup_{t\to\infty}(d(x, \gamma(t)) - t)$$

for any geodesic ray $\gamma : [0, \infty) \to X$ and

$$b_\infty(x, y) = \sup\{b(x, \gamma); \gamma \in R(X, y), [\gamma]_\sim = x_\infty\}.$$

The proof of the fact that d is a good distance function is quite technical and omitted here. An interested reader is referred to [**GH**], Chapter 8.

Theorem 1.5.15. *If f is an hyperbolic isometry, then*

(i) *∂X contains exactly two points, say x_∞ and y_∞, fixed by f,*

(ii) *if $\Phi(f, x_\infty) < 1$, U and V are arbitrary open neighbourhoods of x_∞ and y_∞, respectively, then $f^n(\partial X \setminus U) \subset V$ and $f^{-n}(\partial X \setminus V) \subset U$ for all $n \in \mathbb{N}$ sufficiently large,*

(iii) *$\Phi(f, x_\infty) < 1$ whenever $\Phi(f, y_\infty) > 1$,*

(iv) *f^n is hyperbolic for all $n \in \mathbb{N} \setminus \{0\}$.*

Proof. If x_∞ is fixed by f and $\Phi(f, x_\infty) \neq 1$, say $\Phi(f, x_\infty) < 1$, then the formula

$$d_f(u_\infty, w_\infty) = \lim_{n\to\pm\infty} \frac{1}{\Phi(f, x_\infty)^n} \cdot d(f^n(u_\infty), f^n(w_\infty))$$

defines a complete metric on $\partial X \setminus \{x_\infty\}$ such that

$$d_f(f(u_\infty), f(w_\infty)) = \Phi(f, x_\infty) d_f(u_\infty, w_\infty). \tag{1.5.27}$$

So, f is a contraction of a complete metric space $\partial X \setminus \{x_\infty\}$ and, by the classical Banach theorem, it has a unique fix point y_∞. Moreover, if U and V are neighbourhoods of x_∞ and y_∞, then there exist positive numbers r and R such that $\partial X \setminus U \subset B_f(y_\infty, R)$ and $V \subset B_f(y_\infty, r)$, where B_f denotes a ball in the metric d_f. Condition (ii) follows immediately.

To prove (iii) one has just to observe that f cannot have two attracting fix points. Part (iv) becomes evident after one can notice that the equality

$$\Phi(f^n, x_\infty) = (\Phi(f, x_\infty))^n \tag{1.5.28}$$

follows directly from the definition (1.5.26) of the force. □

In classical hyperbolic geometry, the isometries of all three types (elliptic, hyperbolic and parabolic) may exist. For instance, the shift $z \to z+1$ in $\{z; \operatorname{re} z > 0\} \subset \mathbb{C}$ represents a parabolic isometry of the hyperbolic plane in the half-plane model. Hyperbolic groups, however, are very special from this point of view.

Theorem 1.5.16. *Hyperbolic groups contain no parabolic elements.*

Proof. If G is a hyperbolic group, X is its Cayley graph and $g \in G$ is non-elliptic, then $g^{n_i} \to x_\infty$ for some $n_1 < n_2 < \cdots$ and $x_\infty \in \partial X$. Obviously, $(g^{n_i}) \sim (g^{n_i+1})$, that is $g(x_\infty) = x_\infty$.

We claim that, given $a > 0$, there exist at most finitely many elements $h \in G$ such that

$$h(x_\infty) = x_\infty \quad \text{and} \quad e^{-a} \leq \Phi(h, x_\infty) \leq e^a. \tag{1.5.29}$$

To prove the claim take elements $h_1, \ldots, h_k$ satisfying (1.5.29) and a geodesic ray $\gamma : [0, \infty) \to X$ such that $\gamma(0) = e$ and $[\gamma]_\sim = x_\infty$. Put $\gamma_i = h_i \circ \gamma$ and find numbers $s_1, \ldots, s_k$ such that $d(\gamma(t), \gamma_i(t - s_i)) \leq 16\delta$ for all $t \geq t_0$ (see (1.5.8)). The definitions of the force, distances d'_ε and Busemann functions imply that there exists a constant $k_\varepsilon > 1$ such that

$$\frac{1}{k_\varepsilon} \cdot e^{\varepsilon\beta_\infty(x, f(x))} \leq \Phi(f, x_\infty) \leq k_\varepsilon \cdot e^{\varepsilon\beta_\infty(x, f(x))} \tag{1.5.30}$$

whenever f is an isometry of X fixing x_∞ at the boundary. Inequalities (1.5.29) and (1.5.30) imply that

$$|\beta_\infty(x, h_i(x))| \leq A, \quad i = 1, \ldots, k, \tag{1.5.31}$$

where $A = \varepsilon^{-1}(a + \log k_\varepsilon)$. Applying (1.5.31) to γ and γ_i we get the inequality $d(\gamma(t), \gamma_i(s)) \geq s - t - 2A$ for all $s \geq 0$. This shows that $|s_i| \leq b$, where $b > 0$ is a constant which depends only on δ, ε and a. Next, choose $m \in \mathbb{N}$ large enough, put $x = \gamma(m)$ and observe that $d(x, h_i(x)) \leq b + 16\delta$ for all $i \leq k$. It follows that $k \leq \#G_N$, where N is any natural number greater than $b + 16\delta$.

Now, if our element g were not hyperbolic, we would have $g^n(x_\infty) = x_\infty$ and $\Phi(g^n, x_\infty) = 1$ for all $n \in \mathbb{N}$. If so, $g^k = e$ for some k and g would be elliptic, a contradiction. □

We shall complete this section by

Theorem 1.5.17. *Any subgroup H of a hyperbolic group G either is finite, or contains an infinite cyclic subgroup of finite index, or contains a free group with two generators.*

To prove the Theorem we need the following classical fact known as the Ping-pong Lemma ([**Har**].

Lemma 1.5.18. *If f_1 and f_2 are two transformations of a set S which contains two nonempty subsets S_1 and S_2 such that $S_2 \setminus S_1 \neq \emptyset$, $f_1^n(S_2) \subset S_1$ and $f_2^n(S_1) \subset S_2$ for all $n \neq 0$, then the group generated by f_1 and f_2 is free.*

Proof. If $f = f_1^m \circ f_2^{m_1} \circ f_1^{m_2} \circ \cdots \circ f_2^{m_k} \circ f_1^n$ and $m, m_1, \ldots, m_k, n \neq 0$, then $f(S_2) \subset S_1$ and $f \neq \mathrm{id}$. If $f = f_2^m \circ f_1^{m_1} \circ f_2^{m_2} \circ \cdots \circ f_1^{m_k} \circ f_2^n$, then $f' = f_1 \circ f \circ f_1^{-1}$ maps S_2 to S_1, $f' \neq \mathrm{id}$ and $f = f_1^{-1} \circ f' \circ f_1 \neq \mathrm{id}$. If $f = f_1^m \circ f_2^{m_1} \circ f_1^{m_2} \circ \cdots \circ f_1^{m_k} \circ f_2^n$, then $f' = f_1^p \circ f \circ f_1^{-p}$, $p \neq 0, -m$, maps S_2 to S_1, and again $f' \neq \mathrm{id}$ and $f \neq \mathrm{id}$. The same holds when $f = f_2^m \circ f_1^{m_1} \circ f_2^{m_2} \circ \cdots \circ f_2^{m_k} \circ f_1^n$. □

To prove the theorem assume that H is infinite. Then H contains a hyperbolic element f_1. Let x_∞ and y_∞ be the points of ∂X, X being the Cayley graph of G, fixed by f_1. If all the elements of H fix both, x_∞ and y_∞, then the subgroup generated by f_1 is cyclic, infinite and has finite index in H. The same holds when all the elements of H leave the set $\{x_\infty, y_\infty\}$ invariant. If so, the subgroup of H consisting of all $f \in H$ for which $f(x_\infty) = x_\infty$ has index 2 and contains an infinite cyclic subgroup of finite index.

Therefore, if H contains no cyclic infinite subgroups of finite index, then there exists another hyperbolic element $f \in H$ such that $\{f(x_\infty), f(y_\infty)\} \cap \{x_\infty, y_\infty\} = \emptyset$. The element $f_2 = f f_1 f^{-1}$ leaves the points $f(x_\infty)$ and $f(y_\infty)$ fixed. Applying Lemma 1.5.18 to $S = \partial X$ and the unions S_1 and S_2 of pairwise disjoint neighbourhoods of the points x_∞, y_∞ and $f(x_\infty), f(y_\infty)$ we can see that the group generated by f_1^n and f_2^n, $n \in \mathbb{N}$ being sufficiently large, is free. □

Chapter 2

Growth

The type of growth of a non-decreasing function $f : \mathbb{N} \to [0, +\infty)$ (or, of a non-decreasing sequence of such functions) describes its asymptotic behaviour at infinity. The formal definitions which can be found in Section 2.1 of this chapter are consistent to some extent with our intuition: linear functions grow slower than quadratic ones, all the polynomials grow slower than exponential functions $n \mapsto a^n$, $a > 1$. However, all the exponential functions mentioned above have the same type of growth. This could be considered by a reader as a bit strange, but suits our purposes properly.

The type of growth of a finitely generated group is defined as the type of growth of functions counting the elements of the group obtained from words of a given length built of given generating elements. Similarly, the type of growth of an orbit $G(x)$ of a group (or, a pseudogroup) G acting on a space X ($x \in X$) is defined as the type of growth of the number of points of $G(x)$ which can be obtained from x by action of elements of G of a given length. It was shown by Milnor [**Mi1**] (compare also [**Sv**]) that the type of growth of the fundamental group of a compact Riemannian manifold M determines, to some extent, the geometry of M. In fact, this growth type is the same as that of volumes of balls in the universal covering $\tilde{M}$ of M. A similar phenomenon occurs in the theory of foliations: the growth type of volumes of balls on a leaf L of a foliation $\mathcal{F}$ of a compact Riemannian manifold M is the same as that of an orbit $\mathcal{H}(x)$, $x \in L$, of a holonomy pseudogroup $\mathcal{H}$ of $\mathcal{F}$ [**Pl**]. In dynamical systems and ergodic theory [**Wa**], so called expansion growth (i.e., the type of growth of a sequence of functions s_k counting the maximal numbers of points separated to the distance $\frac{1}{k}$, $k = 1, 2, \dots$) of a system is related to the well-known invariant, the entropy of the system.

In this chapter, we discuss the growth types mentioned above for groups, pseudogroups and foliations.

2.1 Growth types

Let us begin with two non-decreasing sequences (a_n) and (b_n) of non-negative numbers. We shall say that (a_n) "grows slower" than (b_n) ($(a_n) \preceq (b_n)$) whenever there exist positive constants A and B such that the inequalities

$$a_n \leq b_{Bn} \tag{2.1.1}$$

hold for all $n \in \mathbb{N}$. We say that *types of growth* of our sequences (a_n) and (b_n) are the same whenever

$$(a_n) \preceq (b_n) \preceq (a_n). \tag{2.1.2}$$

Such a definition would be enough for those who want to talk about growth of groups or volumes. Here, we need a bit more. In Chapter 3, we consider entropies of different systems which are defined as suitable limits of quantities indexed by two parameters, $n \in N$ and $\varepsilon \in \mathbb{R}_+$, as $n \to \infty$ and $\varepsilon \to 0$. Roughly speaking, such entropy is equal to the rate of exponential growth of these quantities. From this point of view, complexity of systems with zero entropy cannot be distinguished, one from another. However, we shall see in Section 2.4 that these quantities may grow differently even if entropies are the same (and equal to 0). For the purpose of more detailed distinction of type of growth of such quantities, we generalize the above notion of growth types to that for "sequences of sequences". This may seem to be a bit too technical and superfluously complicated but is useful as we shall see later.

Let now $\mathcal{T}$ be the set of non-negative increasing functions defined on $\mathbb{N}$:

$$\mathcal{T} = \{t : \mathbb{N} \to [0, \infty); t(n) \leq t(n+1) \text{ for all } n \in \mathbb{N}\}, \tag{2.1.3}$$

and $\hat{\mathcal{T}}$ the set of increasing sequences with entries in $\mathcal{T}$:

$$\hat{\mathcal{T}} = \{(t_j)_{j\in\mathbb{N}}; t_j \in \mathcal{T} \text{ and } t_j(n) \leq t_{j+1}(n) \text{ for all } j \text{ and } n \in \mathbb{N}\}. \tag{2.1.4}$$

Elements t of $\mathcal{T}$ can be identified with sequences $(t, t, \dots)$, so $\mathcal{T}$ can be considered as a subset of $\hat{\mathcal{T}}$.

A preorder $\preceq$ defined by the condition

(*) $(t_j)_{j\in\mathbb{N}} \preceq (\tau_k)_{k\in\mathbb{N}}$ if and only if there exists $b \in \mathbb{N}$ such that for all $j \in \mathbb{N}$ the inequalities

$$t_j(n) \leq a\tau_k(bn), \quad n \in \mathbb{N}, \tag{2.1.5}$$

hold for some $k \in \mathbb{N}$ and $a > 0$,

induces an equivalence relation $\simeq$ in $\hat{\mathcal{T}}$:

$$(t_j) \simeq (\tau_k) \text{ if and only if } (t_j) \preceq (\tau_k) \text{ and } (\tau_k) \preceq (t_j). \tag{2.1.6}$$

In particular, if t and $\tau \in \mathcal{T}$, then

$$t \preceq \tau \Leftrightarrow t(n) \leq a\tau(bn) \tag{2.1.7}$$

and

$$t \simeq \tau \Leftrightarrow a^{-1}\tau([n/b]) \leq t(n) \leq a\tau(bn) \tag{2.1.8}$$

for all $n \in \mathbb{N}$ and some $a, b \in \mathbb{N}$. (Here, $[x]$ is the largest integer which does not exceed x, $x \in \mathbb{R}$.)

Definition 2.1.1. Elements of the quotient $\hat{\mathcal{E}} = \hat{\mathcal{T}}/\simeq$ are called *growth types* [**Eg1**]. The growth type of $(t_j) \in \hat{\mathcal{T}}$ (resp., of $t \in \mathcal{T}$) is denoted by $[(t_j)]$ (resp., by $[t]$). Also, we let $\mathcal{E} = \{[t];\, t \in \mathcal{T}\}$. $\mathcal{E}$ is the set of growth types of monotone functions (in the sense of [**HH**]). The preorder $\preceq$ induces a partial order (denoted again by $\preceq$) in $\hat{\mathcal{E}}$.

Example 2.1.2. $[0] \preceq [1] \preceq [\log n] \preceq [n] \preceq [n^2] \preceq \cdots \preceq [(1, n, n^2, \ldots)] \preceq [2^n] \preceq [(1, 2^n, 3^n, \ldots)]$ and all the growth types listed above are different. The growth type of any polynomial of degree $d \geq 0$ is equal to $[n^d]$ and is called *polynomial* (of degree d). $[a^n] = [e^n]$ for any $a > 1$ and this growth type is called *exponential.* A growth type $[(t_j)]$ is said to be *quasi-exponential* if

$$\lim_{j\to\infty} \limsup_{n\to\infty} \frac{1}{n} \log t_j(n) > 0. \tag{2.1.9}$$

In particular, a growth type $[t] \in \mathcal{E}$ is quasi-exponential when

$$\limsup_{n\to\infty} \frac{1}{n} \log t(n) > 0. \tag{2.1.10}$$

If (2.1.9) (or, (2.1.10)) does not hold, the corresponding growth type is said to be *subexponential* or *quasi-polynomial* . Also, if

$$\lim_{j\to\infty} \liminf_{n\to\infty} \frac{1}{n} \log t_j(n) \leq 0, \tag{2.1.11}$$

then the growth type $[(t_j)]$ is called *non-exponential.*

It is easy to see that the following definitions of the sum and the product of growth types are correct:

$$[(t_j)] + [(\tau_j)] = [(t_j + \tau_j)], \quad [(t_j)] \cdot [(\tau_j)] = [(t_j\tau_j)]. \tag{2.1.12}$$

Also, the following relations are easy to check:

$$\theta \cdot [0] = [0] \text{ and } \theta + [0] = \theta \cdot [1] = \theta \text{ for any } \theta \in \hat{\mathcal{E}}, \tag{2.1.13}$$

$$\text{if } \theta_1 \preceq \xi_1 \text{ and } \theta_2 \preceq \xi_2, \text{ then } \theta_1 + \theta_2 \preceq \xi_1 + \xi_2 \text{ and } \theta_1\theta_2 \preceq \xi_1\xi_2, \tag{2.1.14}$$

$$\text{if } \theta_1 \leq \theta_2, \text{ then } \theta_1 + \theta_2 = \theta_2. \tag{2.1.15}$$

In particular,

$$\theta + [1] = \theta \text{ for any } \theta \in \hat{\mathcal{E}},\ \theta \neq [0]. \tag{2.1.16}$$

Certainly, the world of non-decreasing functions (and sequences of functions) under consideration is very large and one can find there elements with growth rates which cannot be compared in any reasonable way. However, the partial order $\preceq$ defined above is dense in the sense that if you have arbitrary (non-trivial) different types of growth which can be compared by this relation, then you can find the third one, different from and comparable with both of them, and located between them [**Bad**].

2.2 Growth in groups

Let G be a finitely generated group and G_1 a finite *symmetric* (i.e., such that $e \in G_1$ and $G_1^{-1} = \{g^{-1};\, g \in G_1\} \subset G_1$) set generating it. For any $n \in \mathbb{N}$ let

$$G_n = \{g_1 \cdot \dots \cdot g_n;\ g_1, \dots, g_n \in G_1\} \tag{2.2.1}$$

and

$$t_G(n) = \# G_n. \tag{2.2.2}$$

Obviously, $G_n \subset G_{n+1}$ for all n and $t_G \in \mathcal{T}$. If G_1' is another finite symmetric set of generators for G, then there exists $b \in \mathbb{N}$ such that $G_1' \subset G_b$ and $G_1 \subset G_b'$. For this b and all $n \in \mathbb{N}$ we have

$$G_n' \subset G_{bn},\ G_n \subset G_{bn}',\ t_G'(n) \le t_G(bn) \text{ and } t_G(n) \le t_G'(bn), \tag{2.2.3}$$

where $t_G'(k) = \# G_k'$. Consequently, $[t_G] = [t_G']$ and the following definition is correct.

Definition 2.2.1. The *growth type* $\mathrm{gr}(G)$ of G is defined as $[t_G]$, t_G being defined by (2.2.2) for any finite symmetric generating set G_1. If G acts on a space X and $x \in X$, then the *growth type* $\mathrm{gr}(G, x)$ of G at x is defined in a similar way: $\mathrm{gr}(G, x) = [t_x]$, where

$$t_x(n) = \#\{g(x);\ g \in G_n\} \tag{2.2.4}$$

for any fixed finite symmetric generating set G_1.

Clearly,

$$t_x(n) \le t_G(n) \tag{2.2.5}$$

for all n and x, and therefore

$$\mathrm{gr}(G, x) \preceq \mathrm{gr}(G) \tag{2.2.6}$$

for any x of X.

Example 2.2.2. A finite group has the growth type $[1]$ while the abelian group $\mathbb{Z}^d$ has the polynomial growth of degree d. Any free (non-abelian) group has the exponential growth $[e^n]$.

Proposition 2.2.3. *For any finitely generated group G and any normal subgroup H of G we have*

$$\mathrm{gr}(G/H) \preceq \mathrm{gr}(G) \preceq [e^n]. \tag{2.2.7}$$

Proof. If G_1 generates G, then $K_1 = \{gH;\, g \in G_1\}$ generates $K = G/H$ and $t_K(n) \leq t_G(n)$ for all n. This proves the first part of (2.2.7). Also, since $G_{n+m} \subset \bigcup_{g \in G_n} g(G_m)$, we can observe that

$$t_G(n+m) \leq t_G(n) \cdot t_G(n) \quad (n, m \in \mathbb{N}) \tag{2.2.8}$$

and therefore $t_G(n) \leq t_G(1)^n$. This proves the second part of (2.2.7). □

This proposition can be generalized as follows. Let

$$\begin{array}{ccccccccc} 1 & \longrightarrow & H' & \longrightarrow & H & \longrightarrow & H/H' & \longrightarrow & 1 \\ & & \downarrow & & \downarrow{\scriptstyle \iota} & & \downarrow & & \\ 1 & \longrightarrow & G' & \longrightarrow & G & \longrightarrow & G/G' & \longrightarrow & 1 \end{array}$$

be a commutative diagram of finitely generated groups with $G' = \iota(H')$, $\iota : H \to G$ being a homomorphism.

Proposition 2.2.4. *If ι is a monomorphism, then $\mathrm{gr}(H/H') \preceq \mathrm{gr}(G/G')$. If, moreover, $\iota(H)$ has finite index in G, then $\mathrm{gr}(H/H') = \mathrm{gr}(G/G')$.*

Proof. Choose a finite generating set H_1 of H and extend $\iota(H_1)$ to a finite generating set G_1 of G. The cosets hH' with $h \in H_1$ (resp., gG' with $g \in G_1$) generate H/H' (resp., G/G') while ι induces a monomorphism $\hat{\iota} : H/H' \to G/G'$ such that $\hat{\iota}(hH') = \iota(h)G'$ for all $h \in H$. Therefore, $t_{H/H'}(n) \leq t_{G/G'}(n)$ for all n and this proves the first part of the statement.

To get the second part, denote by d the index of $\iota(H)$ in G and choose elements $r_1 = e, r_2, \ldots, r_d$ representing all the cosets of G/H. For any $g \in G$ define $\phi(g) \in H$ by $\phi(g) = r_i^{-1} g$, where $i \leq d$ is the only index for which $r_i H = gH$. Then

$$\#\phi^{-1}(\{h\}) = d \quad \text{for any} \quad h \in H.$$

Next, choose a generating set G_1 of G and define H_1 by

$$H_1 = \{\phi(g \cdot r_j);\, g \in G_1, j \leq d\}.$$

H_1 generates H. In fact, if $g = g_1 \ldots g_k \in H$ and $g_1, \ldots, g_k \in G_1$, then g equals the product of all the elements of the form $\phi(g_j \cdot r_{\sigma(j)})$, $j = 1, \ldots, k$, where $\sigma(j) \leq d$ satisfies $g_{j+1} \cdot \cdots \cdot g_k H = r_{\sigma(j)} H$. (The above method of writing elements of H in terms of generators of H_1 is called the Reidemeister rewriting procedure, see [**MKS**], Theorem 2.7 and Corollary 2.7.2.) Moreover, the above shows that $\phi(G_n) \subset H_{n+n_0}$, where n_0 is chosen in such a way that $r_i \in G_{n_0}$ for all $i \leq d$. Therefore,

$$t_G(n) \leq d \cdot t_H(n + n_0)$$

and

$$\mathrm{gr}(H) = \mathrm{gr}(G).$$

Passing as before to the quotients G/G' and H/H' we obtain the required equality of growth types. □

Proposition 2.2.5. *If ι is an epimorphism, then $\mathrm{gr}(G/G') \preceq \mathrm{gr}(H/H')$. If, moreover, ι has finite kernel, then $\mathrm{gr}(G/G') = \mathrm{gr}(H/H')$.*

Proof. If H_1 generates H, then $G_1 = \iota(H_1)$ generates G. As before, the cosets hH' with $h \in H_1$ generate H/H', the cosets gG' with $g \in G_1$ generate G/G' and ι induces an epimorphism $\hat{\iota} : H/H' \to G/G'$. Since $\iota(H_n) = G_n$, $t_{G/G'}(n) \le t_{H/H'}(n)$ for all n. This proves the first part of the statement.

Now, if $m = \#\ker\iota < \infty$, then $\#\iota^{-1}(\{g\}) = m$ and $\#\hat{\iota}^{-1}(\{gG'\}) \le m$ for any $g \in G$. Therefore, $t_{H/H'}(n) \le m \cdot t_{G/G'}(n)$ for all n. This implies the relation $\mathrm{gr}(H/H') \preceq \mathrm{gr}(G/G')$ which, together with the first part of the statement, yields the required equality of growth types. □

Since any finitely generated abelian group G of rank d is a finite extension of $\mathbb{Z}^d$, Proposition 2.2.4 applied to the diagram

$$\begin{array}{ccccccccc} 0 & \longrightarrow & 0 & \longrightarrow & \mathbb{Z}^d & \longrightarrow & \mathbb{Z}^d & \longrightarrow & 0 \\ & & \downarrow & & \downarrow & & \downarrow & & \\ 0 & \longrightarrow & 0 & \longrightarrow & G & \longrightarrow & G & \longrightarrow & 0 \end{array}$$

yields the following.

Corollary 2.2.6. *Any finitely generated abelian group of rank d has the growth type $[n^d]$.*

Another consequence of the above concerns nilpotent groups.

Corollary 2.2.7. [**Wo2**] *Any finitely generated nilpotent group is of polynomial type of growth.*

Proof. Let G be finitely generated and nilpotent. Let $G(0) = G$ and $G(k+1) = [G, G(k)]$ for $k \ge 0$. Then, $G(m+1) = \{e\}$ for some $m \in \mathbb{N}$, each $G(k+1)$ is a normal subgroup of $G(k)$ and the quotients $G(k)/G(k+1)$ are abelian and finitely generated.

Choose a finite symmetric generating set $S_m \subset G(m)$. For each $k < m$ choose a finite set $\tilde{S}_k$ generating $G(k)/G(k+1)$ and for any $z \in \tilde{S}_k$ choose an element $g_z \in G_k$ representing z. Set $S_k = \{g_z; z \in \tilde{S}_k\}$ and $S = S_0 \cup S_1 \cup \cdots \cup S_m$.

The set S generates G. In fact, if $g \in G$, then $gG(1) = h_1 \cdot \cdots \cdot h_{k_1} \cdot G(1)$ for some $h_1, \ldots, h_{k_1} \in S_0 \subset S$. This means that $g = h_1 \cdot \cdots \cdot h_{k_1} g_1$ for some $g_1 \in G(1)$. Again, $g_1 G(2) = h_{k_1+1} \cdot \cdots \cdot h_{k_2} G(2)$ and $g_1 = h_{k_1+1} \cdot \cdots \cdot h_{k_2} g_2$ for some $h_{k_1+1}, \ldots, h_{k_2} \in S_1 \subset S$ and $g_2 \in G(2)$. Continuing by induction one can write g in the form $g = h_1 \cdot \cdots \cdot h_{k_m}$, where $h_j \in S$ for all j.

Moreover, an argument similar to that above shows that any element $g \in G$ of length (in the alphabet S) $\leq n$ can be written in the form

$$g = h_1 \cdot \cdots \cdot h_{k_2} h_{k_1+1} \cdot \cdots \cdot h_{k_2} \cdot \cdots \cdot h_{k_{m-1}+1} \cdot \cdots \cdot h_{k_m}$$

for some h_j and k_i such that $k_1 \leq k_2 \leq \cdots \leq k_m$, $k_{i+1} - k_i \leq n$ and $h_j \in S_i$ when $k_i + 1 \leq j \leq k_{i+1}$. Therefore,

$$t_G(n) \leq t_{G/G(1)}(n) \cdot \cdots \cdot t_{G(m-1)/G(m)}(n) \cdot t_{G(m)}(n)$$

and the result follows since by Corollary 2.2.6 all the functions $t_{G(k)/G(k+1)}$ (as well as $t_{G(m)}$) grow polynomially. □

The same reasoning allows us to improve the above to obtain

Corollary 2.2.8. *If a finitely generated group contains a nilpotent group of finite index, then it has a polynomial type of growth.* □

The converse is also true: in the remarkable paper [**Gro1**], Gromov proved that any finitely generated group of polynomial type of growth contains a nilpotent subgroup of finite index. Also, Wolf [**Wo2**] and Milnor [**Mi2**] proved that any finitely generated solvable group which contains no nilpotent subgroups of finite index has exponential growth.

Some very natural examples of finitely generated groups are provided by fundamental groups of compact manifolds. Below we show how the growth type of the fundamental group of a compact Riemannian manifold is related to its geometry. To this end let us consider a complete Riemannian manifold $(M, \langle\cdot,\cdot\rangle)$, fix a point $x \in M$ and let

$$t_M(n) = \operatorname{vol} B(x, n), \tag{2.2.9}$$

where $B(x, r)$ is the ball on M of radius r and centre x while vol is the measure on M induced by the Riemannian structure: if A is a subset of the domain U of a chart $\phi = (\phi_1, \ldots, \phi_m) : U \to \mathbb{R}^m$, $m = \dim M$, and (g_{ij}) is the matrix of components of $\langle\cdot,\cdot\rangle$ in the frame $(d\phi_1, \ldots, d\phi_m)$, $\langle\cdot,\cdot\rangle = \sum_{i,j} g_{ij} d\phi_i \otimes d\phi_j$ on U, then — by classical results of Riemannian geometry —

$$\operatorname{vol} A = \int_{\phi(A)} \sqrt{\det(g_{ij})} dx_1 \ldots dx_m. \tag{2.2.10}$$

If x' is another point of M, then $B(x', r) \subset B(x, r + r_0)$, where $r_0 = d(x, x')$ and d is the distance function on $(M, \langle\cdot,\cdot\rangle)$. Therefore, the function t'_M defined by (2.2.9) with x replaced by x' satisfies $t'_M(n) \leq t_M(n + n_0) \leq t_M(2n)$ when $n > n_0 \geq r_0$ and the following definition is correct.

Definition 2.2.9. $\operatorname{gr}(M) = [t_M]$ is called the *growth type* of a Riemannian manifold $(M, \langle\cdot,\cdot\rangle)$.

One of the very first and most important results relating growth types of Riemannian manifolds and groups is the following theorem due to Milnor [**Mi1**] and Svarc [**Sv**].

Theorem 2.2.10. *If M is compact, then the fundamental group $\pi_1(M)$ and the universal covering $\tilde{M}$ of M have the same growth type.*

Proof. First, fix a base point $x_0 \in M$ and a neighbourhood U of x covered regularly by $\pi : \tilde{M} \to M$. Choose $\tilde{x}_0 \in \pi^{-1}(\{x_0\})$ and let

$$l_0 = \max\{|g|;\ g \in G_1\}, \tag{2.2.11}$$

where G_1 is a finite symmetric set generating $G = \pi_1(M)$,

$$|g| = \inf\{l(\gamma);\ \gamma \in g\}, \tag{2.2.12}$$

and $l(\gamma)$ is the length of a curve γ on M. For any $g \in G_n$, $g \neq e$, the distance of $\tilde{x}_0$ and $g(\tilde{x}_0)$ does not exceed $l_0 n$ and the connected components of $\pi^{-1}(U)$ containing $\tilde{x}_0$ and $g(\tilde{x}_0)$ are disjoint. Therefore, if l_1 is any natural number greater than l_0, then

$$t_{\tilde{M}}(l_1 n) \geq t_G(n) \cdot \operatorname{vol} U \quad (n \in \mathbb{N}) \tag{2.2.13}$$

and

$$\operatorname{gr}(G) = [t_G] \preceq [t_{\tilde{M}}] = \operatorname{gr}(\tilde{M}). \tag{2.2.14}$$

Second, let $\tilde{M}_0 = B(\tilde{x}_0, \delta) \subset \tilde{M}$ be the closed ball of radius $\delta = \operatorname{diam} M$. All the sets $g(\tilde{M}_0)$, $g \in G$, cover $\tilde{M}$ and this covering is locally finite.

In fact, if $\tilde{B}(\tilde{x}_0, r)$, $r > 0$, were to intersect infinitely many sets of the form $g(\tilde{M}_0)$, then $\tilde{B}(\tilde{x}_0, \delta + r + \varepsilon)$ would contain infinitely many disjoint copies of the ball $\tilde{B}(\tilde{x}_0, \varepsilon)$, each of them of the same volume $V(\varepsilon) > 0$. Hence, $\tilde{B}(\tilde{x}_0, \delta + r + \varepsilon)$ would have infinite volume. Here, $\varepsilon > 0$ is so small that the ball $\tilde{B}(\tilde{x}_0, \varepsilon)$ is covered regularly by π.

Let $H = \{g \in G;\ g(\tilde{M}_0) \cap \tilde{M}_0 \neq \emptyset\}$ and $\eta = \min\{\tilde{d}(\tilde{M}_0, g(\tilde{M}_0));\ g \notin H\}$. Assume that $\tilde{d}(\tilde{x}_0, g(\tilde{M}_0)) < \eta n + \delta$ for some $n \in \mathbb{N}$. Choose $\tilde{x} \in g(\tilde{M}_0)$ with $\tilde{d}(\tilde{x}_o, \tilde{x}) < \eta n + \delta$ and such points $\tilde{x}_0 = y_0, y_1, \ldots, y_{n+1} = \tilde{x}$ along the minimal geodesic joining $\tilde{x}$ to $\tilde{x}_0$ that $\tilde{d}(y_i, y_{i+1}) < \eta$ for $i = 1, \ldots, n$. For any i, $y_i \in h_i(\tilde{M}_0)$, where $h_i \in G$, $h_0 = e$ and $h_{n+1} = g$. If $g_i = h_i^{-1} h_{i+1}$, then $g = g_1 \cdot \cdots \cdot g_n$ and $g_i \in H$ since both $h_i^{-1}(y_i)$ and $h_i^{-1}(y_{i+1})$ lie in $\tilde{M}_0 \cap h_i(\tilde{M}_0)$ while $\tilde{d}(h_i^{-1}(y_i), h_i^{-1}(y_{i+1})) < \eta$. Consequently,

$$\operatorname{vol} \tilde{B}(\tilde{x}_0, \eta n + \delta) \leq t_G(kn) \cdot \operatorname{vol} \tilde{B}(x_0, \delta), \tag{2.2.15}$$

where $k \in \mathbb{N}$ is such that $H \subset G_k$. In other words,

$$t_{\tilde{M}}(n) \leq \operatorname{Vol} \tilde{B}(\tilde{x}_0, \delta) \cdot t_G(([k/2\eta] + 1)\, n) \tag{2.2.16}$$

for n large enough, $n > 2\delta$, and finally

$$\operatorname{gr}(\tilde{M}) = [t_{\tilde{M}}] \leq [t_G] = \operatorname{gr}(G). \tag{2.2.17}$$

□

The argument in the proof above shows in fact the following.

Corollary 2.2.11. *The universal covering $\tilde{M}$ of any compact Riemannian manifold M is quasi-isometric to the fundamental group $\pi_1(M)$ (i.e., to its Cayley graph).* □

Moreover, well-known estimates of volume on Riemannian manifolds of negative curvature (see, for example, [**BC**]) yield

Corollary 2.2.12. *If M is a compact Riemannian manifold of negative sectional curvature, then both $\tilde{M}$ and $\pi_1(M)$ have exponential growth.*

Proof. If the sectional curvature K of M satisfies

$$-b^2 \leq K \leq -a^2 < 0,$$

then the volume $V(r)$ of any ball of radius r in the universal covering $\tilde{M}$ of M satisfies

$$V_a(r) \leq V(r) \leq V_b(r), \tag{2.2.18}$$

where $V_c(r)$ is the volume of a similar ball in the space form of constant curvature $-c^2$:

$$V_c(r) = nc^{1-n}\omega_n \cdot \int_0^r (\sinh ct)^{n-1}dt, \tag{2.2.19}$$

ω_n being the volume of the unit ball in $\mathbb{R}^n$. If r is large enough, then

$$\int_0^r (\sinh at)^{n-1}dt \geq \frac{1}{2^n(n-1)a} \cdot e^{(n-1)ar}. \tag{2.2.20}$$

Also,

$$\int_0^r (\sinh bt)^{n-1}dt \leq \frac{1}{(n-1)b} \cdot e^{(n-1)br}. \tag{2.2.21}$$

From (2.2.18)–(2.2.21) we get the estimates

$$\frac{\omega_n}{(2a)^n(n-1)} \cdot e^{(n-1)ar} \leq V(r) \leq \frac{\omega_n}{b^n(n-1)} \cdot e^{(n-1)br} \tag{2.2.22}$$

which imply that

$$\mathrm{gr}(\pi_1(M)) = \mathrm{gr}(\tilde{M}) = [e^n].$$ □

Using the results of Section 1.5 we can generalize the above as follows.

Corollary 2.2.13. *Any non-elementary hyperbolic group has exponential growth.* □

Definition 2.2.14. Riemannian manifolds M and M' are *Lipschitz equivalent* if there exists a diffeomorphism f of M onto M' and a constant $c \geq 1$ such that

$$\frac{1}{c}|v| \leq |df(v)| \leq c|v| \tag{2.2.23}$$

for all $v \in TM$.

Inequality (2.2.23) is equivalent to

$$\frac{1}{c}d(x,y) \leq d'(f(x), f(y)) \leq cd(x,y) \quad (x, y \in M) \tag{2.2.24}$$

where d and d' denote the Riemannian distance functions on M and M', respectively. Therefore, Lipschitz equivalent Riemannian manifolds are quasi-isometric in the sense of Definition 1.5.5.

Proposition 2.2.15. *If M and M' are Lipschitz equivalent, then* $\operatorname{gr}(M) = \operatorname{gr}(M')$.

Proof. Passing to two-fold coverings, if necessary, we may assume that both M and M' are orientable. Let ω and ω' be the Riemannian volume forms on M and M', respectively. Then $f^*\omega' = \mu\omega$ for some $\mu : M \to \mathbb{R}$. If $(e_1, \ldots, e_m)$, $m = \dim M = \dim M'$, is an orthonormal frame of T_xM, then

$$|\mu(x)| = |f^*\omega'(e_1, \ldots, e_m)| = |\omega'(df(e_1), \ldots, df(e_m)| \leq c^m. \tag{2.2.25}$$

Moreover,

$$B_{M'}(f(x), c^{-1}r) \subset f(B_M(x, r)) \tag{2.2.26}$$

for all $x \in M$ and $r > 0$. Consequently,

$$\operatorname{vol} B_{M'}(f(x), c^{-1}r) \leq c^m \cdot \operatorname{vol} B_M(x, r). \tag{2.2.27}$$

This implies the relation $\operatorname{gr}(M') \preceq \operatorname{gr}(M)$. The converse holds because of the same reason. □

Let us recall now that a Riemannian manifold M has *bounded geometry* when its sectional curvature is bounded and the injectivity radius of its exponential map has a strictly positive lower bound all over M.

Proposition 2.2.16. *If complete Riemannian manifolds M and M' have bounded geometry and are quasi-isometric, then* $\operatorname{gr}(M) = \operatorname{gr}(M')$.

Proof. Since M and M' have bounded geometry, for any $r > 0$ there exist positive constants $v(r)$ and $V(r)$ such that the volumes of all the balls of radius r on M and M' are contained in the interval $[v(r), V(r)]$.

Let $f : M \to M'$ and $f' : M \to M'$ be quasi-isometries satisfying all the conditions of Definition 1.5.5 with $c = \lambda = \varepsilon$. Let $x \in M$ and $x' = f(x)$. Choose $R > 0$ arbitrarily large and a maximal (2ε)-separated subset $X = \{x'_1, \ldots, x'_m\}$ of the ball $B_{M'}(x', R)$. Then, the balls $B_{M'}(x'_i, \varepsilon)$, $i = 1, \ldots, m$, are pairwise disjoint while the balls $B_{M'}(x'_i, 2\varepsilon)$, $i = 1, \ldots, m$ cover the ball $B_{M'}(x', R)$. Therefore

$$\operatorname{vol} B_{M'}(x', R) \leq m \cdot V(2\varepsilon). \tag{2.2.28}$$

Let $x_i = f'(x'_i)$ for all $i \leq m$. Since f' is ε-quasi-isometric, the balls $B_M(x_i, 1)$ are pairwise disjoint. Also,

$$d(x, x_i) \leq d(x, f'(x')) + d(f'(x'), f'(x'_i)) \leq \varepsilon(R + 2),$$

i.e., $x_i \in B_M(x, \varepsilon(R+2))$. Therefore,

$$\operatorname{vol} B_M(x, \varepsilon(R+2)) \geq m \cdot v(1). \tag{2.2.29}$$

Comparing (2.2.28) and (2.2.29) we obtain

$$\operatorname{vol} B_{M'}(x', R) \leq \frac{V(2\varepsilon)}{v(1)} \cdot \operatorname{vol} B_M(x, 2\varepsilon R)$$

for any R sufficiently large. This yields the relation $\operatorname{gr}(M') \preceq \operatorname{gr}(M)$. Similarly, $\operatorname{gr}(M) \preceq \operatorname{gr}(M')$. □

In the same way, if two finitely generated groups G and G' are quasi-isometric, then they have the same growth types. It is easy to guess that the converse is not true.

Example 2.2.17. Let G be the Heisenberg group, i.e., the group of all the matrices of the form

$$\begin{pmatrix} 1 & k & l \\ 0 & 1 & m \\ 0 & 0 & 1 \end{pmatrix} \tag{2.2.30}$$

with k, l and $m \in \mathbb{Z}$. G is obviously isomorphic to $\mathbb{Z}^3$ equipped with the multiplication given by

$$(k, l, m) \cdot (k', l', m') = (k + k', l + l' + km', m + m'). \tag{2.2.31}$$

Also, it is generated by the elements $a = (1, 0, 0)$, $b = (0, 0, 1)$ and $c = (0, 1, 0)$ which satisfy the relations

$$ac = ca, \; bc = cb \;\text{ and }\; c = aba^{-1}b^{-1}. \tag{2.2.32}$$

Let G_1 be the smallest symmetric generating set containing a, b and c. It is easy to see from (2.2.31) and (2.2.32) that if $g = (k, l, m) \in G_n$, then $|k| \leq n$, $|m| \leq n$ and $|l| \leq n^2$. Therefore, if $a^k b^m c^l \in G_n$, then $|k| \leq n$, $|m| \leq n$ and $|l| \leq 2n^2$. It follows that, $\#G_n \leq (2n+1)^2(2n^2+1) \leq (2n+1)^4$. Also, (2.2.31) implies the relation

$$c^{n^2} = a^n b^n a^{-n} b^{-n} \in G_{4n}.$$

Therefore, G_{8n} contains all the elements of the form $a^k b^m c^l$ with $|k| \leq n$, $|m| \leq n$ and $|l| \leq n^2$, and consequently $\#G_{8n} \geq (2n+1)^2(2n^2+1) > 8n^4$. This shows that G has polynomial growth of degree 4.

We encourage the reader to prove directly that G is not quasi-isometric to $\mathbb{Z}^4$. This fact can be extracted also from the Pansu results: G and $\mathbb{Z}^4$ are nilpotent of finite type, to each group Γ of this sort one can associate a nilpotent simply connected Lie group $\tilde{\Gamma}$ which depends only on the quasi-isometry class of Γ ([**Pan**], Theorem 3). Clearly, $\tilde{\mathbb{Z}^4} = \mathbb{R}^4$ while

$$\tilde{G} = \left\{ \begin{pmatrix} 1 & x & y \\ 0 & 1 & z \\ 0 & 0 & 1 \end{pmatrix}, \ x, y, x \in \mathbb{R} \right\}.$$

Let us note that related problems for lattices in Lie groups are situated among the hottest areas in geometric group theory. Interested readers can consult for example the survey [**Es**] and references there.

Remark 2.2.18. Growth of a finitely generated group G can be described more precisely in terms of the *growth function* f (with respect to a given finite generating set G_1) defined by the power series

$$f(z) = \sum_{n=0}^{\infty} a_n z^n, \tag{2.2.33}$$

where a_n is the number of elements g of G satisfying $|g| = n$ and $|g| = \min\{n; g \in G_n\}$ is the length of g in the alphabet G_1. It is known that f is rational when G is either hyperbolic ([**GH**], p. 174) or virtually abelian [**Ben1**]. The growth function f has been computed explicitly for several groups like Coxeter groups ([**Bou**], [**Par**] etc.), Fuchsian groups ([**CW**], [**FP**] etc.) and some nilpotent groups ([**Ben2**], [**Sh**] etc.). Recently, the *complete growth function* F,

$$F(z) = \sum_{g \in G} g z^{|g|}, \tag{2.2.34}$$

was defined as an element of $\mathbb{Z}[G][z]$, the ring of formal series with coefficients in the group ring $\mathbb{Z}[G]$, and has been shown to be rational (in a natural sense) when G is hyperbolic [**GrN**]. From (2.2.33) and (2.2.34) it follows immediately that the homomorphism $h : \mathbb{Z}[G] \to \mathbb{Z}$ induced by the map $G \ni g \mapsto 1$ gives rise to the homomorphism $\tilde{h} : \mathbb{Z}[G][z] \to \mathbb{Z}[z]$ such that $\tilde{h} \circ F = f$. Therefore, rationality of F yields immediately rationality of f. All these results belong to geometric group theory. An interested reader is referred to [**GrH**] for a recent review of related problems.

2.3 Orbit growth for pseudogroups

If $\mathcal{G}$ is a finitely generated pseudogroup and $\mathcal{G}_1 \subset \mathcal{G}$ is a finite symmetric generating set, then one can define the growth function $t_{\mathcal{G}_1,\mathcal{G}}$ by

$$t_{\mathcal{G}_1,\mathcal{G}}(n) = \#\mathcal{G}_n, \tag{2.3.1}$$

where, similarly to (2.2.1),

$$\mathcal{G}_n = \{g_1 \circ \cdots \circ g_n; \, g_1, \ldots, g_n \in \mathcal{G}_1\}. \tag{2.3.2}$$

The example below shows that the growth type $[t_{\mathcal{G}_1,\mathcal{G}}]$ depends strongly on the choice of $\mathcal{G}_1$. Therefore, talking about the growth type of a pseudogroup might be deceiving.

Example 2.3.1. Let $\mathcal{G} = \mathcal{G}(f) = \mathcal{G}(f_0, f_1)$, where $f : X \to X$ is a homeomorphism, $f_i = f|U_i$, U_0 and U_1 are open subsets covering X. If $\mathcal{G}_1 = \{\mathrm{id}, f, f^{-1}\}$ and $\mathcal{G}_1' = \{\mathrm{id}, f_0, f_0^{-1}, f_1, f_1^{-1}\}$, then it might happen that $[t_{\mathcal{G},\mathcal{G}_1}] = [n]$ and $[t_{\mathcal{G},\mathcal{G}_1'}] = [2^n] = [e^n]$. In fact, $f_{i_1} \circ \cdots \circ f_{i_n} = f^n|U_{i_1,\dots,i_n}$, $U_{i_1,\dots,i_n} = f^{-(n-1)}(U_{i_1}) \cap \cdots \cap f^{-1}(U_{i_{n-1}}) \cap U_{i_n}$, and all the sets $U_{i_1,\dots,i_n}$, $i_k \in \{0,1\}$, might be different. This happens, for example, when $X = S^1$, $f = R_\alpha$ is a minimal rotation, $U_i = S^1 \setminus \{x_i\}$, and the points x_0 and x_1 are independent over α (i.e., $x_1 - x_0 \notin \alpha\mathbb{Z}$).

Also, the type of growth of orbits of general pseudogroups may be not well defined, i.e., it may depend on the choice of generators. This is due mostly to the possibility of "gluing" maps in pseudogroups (property (iv) in Definition 1.1.1). So, if we would try to define it as the type of growth of the function $n \mapsto \mathcal{G}_n(x)$, then one could observe that, due to the fact that domains of maps of $\mathcal{G}$ are open in X, it might happen that, given $g \in \mathcal{G}$, the length of the shortest word in alphabet $\mathcal{G}_1$ expressing g in a neighbourhood of $x \in D_g$ escapes to infinity when x approaches the boundary of D_g. Consequently, the types of growth of functions $x \mapsto \mathcal{G}_n(x)$ and $n \mapsto \mathcal{G}_n'(x)$ could be *a priori* different for different sets $\mathcal{G}_1$ and $\mathcal{G}_1'$ generating $\mathcal{G}$. Therefore, to define orbit growth type properly we need some special assumption about either $\mathcal{G}$ or a generating set.

In some particular cases this type is well defined but the proof of this fact may be surprisingly complicated and technical. This happens in the following

Proposition 2.3.2. [Le] *If $\mathcal{G}$ is a pseudogroup of local isometries of the circle S^1 generated by a finite set $\mathcal{G}_1$ of rotations $h_i : (x_i, y_i) \to (x_i + \alpha_i, y_i + \alpha_i)$, $i = 1, \dots, m$, and $x \in S^1$, then the type of the growth of the orbit $\mathcal{G}(x)$ is independent of the choice of such a generating system $\mathcal{G}_1$.*

To prove the above we need the following.

Lemma 2.3.3. *If x and y are points of S^1 and ε is a positive number such that $y+t \in \mathcal{G}(x+t)$ for all $t \in (0,\varepsilon)$, then there exists $\delta > 0$ and an element $g \in \bigcup_n \mathcal{G}_n$ such that $(x, x+\delta) \subset D_g$ and $g((x, x+\delta)) = (y, y+\delta)$.*

Proof. If $y \in \mathcal{G}(x)$, the statement is obvious. So, assume that

$$y \notin \mathcal{G}(x). \tag{2.3.3}$$

First, consider the set $A = \{x_i, x_i + \alpha_i; i = 1, \dots, m\}$. Since A is finite, there exists $\eta \in (0, \varepsilon)$ such that for any points u and v of A with $v \in \mathcal{G}(u)$ there exists $\gamma \in \bigcup_n \mathcal{G}_n$ such that

$$[u - \eta, u] \subset D_\gamma \quad \text{and} \quad \gamma(u) = v. \tag{2.3.4}$$

Next, consider the set $\tilde{\mathcal{G}}$ of all $g \in \bigcup_n \mathcal{G}_n$ for which there exist numbers $t = t_g$ and $t' = t'_g$ such that $0 \le t < t' < \eta$, $(x+t, x+t') \subset D_g$, $x + t \notin D_g$ and $g((x+t, x+t')) = (y+t, y+t')$. By (2.3.3), $\tilde{\mathcal{G}} \neq \emptyset$.

For any $g \in \tilde{\mathcal{G}}$, $g = f_1 \circ \cdots \circ f_p \in \mathcal{G}_p$ for some $f_j = h_{i_j} \in \mathcal{G}_1$, let $c(g) = \#\{j; (x, x+t'_g) \not\subset D_{f_j \circ \cdots \circ f_p}\}$. We claim that

$$\min\{c(g); g \in \tilde{\mathcal{G}}\} = 0. \tag{2.3.5}$$

In fact, if $g \in \tilde{\mathcal{G}}$, $c(g) > 0$,

$$r = \max\{j; f_{j+1} \circ \cdots \circ f_p(x + t_g) \notin D_{f_j}\}$$

and

$$s = \min\{j; f_{j-1}^{-1} \circ \cdots \circ f_1^{-1}(y + t_g) \notin D_{f_j^{-1}}\},$$

then $t_g > 0$ (obviously), $r \ge s$ (otherwise, $f_{s+1} \circ \cdots \circ f_p(x+t_g) \in D_{f_s}$, $x + t_g \in D_{f_s \circ \cdots \circ f_p}$ and $y + t_g \in D_{(f_1 \circ \cdots \circ f_{r-1})^{-1}} \subset D_{(f_1 \circ \cdots \circ f_{s-1})^{-1}}$, so $x + t_g \in D_{f_1 \circ \cdots \circ f_p}$ contradicting the definition of $\tilde{\mathcal{G}}$) and $c(g) \ge r$ (by similar arguments). Let now

$$u = f_{r+1} \circ \cdots \circ f_p(x + t_g) \quad \text{and} \quad v = f_{s-1}^{-1} \circ \cdots \circ f_1^{-1}(y + t_g).$$

Then, $v \in \mathcal{G}(u)$ (since $y + t \in \mathcal{G}(x+t)$). Let γ be an element of $\bigcup_n \mathcal{G}_n$ satisfying (2.3.4) and $g' = f_1 \circ \cdots \circ f_{s-1} \circ \gamma \circ f_{r+1} \circ \cdots \circ f_p \in \tilde{\mathcal{G}}$. Now, $[x+t_g, x+t'_g) \subset D_{f_{r+1} \circ \cdots \circ f_p}$ and, since $0 < t' < \eta$, $(x, x+t'') \subset D_{\gamma \circ f_{r+1} \circ \cdots \circ f_p}$ for some $t'' > 0$. Consequently, $c(g') \le s - 1 < c(g)$ and this proves (2.3.5).

Obviously, (2.3.5) implies the statement of our lemma. □

Let now $\mathcal{G}'_1$ be another generating system for $\mathcal{G}$ and let $g' \in \mathcal{G}'_1$. Lemma 2.3.3 implies that there exists a finite covering $U_1, \ldots, U_m$ of $D_{g'}$ such that $g'|U_j = \gamma_j|U_j$ for some $\gamma_j \in \mathcal{G}_{n(j)}$, $j = 1, \ldots, m$. Let $n(g') = \max\{n(j); j \le m\}$ and $n_0 = \max\{n(g'); g' \in \mathcal{G}'_1\}$. Then, for any $x \in S^1$ and $n \in \mathbb{N}$,

$$\mathcal{G}'_n(x) \subset \mathcal{G}_{n_0 n}(x). \tag{2.3.6}$$

Relation (2.3.6) shows that the growth of $\mathcal{G}(x)$ defined by $\mathcal{G}'_1$ does not exceed that defined by $\mathcal{G}_1$. The opposite relation holds accordingly. This completes the proof of Proposition 2.3.2. □

Note that the above result is valid for pseudogroups of local isometries of S^1 only. In the proof of Lemma 2.3.3, we strongly use the order of points along the circle as well as the fact that the maps of $\mathcal{G}$ preserve the length of arcs. As far as we know, the analogous problem for pseudogroups of local affine maps of S^1 is still open (and of some interest). In the case of arbitrary pseudogroups, the situation is even more complicated. Therefore, we shall accept the following definitions.

Definition 2.3.4. As before, let $\mathcal{G}_1$ be a finite symmetric set generating a pseudogroup $\mathcal{G}$ on X. For any $g \in \mathcal{G}_1$ choose a compact set $K_g \subset D_g$ in such a way that the maps $g|\operatorname{int} K_g$, $g \in \mathcal{G}_1$ still generate $\mathcal{G}$. If this is possible, $\mathcal{G}_1$ is said to be a *good* set of generators. Pseudogroups which admit good generating sets are also called *good*.

Clearly, any finitely generated group of transformations of a compact space is good. Also, the holonomy pseudogroup $\mathcal{H}$ corresponding to a nice covering $\mathcal{U}$ of a compact foliated manifold $(M, \mathcal{F})$ is good.

Let

$$\mathcal{G}_n^c = \{(g_1|K_{g_1}) \circ \cdots \circ (g_n|K_{g_n});\ g_i \in \mathcal{G}_1\} \tag{2.3.7}$$

and

$$t_{\mathcal{G},x}(n) = \#\{g(x);\ g \in \mathcal{G}_n^c \text{ and } x \in D_g\} \tag{2.3.8}$$

for $n = 1, 2, \ldots$ and $x \in X$. If $\mathcal{G}_1'$ is another good generating set, $g \in \mathcal{G}_1$ and $x \in K_g$, then $g|U(x) = g'|U(x)$, where $U(x)$ is a neighbourhood of x and $g' \in \mathcal{G}_{k(x)}'$. Covering K_g by finitely many such neighbourhoods $U(x_1), \ldots, U(x_m)$, putting $k_g = \max\{k(x_1), \ldots, k(x_m)\}$ and $k = \max\{k_g;\ g \in \mathcal{G}_1\}$ we can see that

$$\{g(x);\ g \in \mathcal{G}_n^c \text{ and } x \in D_g\} \subset \{g'(x);\ g' \in \mathcal{G}_{kn}'^c \text{ and } x \in D_{g'}\} \tag{2.3.9}$$

for all $n \in \mathbb{N}$. Therefore,

$$t_{\mathcal{G},x}(n) \leq t'_{\mathcal{G},x}(kn) \quad \text{and} \quad [t_{\mathcal{G},x}] \preceq [t'_{\mathcal{G},x}], \tag{2.3.10}$$

where $t'_{\mathcal{G},x}$ is defined by (2.3.8) with $\mathcal{G}_n^c$ replaced by $\mathcal{G}_n'^c$. Consequently, the following definition is correct.

Definition 2.3.5. The *growth type* of $\mathcal{G}$ at x is defined by

$$\operatorname{gr}(\mathcal{G}, x) = [t_{\mathcal{G},x}], \tag{2.3.11}$$

where $t_{\mathcal{G},x}$ is given by (2.3.7) and (2.3.8) for any good generating set $\mathcal{G}_1$.

Analogously to (2.2.8) and (2.2.7) we have

$$t_{\mathcal{G},x}(n+m) \leq t_{\mathcal{G},x}(n) \cdot t_{\mathcal{G},x}(m) \tag{2.3.12}$$

and

$$\operatorname{gr}(\mathcal{G}, x) \preceq [e^n]. \tag{2.3.13}$$

Also, if $y = g(x)$ for some $g \in \mathcal{G}$, then

$$\operatorname{gr}(\mathcal{G}, y) = \operatorname{gr}(\mathcal{G}, x). \tag{2.3.14}$$

In fact, $g|U = g'|U$ for some neighbourhood U of x and $g' \in \mathcal{G}_k^c$; therefore, $t_{\mathcal{G},x}(n) \geq t_{\mathcal{G},y}(n+k)$, $n = 1, 2, \ldots$, and $\operatorname{gr}(\mathcal{G}, y) \preceq \operatorname{gr}(\mathcal{G}, x)$. The converse inequality holds as well. Therefore, we can speak about the type of growth of the orbits of $\mathcal{G}$ instead of the growth type of $\mathcal{G}$ at the points of X.

Remark 2.3.6. In [**Hae4**] Haefliger discussed the notion of compact generation. A pseudogroup $\mathcal{G}$ on X is *compactly generated* when (1) X is locally compact and contains a relatively compact open subset X_0 meeting all the orbits of $\mathcal{G}$, and (2) $\mathcal{G}$ contains a finite subset $\mathcal{G}_0$ generating $\mathcal{G}|X_0$ and such that any element $g \in \mathcal{G}_0$ can be extended to a map $\tilde{g} \in \mathcal{G}$ defined on the domain $D_{\tilde{g}}$ which contains the closure of D_g. Evidently, good pseudogroups are compactly generated (take $X_0 = \bigcup\{K_g;\, g \in \mathcal{G}_1\}$ and $\mathcal{G}_0 = \{g|\operatorname{int} K_g;\, g \in \mathcal{G}_1\}$, where $\mathcal{G}_1$ is a good generating set) but the converse is not true (for example, the pseudogroup $\mathcal{G}(f)$, where $f(t) = t^3$ for all $t \in \mathbb{R}$, is compactly generated but it is not good). As was observed in [**Hae4**], the growth type of orbits can be defined for compactly generated pseudogroups as the growth type of orbits of $\mathcal{G}|X_0$ (with respect to the generating set $\mathcal{G}_0$). The argument showing that such a type of growth is well defined is essentially the same as that above. In fact, several notions and results discussed hereafter could be formulated in the case of compact generation.

For the holonomy pseudogroup $\mathcal{H}$ of a compact foliated manifold $(M, \mathcal{F})$ without boundary we have the following, analogous to Theorem 2.2.10, result.

Proposition 2.3.7. $\operatorname{gr}(\mathcal{H}, x) = \operatorname{gr}(L)$, *where L is the leaf of $\mathcal{F}$ passing through x equipped with any Riemannian metric inherited from M.*

Proof. Let $\mathcal{U}$ be the nice covering defining $\mathcal{H}$, $\langle\cdot,\cdot\rangle$ a Riemannian metric on M and T the complete transversal with $\mathcal{H}$ acting on it. Since M is compact, there exists $\varepsilon \in (0,1)$ such that the leaf distance $d_L(x,y)$ (defined by the Riemannian structure on L induced from $\langle\cdot,\cdot\rangle$) of arbitrary points x and y in $L \cap T$, $x \neq y$, satisfies the inequality $d_L(x,y) > 2\varepsilon$. Also, $v(\varepsilon) = \min\{\operatorname{vol} B_{L'}(x,\varepsilon);\, x \in L'$ and L' is a leaf of $\mathcal{F}\}$ is positive and $\delta = \max\{\operatorname{diam} P;\ P$ is a plaque of $\mathcal{U}\}$ is finite. If $x \in L$, $n \in \mathbb{N}$ and $h \in \mathcal{H}_n$, then $d_L(x, h(x)) \leq n\delta$ and therefore

$$\begin{aligned} v(\varepsilon)\cdot t_{\mathcal{H},x}(n) &\leq \sum_{h\in\mathcal{H}_n\cap\{h;\, x\in D_h\}} \operatorname{vol} B_L(h(x),\varepsilon) \\ &\leq \operatorname{vol} B_L(x, n\delta+\varepsilon) \leq t_L(n([\delta+1]+1)). \end{aligned} \tag{2.3.15}$$

Now, let $\lambda > 0$ be a Lebesgue number of the covering of M by the plaques: if $x, y \in L'$, L' is a leaf of $\mathcal{F}$ and $d_{L'}(x,y) < \lambda$, then there exists a chart $U \in \mathcal{U}$ and a plaque $P \subset U$ such that x and y lie in P. Also, let $V(\delta) = \max\{\operatorname{vol} B_{L'}(z,\delta);\, z \in L'$ and L' is a leaf$\}$. Since M is compact, $V(\delta)$ is finite. For any $x \in L$ and any $n \in \mathbb{N}$ we have

$$\operatorname{vol} B_L(x,n) \leq \sum_{h\in\mathcal{H}_{k(n)}\cap\{h;\, x\in D_h\}} \operatorname{vol} B_L(h(x),\delta) \leq V(\delta)\cdot t_{\mathcal{H},x}(k(n)), \tag{2.3.16}$$

where $k(n) = [n/\lambda] + 1$.

Inequalities (2.3.15) and (2.3.16) prove our statement. □

Corollary 2.3.8. *If a Riemannian manifold N is quasi-isometric to a leaf L of a foliation $\mathcal{F}$ of a compact manifold M, then*

$$\mathrm{gr}(N) \preceq [\exp]. \tag{2.3.17}$$

□

Equality in (2.3.17) may hold. To see this let us recall the following

Definition 2.3.9. A leaf L of a codimension-one foliation $\mathcal{F}$ is said to be *resilient* if there exists a holonomy map $f \in \mathcal{H}$ defined in a neighbourhood $U \subset T$ of a point $x \in L \cap T$ and another point $y \in L \cap U$ such that $y \neq x$,

$$f(x) = x \text{ and } f^n(y) \to x \text{ when } n \to \infty. \tag{2.3.18}$$

In other words, L is resilient whenever it contains a loop Γ and two points x and y such that $x \in \Gamma \cap T$, $y \in T$, $x \neq y$ and $y_n = h_\gamma^n(y) \to x$ as $n \to \infty$. The term "resilient" comes from the French word "*ressort*" which means that our leaf has the shape (not the properties) of a spring (Figure 2.1). This translation is not perfect, some authors (see, for instance, [**CaC1**]) call it "unfortunate".

Recall also that, due to a classical Sacksteder result ([**Sa**], see also [**HH**]), any exceptional minimal set on a compact C^2-foliated manifold $(M, \mathcal{F})$ contains a resilient leaf. Therefore, resilient leaves do exist.

If $\mathcal{F}$ is transversely oriented and L is resilient, $\mathcal{H}$ contains a subpseudogroup $\mathcal{H}'$ isomorphic to a pseudogroup generated by two local homeomorphisms g and h of $\mathbb{R}$ such that

$$h : [a, b] \to (b, c], \;\; g : [a, c] \to [a, b) \text{ and } g(a) = a \tag{2.3.19}$$

for some a, b and $c \in \mathbb{R}$, $a < b < c$. In fact, since $\operatorname{codim} \mathcal{F} = 1$, we can identify a transversal T with an interval contained in $\mathbb{R}$. Assume that f, x and $y > x$ have the properties listed in Definition 2.3.9. Put $a = x$ and let h be any holonomy map satisfying $h(a) = f^k(y) < y$ with some $k \in \mathbb{N}$ (such maps exist since a and $f^k(y)$ belong to the same leaf L). Fix an element c of the interval $(f^k(y), y)$ and find b such that $b > a$, $[a, b] \subset D_f \cap D_h$ and $h(b) \leq c$. Finally, take $m \in \mathbb{N}$ as large that $f^m(y) < b$ and put $g = f^m$. Certainly, these a, b, c, h and g satisfy all the conditions of (2.3.19).

Now, if $x_0 \in [a, b]$ and $n \in \mathbb{N}$, then all the points

$$(g^{\varepsilon_n} \circ h \circ g^{\varepsilon_{n-1}} \circ h \cdots \circ h \circ g^{\varepsilon_1} \circ h)(x_0), \;\; \varepsilon_i \in \{1, 2\}, \tag{2.3.20}$$

are different because $g([a, b]) \cap h([a, b]) = \emptyset$. (Note that all the maps which appeared in (2.3.20) are defined on the whole interval $[a, b]$!) Consequently,

$$t_{\mathcal{H}, x_0}(3n) \geq t_{\mathcal{H}', x_0}(3n) \geq 2^n \tag{2.3.21}$$

and

$$[\exp] \preceq \mathrm{gr}(L). \tag{2.3.22}$$

From (2.3.22) and Corollary 2.3.8 we obtain the following.

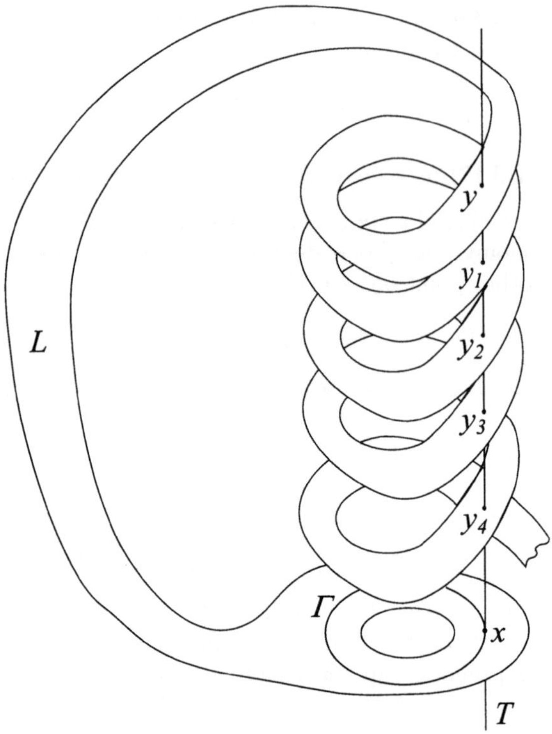

Figure 2.1: A piece of a resilient leaf.

Proposition 2.3.10. *If a Riemannian manifold N is quasi-isometric to a resilient leaf L on a compact foliated manifold $(M, \mathcal{F})$, then*

$$\mathrm{gr}(N) = [\exp]. \tag{2.3.23}$$

□

In the same way, one can define resilient orbits of arbitrary pseudogroups on $\mathbb{R}$. An orbit $\mathcal{G}(x)$ of x is said to be *resilient* if there exist elements g and h of $\mathcal{G}$ such that $x \in D_g \cap D_h$, $g(x) = x$, $y = h(x) \in D_g$ and $g^n(y) \to x$ while $n \to \infty$. Of course, resilient leaves of a foliation $\mathcal{F}$ correspond exactly to resilient orbits of its holonomy pseudogroup. Analogously to Proposition 2.3.10 we have

Proposition 2.3.11. *If an orbit $\mathcal{G}(x)$ is resilient, then*

$$\mathrm{gr}(\mathcal{G}, x) = [\exp]. \tag{2.3.24}$$

□

Example 2.3.12. (i) The pseudogroup $\mathcal{G}$ on $X = [0,1]$ generated by the maps h_1 and h_2 considered in Example 1.4.4 (b) admits resilient orbits. For example, $\mathcal{G}(0)$

is resilient: $h_1(0) = 0$, $h_2(0) = \frac{2}{3}$ and $h_1^n(\frac{2}{3}) = 2 \cdot 3^{-(n+1)} \to 0$ with $n \to \infty$. This orbit is contained and dense in Z_0, an exceptional minimal set. The reader can check that $\mathcal{G}$ admits more resilient orbits: The orbit $\mathcal{G}(x)$ of a point $x \in Z_0$ is resilient if and only if

$$x = \sum_{n=1}^{\infty} \frac{x_n}{3^n}, \quad x_n \in \{0, 2\},$$

and the sequence (x_n) is eventually periodic, i.e., becomes periodic after deleting terms $x_1, \dots, x_m$ for some $m \in \mathbb{N}$. (ii) The pseudogroup $\mathcal{G}(f)$ generated by the Denjoy diffeomorphism f of S^1 (Example 1.2.2) admits no resilient orbits because the only element of $\mathcal{G}(f)$ which has fix points is id. Therefore, the suspension of such an f provides a C^1-foliation with an exceptional minimal set but without resilient leaves.

2.4 Expansion growth

Let $\mathcal{G}_1$ be a good finite symmetric set generating a pseudogroup $\mathcal{G}$ on a compact metric space (X, d). Let $K_g \subset D_g$ be compact sets such that the maps $g|\mathrm{Int}\,(K_g)$, $g \in \mathcal{G}_1$ still generate $\mathcal{G}$. Following [**GLW2**] we shall accept the following.

Definition 2.4.1. Points x and y of X are *(n, ε)-separated* ($n \in \mathbb{N}, \varepsilon > 0$) if there exists $g \in \mathcal{G}_n^c$ such that

$$\{x, y\} \subset D_g \quad \text{and} \quad d(g(x), g(y)) \geq \varepsilon. \tag{2.4.1}$$

A set $A \subset X$ is *(n, ε)-separated* when any two points x and y of A, $x \neq y$, have this property.

Since X is compact, any (n, ε)-separated set is finite and we may put

$$s(n, \varepsilon, \mathcal{G}_1) = \max\{\#A;\ A \subset X \text{ is } (n, \varepsilon)\text{-separated}\}. \tag{2.4.2}$$

Definition 2.4.2. [**Eg1**] The type of growth of the sequence $(s(\cdot, \frac{1}{k}, \mathcal{G}_1))_{k \in \mathbb{N}}$ is called the *expansion growth* of $\mathcal{G}$.

This definition is correct: the type of growth of the sequence $(s(n, \frac{1}{k}))$ does not depend on the choices involved in its definition. Indeed, we have the following.

Lemma 2.4.3. *If $\mathcal{G}_1'$ is another good finite symmetric set generating $\mathcal{G}$, then*

$$[(s(\cdot, \frac{1}{k}, \mathcal{G}_1'))] = [(s(\cdot, \frac{1}{k}, \mathcal{G}_1))]. \tag{2.4.3}$$

Proof. For any $g \in \mathcal{G}_1$ and $x \in K_g$ find a neighbourhood $U_g(x)$ and a map $g' \in {\mathcal{G}'}^c_{m(x,g)}$ such that $g = g'$ on $U_g(x)$. Cover K_g with sets $U_g(x_1), \dots, U_g(x_{r(g)})$, denote by λ_g a Lebesgue number of this covering, and put $\lambda = \min\{\lambda_g;\ g \in \mathcal{G}_1\}$ and $m = \max\{m(x_i, g);\ g \in \mathcal{G}_1 \text{ and } i = 1, \dots, r(g)\}$. Thus, if $x, y \in X$ and $d(x, y) < \lambda$, then there exist maps $g \in \mathcal{G}_1$ and $g' \in \mathcal{G}'^c_m$ such that $\{x, y\} \subset K_g \cap D_{g'}$, $g(x) = g'(x)$ and $g(y) = g'(y)$. It follows that, for any $\varepsilon < \lambda$ and $n \in \mathbb{N}$, all $(n, \varepsilon, \mathcal{G}_1)$-separated sets are $(mn, \varepsilon, \mathcal{G}'_1)$-separated. Consequently,

$$s(n, \varepsilon, \mathcal{G}_1) \leq s(mn, \varepsilon, \mathcal{G}'_1) \quad (n \in \mathbb{N}, \varepsilon < \lambda). \tag{2.4.4}$$

This implies (2.4.3) immediately. □

Also, it is easy to show that isomorphic good pseudogroups (in particular, holonomy pseudogroups corresponding to different nice coverings of a compact foliated manifold $(M, \mathcal{F})$) have the same expansion growth. The proof of this fact is similar to that of the lemma above. Therefore, we may accept the following.

Definition 2.4.4. The *expansion growth* of a foliation $\mathcal{F}$ of a compact manifold M is defined as the expansion growth of the holonomy pseudogroup of $\mathcal{F}$.

Remark 2.4.5. The expansion growth of a pseudogroup could be defined also with the aid of (n, ε)-spanning sets. A subset B of X is said to be $(n, \varepsilon, \mathcal{G}_1)$-*spanning*, $\mathcal{G}_1$ being — as before — a good symmetric finite set generating a pseudogroup $\mathcal{G}$, if for any x of X there exists $y \in B$ such that for any $g \in \mathcal{G}^c_n$,

$$\{x, y\} \subset D_g \quad \text{implies} \quad d(g(x), g(y)) < \varepsilon. \tag{2.4.5}$$

Since X is compact, finite $(n, \varepsilon, \mathcal{G}_1)$-spanning sets exist for all $n \in \mathbb{N}$ and $\varepsilon > 0$, and we may put

$$r(n, \varepsilon, \mathcal{G}_1) = \min\{\#B;\ B \subset X \text{ is } (n, \varepsilon, \mathcal{G}_1)\text{-spanning}\}. \tag{2.4.6}$$

Again, the type of growth of the sequence $(r(\cdot, \frac{1}{k}))_{k \in \mathbb{N}}$ is independent of the choice of $\mathcal{G}_1$. Moreover (see [**Eg1**] for a detailed proof)

$$[(r(\cdot, \frac{1}{k}))] = [s(\cdot, \frac{1}{k}))]. \tag{2.4.7}$$

Hereafter, the expansion growth type of a pseudogroup $\mathcal{G}$ (resp., of a foliation $\mathcal{F}$) will be denoted by $\operatorname{egr}(\mathcal{G})$ (resp., by $\operatorname{egr}(\mathcal{F})$).

Example 2.4.6. (i) Let $\mathcal{G} \subset \mathrm{Iso}\ (X)$ be a pseudogroup of local isometries. Then for any $g \in \mathcal{G}$ and $x \in D_g$ there exists a neighbourhood U of x such that g maps U isometrically onto $g(U)$. Recall (Section 1.2) that, in general, the elements g of $\mathcal{G}$ need not map D_g isometrically onto R_g. However, if $\mathcal{G}$ is finitely generated (and good), then there exists a finite symmetric (and good) set $\mathcal{G}_1$ generating $\mathcal{G}$ and such that any map $g \in \mathcal{G}_1$ is an isometry of D_g (equipped with the distance function induced from X) onto R_g. For such $\mathcal{G}_1$, $d(g(x), g(y)) = d(x, y)$ for all $g \in \mathcal{G}^c_n$, $n \in \mathbb{N}$, x and $y \in D_g$. Therefore, $s(n, \varepsilon, \mathcal{G}_1) = s(0, \varepsilon, \mathcal{G}_1) = s(\varepsilon) > 0$ and

$$\operatorname{egr}(\mathcal{G}) = [(s(1), s(\frac{1}{2}), \dots)] = [1]. \tag{2.4.8}$$

In other words,

1. *any good pseudogroup of local isometries has constant expansion growth.*

Recall that foliations with holonomy pseudogroups consisting of local isometries (of a transversal equipped with a Riemannian structure) are called *Riemannian.* Therefore,

2. *all the Riemannian foliations of compact manifolds have constant expansion growth.*

(ii) If $\mathcal{G} = \mathcal{G}(f)$ is the pseudogroup on S^1 generated by the map $f : z \mapsto z^2$, then $\operatorname{egr}(\mathcal{G})$ is exponential. In fact, if two points x and $y \in S^1$ are close enough, then $d(f(x), f(y)) = 2d(x, y)$. Therefore, if $\varepsilon > 0$ is small enough (say, $\varepsilon < \varepsilon_0 = \pi = \operatorname{diam} S^1$) and $n \in \mathbb{N}$, then on any closed arc $C \subset S^1$ of length ε one can find $2^n + 1$ points which constitute an (n, ε)-separated set. Consequently, $s(n, \varepsilon) > 2^n$ for all n and $\varepsilon < \varepsilon_0$, and $\operatorname{egr}(\mathcal{G}) \succeq [2^n] = [\exp]$. On the other hand, since S^1 is a compact manifold and f is smooth, $\operatorname{egr}(\mathcal{G}) \preceq [\exp]$ (see Proposition 2.4.7 below).

This pseudogroup $\mathcal{G}$ can be realized as the holonomy of a foliation $\mathcal{F}$ of a compact 3-manifold M. Indeed, let $N = D^2 \times S^1$ be the solid torus, $\phi : N \to N$ be given by

$$\phi(z, w) = (\frac{1}{2}w + \frac{1}{4}z, w^2), \quad |z| \leq 1, \ |w| = 1, \tag{2.4.9}$$

and $N_0 = N \setminus \operatorname{Int} \phi(N)$. N_0 is a compact 3-manifold with boundary $\partial N_0 = \partial_0 \cup \partial_1$, ∂_i being diffeomorphic to T^2. N_0 carries the natural foliation $\mathcal{F}_0$ by the surfaces $w =$ const. Let M be the manifold (without boundary) obtained from N_0 by identifying ∂_0 and ∂_1 *via* ϕ and $\mathcal{F}$ — the foliation induced on M from $\mathcal{F}_0$. $\mathcal{F}$ has the required property. It is called *Hirsch foliation.* Therefore,

3. *Hirsch foliation has exponential expansion growth.*

Proposition 2.4.7. *If M is a compact Riemannian manifold and $\mathcal{G}$ is a finitely generated pseudogroup of C^1-diffeomorphisms on M, then*

$$\operatorname{egr}(\mathcal{G}) \preceq [e^n]. \tag{2.4.10}$$

Proof. Let $m = \dim M$ and $\mathcal{U} = \{U_1, \dots, U_k\}$ be a finite covering of M by charts $\phi_i : U_i \to Q$, Q being the unit cube in $\mathbb{R}^m$. Choose a good set $\mathcal{G}_1$ generating $\mathcal{G}$ and let $a = \inf\{\|d(\phi_i \circ g \circ \phi_j^{-1})(x)\|;\ g \in \mathcal{G}_1^c, i, j \leq k \text{ and } x \in D_{\phi g \phi^{-1}}\}$. Certainly, a is a positive constant. Also, there exists another constant $b \geq 1$ such that $b^{-1}|\phi_i(x) - \phi_i(y)| \leq d_M(x, y) \leq b|\phi_i(x) - \phi_i(y)|$ for all $i \leq k$, x and $y \in U_i$.

If a set $A \subset Q$ is (n, ε)-separated with respect to the pseudogroup on Q generated by the maps $\phi_i \circ g \circ \phi_i^{-1}$, then $|x - y| \geq \frac{\varepsilon}{a^n}$ for any x and y of A, and, consequently, $\#A \leq \left(\frac{a^n}{\varepsilon}\right)^m$. Finally, if λ is a Lebesgue number of $\mathcal{U}$, $B \subset M$ is (n, ε)-separated and $\varepsilon < \lambda$, then $\phi_i(B \cap U_i)$ is $(n, b\varepsilon)$-separated and $\#\phi_i(B \cap U_i) \leq \left(\frac{a^n}{b\varepsilon}\right)^m$ for $i = 1, \dots, k$. Therefore,

$$s(n, \varepsilon, \mathcal{G}_1) \leq \frac{k}{(b\varepsilon)^m} \cdot a^{mn}. \tag{2.4.11}$$

□

As an immediate consequence we get the following.

Corollary 2.4.8. $\mathrm{egr}(\mathcal{F}) \preceq [e^n]$ *for any* C^1*-foliation* $\mathcal{F}$ *of any compact manifold* M. □

Remark 2.4.9. In a similar way, one can define the expansion growth $\mathrm{egr}(\mathcal{G}, K)$ of $\mathcal{G}$ on any compact set $K \subset X$. To this end we shall replace $s(n, \varepsilon, \mathcal{G}_1)$ by $s(n, \varepsilon, \mathcal{G}_1, K)$ defined by (2.4.2) modified by considering $(n, \varepsilon, \mathcal{G}_1)$-separated subsets of K only. Obviously, $s(n, \varepsilon, \mathcal{G}_1, K) \leq s(n, \varepsilon, \mathcal{G}_1)$ for all $n, \varepsilon, \mathcal{G}_1$ and K, and

$$\mathrm{egr}(\mathcal{G}, K) \preceq \mathrm{egr}(\mathcal{G}) \tag{2.4.12}$$

for all K. Even if X is non-compact but locally compact only, $\mathrm{egr}(\mathcal{G})$ can be defined by

$$\mathrm{egr}(\mathcal{G}) = \sup_K \mathrm{egr}(\mathcal{G}, K), \tag{2.4.13}$$

where K ranges over all the compact subsets of X. Clearly, this definition is consistent with that for compact spaces.

Now, let $\mathcal{G}$ be a pseudogroup on $\mathbb{R}$ and let the orbit of a point $x \in \mathbb{R}$ be resilient. Then, $\mathcal{G}$ contains a subpseudogroup $\mathcal{G}'$ generated by two elements $f_j : I \to I_j$, $j = 1, 2$, such that I and I_j are closed intervals satisfying $I_j \subset I$ and $I_1 \cap I_2 = \emptyset$. In fact, one can take $f_1 = g^k \circ h$ and $f_2 = g^l$, where g and $h \in \mathcal{G}$ satisfy (2.3.19) while k and $l \in \mathbb{N}$ are sufficiently large.

Let $\varepsilon_0 = \mathrm{dist}(I_1, I_2)$ and $x_0 \in I$. Fix $n \in \mathbb{N}$. For any $i_1, \ldots, i_n \in \{1, 2\}$ let $y(i_1, \ldots, i_n) = f_{i_1} \circ \cdots \circ f_{i_n}(x_0)$. The set A_n of all such points is (n, ε) separated for any $\varepsilon < \varepsilon_0$. In fact, if $y = y(i_1, \ldots, i_k, i_{k+1}, \ldots, i_n)$ and $y' = y(i_1, \ldots, i_k, j_{k+1}, \ldots, j_n)$ with $i_{k+1} \neq j_{k+1}$, then the map $f = f_{i_k}^{-1} \circ \cdots \circ f_{i_1}^{-1}$ is defined on an interval containing both y and y', $f(y) \in I_{i_k}$ and $f(y') \in I_{j_k}$, hence $d(f(y), f(y')) \geq \varepsilon_0 \geq \varepsilon$. Clearly, $\#A_n = 2^n$. Consequently, if $f_1, f_2 \in \mathcal{G}_m$, then

$$s(mn, \varepsilon, \mathcal{G}) \geq s(n, \varepsilon, \mathcal{G}') \geq 2^n \tag{2.4.14}$$

for all n and $\varepsilon \leq \varepsilon_0$.

Combining (2.4.14) with Proposition 2.4.7 we obtain the following.

Proposition 2.4.10. *If a pseudogroup* $\mathcal{G}$ *of* C^1*-diffeomorphisms of a closed interval* $J \subset \mathbb{R}$ *admits a resilient orbit, then*

$$\mathrm{egr}(\mathcal{G}) = [e^n]. \tag{2.4.15}$$

□

Geodesic flows on compact negatively curved manifolds are of great interest in differential geometry. They provide natural examples of Anosov dynamical systems. Anosov systems generate naturally defined foliations. The structure of these foliations determines to some extent the geometry of the underlying manifold. We complete this section by calculation of the type of expansion growth of these foliations in the case of the geodesic flow of a compact hyperbolic (curvature = -1) manifold.

First, let us recall basic notions.

Definition 2.4.11. A non-singular C^1-flow (ϕ_t) on a compact manifold M is *Anosov* ([**An**], [**Ma**], [**Kl**] etc.) if TM splits into the direct sum

$$TM = E^0 \oplus E^s \oplus E^u \tag{2.4.16}$$

of $d\phi_t$-invariant subbundles such that E^0 is one-dimensional spanned by $X = d\phi_t/dt$,

$$\|d\phi_t(v)\| \leq a\lambda^{-t}\|v\| \text{ and } \|d\phi_{-t}(w)\| \leq a\lambda^{-t}\|w\| \tag{2.4.17}$$

for all $t > 0$, $v \in E^s$, $w \in E^u$, some constants $a > 0$ and $\lambda > 1$, and the norm $\|\cdot\|$ induced by some Riemannian metric on M. The similar notion of an *Anosov diffeomorphism* $f : M \to M$ can be introduced by replacing (2.4.16) with

$$TM = E^s \oplus E^u, \tag{2.4.18}$$

and (2.4.17) with

$$\|df^n(v)\| \leq a\lambda^{-n}\|v\| \text{ and } \|df^{-n}(w)\| \leq a\lambda^{-n}\|w\| \quad (n \in \mathbb{N}); \tag{2.4.19}$$

now E^s and E^u become, of course, df-invariant.

The subbundles E^s and E^u are said to be, respectively, *stable* and *unstable* . In the case of a flow, $E^0 \oplus E^s$ and $E^0 \oplus E^u$ are said to be *weakly stable* and *weakly unstable*. All the subbundles E^s, E^u, $E^0 \oplus E^s$ and $E^0 \oplus E^u$ are integrable. The corresponding foliations are called (and denoted by), respectively, *strongly stable* ($\mathcal{F}^{ss}$), *strongly unstable* ($\mathcal{F}^{su}$), *weakly stable* ($\mathcal{F}^{ws}$) and *weakly unstable* ($\mathcal{F}^{wu}$).

In general, stable and unstable foliations are not differentiable. Below, we provide some examples, where these foliations are smooth.

Example 2.4.12. (i) A matrix $A \in SL(m, \mathbb{Z})$ is *hyperbolic* if it has no eigenvalues λ with $|\lambda| = 1$. Any matrix like that provides an Anosov diffeomorphism of the m-dimensional torus $T^m = \mathbb{R}^m/\mathbb{Z}^m$. $\mathcal{F}^{ss}$ and $\mathcal{F}^{su}$ are the linear foliations of T^m by (projections of) parallel affine hyperplanes in $\mathbb{R}^m$ spanned by the eigenvectors corresponding to the eigenvalues λ of A with $|\lambda| < 1$ and $|\lambda| > 1$, respectively. Foliations $\mathcal{F}^{ss}$ and $\mathcal{F}^{su}$ are, obviously, Riemannian and therefore

$$\operatorname{egr}(\mathcal{F}^{ss}) = \operatorname{egr}(\mathcal{F}^{su}) = [1] \tag{2.4.20}$$

in this case.

(ii) Now, let M be a complete Riemannian manifold of constant sectional curvature -1 and (ϕ_t) the *geodesic flow* on the unit tangent bundle SM of M. Recall — compare Figure 2.2 — that the maps ϕ_t are defined by the formula

$$\phi_t(w) = \dot{\gamma}_w(t),$$

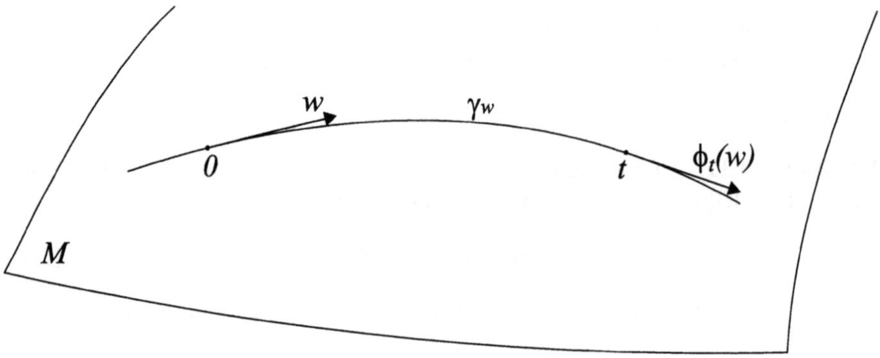

Figure 2.2: Geodesic flow.

where $\gamma_w : \mathbb{R} \to M$ is the unique geodesic such that $\dot{\gamma}_w(0) = w$.

Split TTM into the direct sum of vertical and horizontal (with respect to the Levi-Civita connection ∇ on M) subbundles:

$$TTM = V \oplus H. \tag{2.4.21}$$

For any $w \in T_xM$, V_w and H_w can be identified with the tangent space T_xM *via* π_* and the connection map K. (Here, $\pi : TM \to M$ is the canonical projection, and $K : TTM \to TM$ is determined uniquely by the condition

$$K(Z_*u) = \nabla_u Z \tag{2.4.22}$$

for all $u \in TM$ and any vector field Z on M.)

The vector field $\Phi = d\phi_t/dt$ is horizontal (i.e., $K \circ \Phi = 0$) and $\pi_* \circ \Phi = \mathrm{id}_{TM}$. Moreover, if $\xi = Z_*(w) \in TSM$, $w \in T_xM$, Z is a unit vector field on M and $u = Z(x)$, then

$$\langle K(\xi), u\rangle = \langle KZ_*w, Z(x)\rangle = \frac{1}{2}\langle \nabla_w Z, Z\rangle = \frac{1}{2}w|Z|^2 = 0, \tag{2.4.23}$$

i.e., the space T_uSM consists of all $\xi \in T_uTM$ which have the vertical part orthogonal to u. Let

$$E^s(w) = \{\xi \in T_wSM;\ \xi^v = -\xi^h\} \text{ and } E^u(w) = \{\xi \in T_wSM;\ \xi^v = \xi^h\}, \tag{2.4.24}$$

where ξ^v and ξ^h are, respectively, the (projected to TM) vertical and horizontal part of ξ. We shall show that the geodesic flow (ϕ_t) is Anosov with E^s and E^u being its stable and unstable subbundles.

To this end, recall that the curvature tensor R of M is given by

$$R(u,v)w = \langle u,w\rangle v - \langle v,w\rangle u \quad (u,v,w \in T_xM,\ x \in M) \tag{2.4.25}$$

and that the variation fields Y corresponding to variations among geodesics of a fixed geodesic $\gamma : \mathbb{R} \to M$ satisfy the Jacobi equation

$$Y'' - R(\dot\gamma, Y)\dot\gamma = 0, \tag{2.4.26}$$

where $Y' = \nabla_{\dot\gamma} Y$ is the covariant derivative of Y along γ. (Obviously, $Y'' = (Y')'$.) Therefore, the maps $\phi_{t*} : TSM \to TSM$ satisfy

$$\pi_* \phi_{t*}\xi = Y_\xi(t) \text{ and } K\phi_{t*}\xi = Y'_\xi(t), \tag{2.4.27}$$

where Y_ξ is the unique Jacobi field (i.e., the solution of (2.4.26)) along the geodesic $\gamma = \gamma_w$ ($\xi \in T_w SM$) satisfying the initial conditions $Y_\xi(0) = \pi_*\xi$ and $Y'_\xi(0) = K\xi$. From (2.4.26) and (2.4.25) it follows easily that, in our case of constant sectional curvature, the Jacobi fields Y_ξ, $\xi \in E^s(w) \oplus E^u(w)$, are given by

$$Y_\xi(t) = \cosh t \cdot E_1(t) + \sinh t \cdot E_2(t), \tag{2.4.28}$$

where E_1 and E_2 are the vector fields parallel along γ and such that $E_1(0) = \pi_*\xi$ and $E_2(0) = K\xi$.

Now, equip TM with the *Sasaki metric* which can be defined as the unique Riemannian structure for which the subbundles V and H are orthogonal and all the maps $\pi_*|H_w : H_w \to T_xM$ and $K|V_w : V_w \to T_xM$, $w \in TM$, $x = \pi(w) \in M$, are isometric. Then, from (2.4.27) and (2.4.28) we get

$$\|\phi_{t*}\xi\|^2 = \|\cosh t \cdot \pi_*\xi + \sinh t \cdot K\xi\|^2 + \|\sinh t \cdot \pi_*\xi + \cosh t \cdot K\xi\|^2. \tag{2.4.29}$$

In particular,

$$\|\phi_{t*}\xi\|^2 = 2(\cosh t - \sinh t)^2 \|\pi_*\xi\|^2 = e^{-2t}\|\xi\|^2 \tag{2.4.30}$$

when $\xi \in E^s$ and

$$\|\phi_{t*}\xi\|^2 = 2(\cosh t + \sinh t)^2 \|\pi_*\xi\|^2 = e^{2t}\|\xi\|^2 \tag{2.4.31}$$

when $\xi \in E^u$. Therefore, E^s and E^u are stable and unstable for (ϕ_t) indeed.

A similar but more complicated argument shows that the geodesic flow of any complete negatively curved Riemannian manifold is Anosov (see **[Kl]**, for example).

Theorem 2.4.13. **[Eg2]** *If $\mathcal{F}$ is the strong stable (or, unstable) foliation for the geodesic flow on the unit tangent bundle SM of a compact oriented m-dimensional Riemannian manifold M of constant negative curvature c, then $\mathcal{F}$ has polynomial expansion growth of degree $m-1$, i.e.,*

$$\operatorname{egr}(\mathcal{F}) = [n^{m-1}]. \tag{2.4.32}$$

Proof. We shall assume that $\mathcal{F} = \mathcal{F}^{ss}$ is strongly stable and that $c = -1$. We shall begin with the proof of a technical lemma in dimension 2.

First, recall that the universal covering $\tilde{M}$ of M is isometric to the hyperbolic space $\mathbb{H}^m = \{(x_1, \dots, x_m);\ x_m > 0\}$ equipped with the Poincaré metric $ds^2 = (\sum_{i=1}^m dx_i^2)/x_m^2$. Also, there exists a discrete subgroup Γ of orientation preserving isometries of $\mathbb{H}^m$ such that $M = \Gamma\backslash\mathbb{H}^m$ and, consequently, $SM = \Gamma\backslash S\mathbb{H}^m$.

Next, denote by d the distance function on SM (and, without any misunderstanding, on M) determined by the Sasaki Riemannian structure (recall Example 2.4.12 (ii)) and by d^{ss} (resp., d^{su}) the distance functions on the leaves of $\mathcal{F}^{ss}$ (resp., $\mathcal{F}^{su}$) determined by the induced Riemannian metric.

Now, assume that $m = 2$ and let h^{ss} (resp., h^{su}) be the contracting (resp., expanding) horocycle flow on SM, i.e., the flow determined by the positively oriented unit vector field tangent to $\mathcal{F}^{ss}$ (resp., $\mathcal{F}^{su}$).

Lemma 2.4.14. *If $A > 1$, $\varepsilon > 0$ and $\delta > 0$ are small enough, x and y belong to the same leaf of the foliation $\mathcal{F}^{su}$ on $S\mathbb{H}^2$, and $l = d^{su}(x, y)$, then*

(i) *the points $h^{ss}_{t/(1-tl)}(y)$ and $h^{ss}_t(x)$ belong to the same leaf of the weak unstable foliation $\mathcal{F}^{wu}$ on $S\mathbb{H}^2$ if only $|t|$ is small enough,*

(ii) *for any $n \in \mathbb{N}$ such that $\delta \geq l \geq A\delta/n$ there exists $t \in \mathbb{R}$ such that $|t| \leq n$ and*

$$d(\pi(h^{ss}_t(x)), \pi(h^{ss}_{t/(1-tl)}(y))) = \delta, \tag{2.4.33}$$

(iii) *for any $n \in \mathbb{N}$ such that $l < \varepsilon/An$ the inequality*

$$d(\pi(h^{ss}_t(x)), \pi(h^{ss}_{t/(1-tl)}(y))) < \varepsilon \tag{2.4.34}$$

holds for all $t \in [-n, n]$.

Proof. Under the standard identification of $S\mathbb{H}^2$ with the group $\mathrm{PSL}(2,\mathbb{R})$ the geodesic flow (ϕ_t) and the horocycle flows (h^{ss}_t) and (h^{su}_t) on $S\mathbb{H}^2$ can be represented by the one-parameter subgroups of $\mathrm{PSL}(2,\mathbb{R})$ given, respectively, by

$$\Phi_t = \pm\begin{pmatrix} e^{t/2} & 0 \\ 0 & e^{-t/2} \end{pmatrix},\ H^{ss}_t = \pm\begin{pmatrix} 1 & t \\ 0 & 1 \end{pmatrix} \text{ and } H^{su}_t = \pm\begin{pmatrix} 1 & 0 \\ t & 1 \end{pmatrix} \tag{2.4.35}$$

(see, for instance, [**Ni**], Chapter 10). The continuity argument shows that if v and w are sufficiently close points of a leaf of $\mathcal{F}^{wu}$, then

$$\begin{gathered} v^{-1}w = \pm\begin{pmatrix} a & 0 \\ c & d \end{pmatrix} \in S\mathbb{H}^2 = PSL(2,\mathbb{R}) \quad \text{and} \\ \frac{1}{A}\max\{|1 - |a||, |c|\} < d(\pi(v), \pi(w))) < A\max\{|1 - |a||, |c|\}. \end{gathered} \tag{2.4.36}$$

Coming back to the notation of our lemma we can assume, without losing generality, that $x = \pm I \in PSL(2,\mathbb{R})$, where I is the identity matrix. Then, $y = H_l^{su}$ (or, H_{-l}^{su}, but we focus on the first case only), $h_t^{ss}(x) = H_t^{ss}$, $h_{t/(1-tl)}^{ss}(y) = H_l^{su} \cdot H_{t(1-tl)}^{ss}$ and

$$h_t^{ss}(x)^{-1} \cdot h_{t/(1-tl)(y)}^{ss} = \pm \begin{pmatrix} 1-tl & 0 \\ l & 1/(1-tl) \end{pmatrix} = \Phi_{2\log(1-tl)} \cdot H_{l(1-tl)}^{su} \quad (2.4.37)$$

if only t is small enough. Therefore, $h_t^{ss}(x)$ and $h_{t(1-tl)}^{ss}(y)$ belong to the same leaf of $\mathcal{F}^{wu}$ and, by (2.4.36),

$$\frac{1}{A}\max\{l, |tl|\} < d(\pi(h_t^{ss}(x)), \pi(h_{t/(1-tl)}^{ss}(y))) < A\max\{l, |tl|\}. \quad (2.4.38)$$

The statements (ii) and (iii) of the lemma follow directly from (2.4.38). □

Coming back to the proof of the theorem, let us fix a nice covering $\mathcal{U} = \{(U_i, \phi_i);\ i = 1, \dots p\}$ of SM by charts distinguished simultaneously by both, $\mathcal{F}^{ss}$ and $\mathcal{F}^{wu}$. Let $T \subset U_1$ be an m-dimensional transversal contained in a leaf L^{wu} of $\mathcal{F}^{wu}$ and let $T' = T \cap L^{su}$, L^{su} being a leaf of $\mathcal{F}^{su}$ contained in L^{wu}.

Fix $\varepsilon > 0$ small enough (say, $\varepsilon < \frac{1}{3}\lambda$, λ being a Lebesgue number of the covering $\mathcal{U}$) and $n \in \mathbb{N}$, and take two points x and y of T' such that $d^{su}(x,y) \geq \frac{\varepsilon}{n}$. If $d^{su}(x,y) \geq \varepsilon$, then x and y are (k,ε)-separated for all $k \geq 1$. Assume that $d^{su}(x,y) < \varepsilon$. By Lemma 2.4.14, passing to the universal covering $\tilde{M} = \mathbb{H}^m$ we can find two curves α and $\beta : [0,1] \to S\tilde{M}$ such that $\alpha(0) = x$, $\alpha(1) = y$, $\alpha(t)$ and $\beta(t)$ belong to the same leaf of $\mathcal{F}^{wu}$ and $d(\pi(\alpha(1)), \pi(\beta(1))) \geq \varepsilon$. We can request α and β to have length bounded by $c_1 \cdot n$, where $c_1 \in \mathbb{N}$ is a constant independent of x, y and n. Therefore, since the projection π restricted to T is bi-Lipschitz, there exists another constant $c_2 > 0$, again independent of all the choices involved, and an element $f \in \mathcal{H}_{Knc_1}$ such that x and y lie in D_f, $f(x) = \alpha(1)$, $f(y) = \beta(1)$ and

$$d^{wu}(f(x), f(y)) = d^{wu}(\alpha(1), \beta(1)) \geq c_2\varepsilon. \quad (2.4.39)$$

Here, $\mathcal{H}$ is the holonomy pseudogroup of $\mathcal{F}^{ss}$ generated by the set $\mathcal{H}_1$ of all the holonomy maps corresponding to pairs of intersecting charts of the covering $\mathcal{U}$ and $K \in \mathbb{N}$ is large enough that $K^{-1} \leq d^{ss}(z, h(z)) \leq K$ whenever $h \in \mathcal{H}_1$, $z \in D_h$ and $z \neq h(z)$. Inequality (2.4.39) shows that x and y are $(Knc_1, \varepsilon, \mathcal{H}_1)$-separated.

Moreover, since $\dim T' = m - 1$ and the map $\phi_1|T' : T' \to \mathbb{R}^{m-1}$ is bi-Lipschitz, we can find a finite set $A \subset T'$ of cardinality $\#A \geq c_3 \cdot (\frac{n}{\varepsilon})^{m-1}$, c_3 being a constant independent on n and ε, such that $d^{su}(x,y) \geq \frac{\varepsilon}{n}$ for all x and $y \in A$, $x \neq y$. Such a set A is $(Knc_1, \varepsilon, \mathcal{H}_1)$-separated and, therefore,

$$s(Knc_1, \varepsilon, \mathcal{H}_1) \geq c_3 \cdot \left(\frac{n}{\varepsilon}\right)^{m-1}. \quad (2.4.40)$$

This yields the relation

$$[n^{m-1}] \preceq \operatorname{egr}(\mathcal{F}^{ss}). \quad (2.4.41)$$

To prove the converse,

$$\mathrm{egr}(\mathcal{F}^{ss}) \preceq [n^{m-1}], \tag{2.4.42}$$

choose for all $i \leq p$ transversals T_i' and $T_i \subset U_i$ contained, respectively, in some leaves $L_i^{su} \subset L_i^{wu}$ of $\mathcal{F}^{su}$ and $\mathcal{F}^{wu}$. Next, choose $\frac{\varepsilon}{n}$-dense subsets B_i' of T_i' and shift all the points of B_i' along the orbits of the geodesic flow to the distance $\pm\varepsilon, \pm 2\varepsilon, \ldots$, as long as it is possible within T_i. Let $B \subset T = \bigcup_i T_i$ be the set of all the points obtained in this way. Since the lengths of all the geodesic segments contained in T_i are bounded from above, there exists a constant $c_4 > 0$ such that

$$\#B \leq pc_4 \cdot \frac{1}{\varepsilon} \cdot \left(\frac{n}{\varepsilon}\right)^{m-1}. \tag{2.4.43}$$

Analogously to the proof of (2.4.41) one can see, using Lemma 2.4.14 again, that the set B is $(n, c_5\varepsilon)$-dense (with respect to the holonomy pseudogroup $\mathcal{H}$) for some universal constant $c_5 > 0$. Therefore,

$$r(n, c_5\varepsilon, \mathcal{H}_1) \leq \frac{pc_4}{\varepsilon^m} \cdot n^{m-1}. \tag{2.4.44}$$

This yields (2.4.42) and completes the proof of the theorem. □

Remark 2.4.15. From the variational principle (Thm. 3.1.13 below) and some other classical results of the ergodic theory (see, for instance, [**Ma**], Chapter IV) it follows that the topological entropy (see Section 3.1 below) of any Anosov system is strictly positive. Therefore, any Anosov flow has exponential expansion growth. It is not too hard to show that this implies the equalities

$$\mathrm{egr}(\mathcal{F}^{ws}) = \mathrm{egr}(\mathcal{F}^{wu}) = [\exp] \tag{2.4.45}$$

for the weak stable and weak unstable foliations $\mathcal{F}^{ws}$ and $\mathcal{F}^{wu}$ of any Anosov flow on a compact manifold.

Chapter 3

Entropy

This chapter is devoted to the central topic of this book. Entropy of different systems (groups, pseudogroups, foliations) is considered here. This is an invariant measuring, in some sense, complexity of a system. Roughly speaking, entropy is defined as the exponential rate of growth (as $n \to \infty$) of the maximal number of points which are pairwise separated (by the system under consideration) to the distance ε (ε being arbitrarily small) along pieces of orbits of length $\leq n$.

First, entropy of measure preserving transformations was defined by Kolmogorov and Sinai in the late 1950s. A few years later, a topological counterpart, the topological entropy of continuous transformations (and flows), was introduced by Adler, Konheim and McAndrew [**AKM**]. Bowen [**Bow1**] found a very nice and useful approach to topological entropy of transformations of metric spaces and soon after Dinaburg [**Di**] and Goodman [**Gd**] proved the variational principle which relates measure theoretic and topological entropies.

Since entropy appeared quickly to be the most successful invariant in ergodic theory and dynamical systems, there were several attempts to find its suitable generalization for other systems. For instance, the entropy for commutative groups of transformations was defined and studied by several authors ([**Con**], [**Mis1**], [**Ru1**] etc.) but this entropy vanishes when the group has non-linear growth and the entropies of generating maps are finite, therefore it is almost useless in study of dynamics of foliations (compare Section 4.7 for more extensive discussion). Also, the word "entropy" is often used in Riemannian geometry in the context of dynamics of geodesic flows, but this one is just the rate of exponential growth of volumes of balls in the universal covering (or, of the fundamental group) of the manifold under consideration and has very little in common with entropies considered here.

In [**GLW2**], the authors proposed a definition of a topological entropy for finitely generated groups and pseudogroups of continuous transformations. In the simplest case of the group $\mathbb{Z}$ generated by a single transformation this entropy coincides with the doubled topological entropy of the generator. The entropy of pseudogroups, when applied to holonomy pseudogroups, yields an entropy of foliations. The last one depends on a Riemannian metric (in fact, on the restriction of this metric to the bundle of all the vectors tangent to the leaves of a foliation) chosen for foliated manifold, so it has more geometric than topological nature. In the particular case of an oriented foliation of dimension 1, the geometric entropy coincides with the doubled topological entropy of the flow of a unit vector field tangent to a foliation. It seems that the geometric entropy is a quite useful tool in studying dynamics and topology of foliations. For instance, vanishing entropy of a codimension-one foliation implies vanishing of its Godbillon–Vey class, which is the first known characteristic class discovered in the theory of foliations. Also, vanishing entropy of a pseudogroup implies existence of so-called transverse invariant Borel probability measures (see Chapter 4).

Here, in Section 3.1, we give a brief *résumé* of entropy for classical dynamical systems. This material is standard and can be found in several books on ergodic theory and dynamical systems. A reader interested in a more elaborate and systematic exposition is referred to one of them: [**Ma**] or [**Wa**], for instance. There are two reasons why we devote some pages to this classical theory: one is historical and the other that Bowen's ideas inspired the notions mentioned above of topological and geometric entropies for pseudogroups and foliations which are studied in Sections 3.2–3.5. Finally, in Section 3.6 we prove the result mentioned above on vanishing Godbillon–Vey class (Theorem 3.6.1). This last Section is not self contained. The proof depends on some deep results of the theory of foliations. We describe them briefly without giving proofs here. The reader not familiar with the theory of foliations is referred either to original papers or to books on foliations ([**CL**], [**CaC1**], [**Go2**], [**HH**], for instance).

3.1 Entropy of classical systems

3.1.1 Topological entropy of a transformation

Let $f : X \to X$ be a continuous transformation of a compact Hausdorff space X. For any open covers $\mathcal{U}$ and $\mathcal{V}$ of X let

$$\mathcal{U} \vee \mathcal{V} = \{U \cap V; U \in \mathcal{U} \text{ and } V \in \mathcal{V}\} \text{ and } f^{-1}\mathcal{U} = \{f^{-1}(U); U \in \mathcal{U}\}. \tag{3.1.1}$$

Also, let $N(\mathcal{U})$ be the minimal cardinality of a finite open subcover of $\mathcal{U}$,

$$\mathcal{U}^{(n)} = \mathcal{U} \vee f^{-1}\mathcal{U} \vee \cdots \vee f^{-(n-1)}\mathcal{U} \tag{3.1.2}$$

and

$$N(f,\mathcal{U},n) = N(\mathcal{U}^{(n)}) \tag{3.1.3}$$

for $n = 1, 2, \ldots$. Note that $N(\mathcal{U}\vee\mathcal{V}) \leq N(\mathcal{U})\cdot N(\mathcal{V})$ and, therefore, $N(f,\mathcal{U},n+m) \leq N(f,\mathcal{U},n) \cdot N(f,\mathcal{U},m)$ for all $\mathcal{U}$, $\mathcal{V}$, m and n. Consequently, the limit

$$h(f,\mathcal{U}) = \lim_{n\to\infty} \frac{1}{n} \log N(f,\mathcal{U},n) \tag{3.1.4}$$

exists for any $\mathcal{U}$. Moreover, if $\mathcal{V}$ is a refinement of $\mathcal{U}$, then

$$h(f,\mathcal{U}) \leq h(f,\mathcal{V}). \tag{3.1.5}$$

Definition 3.1.1. The number

$$h_{\text{top}}(f) = \sup_{\mathcal{U}} h(f,\mathcal{U}) \tag{3.1.6}$$

is called the *topological entropy* of a transformation f.

If X is a compact metric space with the distance function d then, following Bowen [**Bow2**], we can define the entropy in a different way.

For any $n \in \mathbb{N}$, x and y of X let

$$d_n(x,y) = \max\{d(f^k(x), f^k(y)); k = 0, 1, \ldots, n-1\}. \tag{3.1.7}$$

It is easy to see that d_n is a new distance function on X.

Definition 3.1.2. Given $n \in \mathbb{N}$ and $\varepsilon > 0$, a subset A of X is said to be (n,ε)-*separated* if $d_n(x,y) \geq \varepsilon$ for all x and y of A, $x \neq y$. A is (n,ε)-*spanning* if for any $x \in X$ there exists $y \in A$ for which $d_n(x,y) < \varepsilon$.

Since X is compact, (n,ε)-separated sets are always finite and there exist finite (n,ε)-spanning sets for all n and ε. Therefore, we may put

$$s(n,\varepsilon) = \max\{\#A; A \subset X \text{ is } (n,\varepsilon)\text{-separated}\}, \tag{3.1.8}$$

$$r(n,\varepsilon) = \min\{\#A; A \subset X \text{ is } (n,\varepsilon)\text{-spanning}\}, \tag{3.1.9}$$

$$s(\varepsilon) = \limsup_{n\to\infty} \frac{1}{n} \log s(n,\varepsilon) \quad \text{and} \quad r(\varepsilon) = \limsup_{n\to\infty} \frac{1}{n} \log r(n,\varepsilon). \tag{3.1.10}$$

Theorem 3.1.3. *The equalities*

$$h_{\text{top}}(f) = \lim_{\varepsilon\to 0+} s(\varepsilon) = \lim_{\varepsilon\to 0+} r(\varepsilon) \tag{3.1.11}$$

hold for any continuous transformation f of a compact metric space X.

Proof. Obviously, $s(n,\varepsilon) \leq s(n,\varepsilon')$, $s(\varepsilon) \leq s(\varepsilon')$, $r(n,\varepsilon) \geq r(n,\varepsilon')$ and $r(\varepsilon) \geq r(\varepsilon')$ whenever $\varepsilon \geq \varepsilon'$. Therefore, the limits in (3.1.11) exist.

If A is a maximal (n,ε)-separated set, then A is (n,ε)-spanning, so $r(n,\varepsilon) \leq s(n,\varepsilon)$ and $r(\varepsilon) \leq s(\varepsilon)$ for all n and ε. Also, if A is (n,ε)-separated and B is $(n,\varepsilon/2)$-spanning, then one can define a map $\phi : A \to B$ such that $d_n(x,\phi(x)) < \frac{\varepsilon}{2}$ for all x. Such a map is injective, so $\#A \leq \#B$, $s(n,\varepsilon) \leq r(n,\varepsilon/2)$ and $s(\varepsilon) \leq r(\varepsilon/2)$ for all n and ε. This proves the second equality in (3.1.11).

To get the first one, let us take a cover of X by balls of radius ε and any (n,ε)-separated set A. Then, any member U of $\mathcal{U}^{(n)}$ contains at most one point of A. Therefore, $N(\mathcal{U}^{(n)}) \geq s(n,\varepsilon)$, $h(f,\mathcal{U}) \geq s(\varepsilon)$ and $h_{\text{top}}(f) \geq s(\varepsilon)$ for all ε.

On the other hand, since X is compact, any open cover $\mathcal{U}$ admits a positive Lebesgue number, i.e., a number $\lambda > 0$ such that any subset of X of diameter less than λ is contained in one of the members of $\mathcal{U}$. If $\varepsilon < \lambda/2$ and A is (n, ε)-spanning, then

$$X = \bigcup_{x\in A} \bigcap_{k=0}^{n-1} f^{-k}(B(f^k(x), \varepsilon))$$

and each ball $B(f^k(x), \varepsilon)$ is contained in a member of $\mathcal{U}$. Therefore, any intersection $\bigcap_{k=0}^{n-1} f^{-k}(B(f^k(x), \varepsilon))$ is contained in a member of $\mathcal{U}^{(n)}$ and $\#A \geq N(\mathcal{U}^{(n)})$. Consequently, $r(n, \varepsilon) \geq N(\mathcal{U}^{(n)})$ and $r(\varepsilon) \geq h(f, \mathcal{U})$. □

The following properties follow easily from the above definitions.

Theorem 3.1.4. (i) *If the family of all the powers f^n, $n = 1, 2, \ldots$, is equicontinuous, then $h_{\text{top}}(f) = 0$. In particular, $h_{\text{top}}(f) = 0$ for any isometry f.*

(ii) *$h_{\text{top}}(f^k) = k \cdot h_{\text{top}}(f)$ for any $k \in \mathbb{N}$, $h_{\text{top}}(f^{-1}) = h_{\text{top}}(f)$ if f is a homeomorphism and, therefore, $h_{\text{top}}(f^k) = |k| \cdot h_{\text{top}}(f)$ for any homeomorphism f and any $k \in \mathbb{Z}$.*

(iii) *If $f_i : X_i \to X_i$, $i = 1, 2$, then $h_{\text{top}}(f_1 \times f_2) = h_{\text{top}}(f_1) + h_{\text{top}}(f_2)$.*

(iv) *If $Y \subset X$ is closed and f-invariant, $f(Y) \subset Y$, then $h_{\text{top}}(f|Y) \leq h_{\text{top}}(f)$.*

(v) *If $f_i : X_i \to X_i$, $i = 1, 2$, are semiconjugated via a continuous surjective map $F : X_1 \to X_2$ (i.e., $f_2 \circ F = F \circ f_1$), then $h_{\text{top}}(f_1) \geq h_{\text{top}}(f_2)$. In particular, $h_{\text{top}}(f_1) = h_{\text{top}}(f_2)$ when f_1 and f_2 are conjugated via a homeomorphism F.* □

Given a flow $\phi = (\phi_t)$ of continuous transformations, its topological entropy $h_{\text{top}}(\phi)$ is defined as that of its time-1 map:

$$h_{\text{top}}(\phi) = h_{\text{top}}(\phi_1). \tag{3.1.12}$$

Some properties of the flow entropy similar to those in Theorem 3.1.4 can be derived directly from this theorem.

Remark 3.1.5. In **[LW2]**, the reader can find a number of equivalent definitions of topological entropy for flows and continuous transformations, some of them involving use of partitions of unity. For example, if f is a continuous transformation of a compact metric polyhedron X, then $h_{\text{top}}(f)$ equals the largest lower bound for the numbers h satisfying the following condition:

For any finite, piecewise continuous partition of unity $\phi = \{\phi_1, \ldots, \phi_m\}$ on X there exist arbitrarily large $k \in \mathbb{N}$ for which

$$\limsup_{n\to\infty} \frac{1}{n} \log N(f^k, \phi, n) \leq hk,$$

where $N(f,\phi,n) = N(\phi^{(n)})$, $\phi^{(n)}$ is the partition of unity consisting of all the functions of the form

$$\phi_{i_0} \cdot (\phi_{i_1} \circ f) \dots (\phi_{i_{n-1}} \circ f^{n-1}), \quad i_0, i_1, \dots, i_{n-1} \in \{1, \dots, m\},$$

and

$$N(\psi) = \sum_i \sup \psi_i$$

for any finite partition of unity $\psi = (\psi_i)$.

3.1.2 Invariant measures

Now, let us consider the space $\mathcal{M}(X)$ of all Borel probability measures on X. Due to the Riesz Representation Theorem [**Rud**], $\mathcal{M}(X)$ can be identified with the space of non-negative ($\phi \geq 0 \Rightarrow \mu(\phi) \geq 0$) normed ($\mu(1) = 1$) linear functionals on the space $C(X)$ of real continuous functions. Equipped with the weak-$*$ topology, $\mathcal{M}(X)$ becomes compact and metrizable. A metric d^* on $\mathcal{M}(X)$ providing this topology can be defined by

$$d^*(\mu,\nu) = \sum_{n=1}^{\infty} \frac{1}{2^n \|\phi_n\|} \cdot \left| \int \phi_n d\mu - \int \phi_n d\nu \right|, \tag{3.1.13}$$

where $\{\phi_n, n \in \mathbb{N}\}$ is a dense subset of $C(X)$ and $\|\cdot\|$ is the supremum norm.

Given a continuous transformation $f : X \to X$, the set $\mathcal{M}(X,f)$ of all f-*invariant* measures can be considered. $\mathcal{M}(X,f)$ consists of all $\mu \in \mathcal{M}(X)$ such that $f^*\mu = \mu$, where

$$f^*\mu(A) = \mu(f^{-1}(A)) \tag{3.1.14}$$

for all Borel sets $A \subset X$. The space $\mathcal{M}(X,f)$ is non-empty. In fact, if (μ_n) is any sequence of Borel measures on X and

$$\nu_n = \frac{1}{n} \sum_{k=0}^{n-1} (f^k)^* \mu_k, \quad n = 1, 2, \dots \tag{3.1.15}$$

are so called *Cesàro means* for measures (μ_n), then (ν_n) contains a convergent subsequence. It is easy to check that its limit $\nu \in \mathcal{M}(X)$ is f-invariant. Moreover, $\mathcal{M}(X,f)$ is closed (hence, compact) and convex: if $\mu_1, \mu_2 \in \mathcal{M}(X,f)$ and $t \in [0,1]$, then $t\mu_1 + (1-t)\mu_2 \in \mathcal{M}(X,f)$.

It is well known that $\mathcal{M}(X,f)$ coincides with the convex hull of the set $\mathcal{E}(X,f)$ of its extreme points: $\mu \in \mathcal{E}(X,f)$ whenever the relation $\mu = t\mu_1 + (1-t)\mu_2$ with $\mu_1, \mu_2 \in \mathcal{M}(X,f)$ and $t \in [0,1]$ implies that either $t = 0$ or $t = 1$.

Recall that $\mu \in \mathcal{E}(X,f)$ if and only if μ is *ergodic*, i.e., all f-invariant subsets A of X satisfy either $\mu(A) = 0$ or $\mu(A) = 1$. (Here, a set $A \subset X$ is f-invariant when $f^{-1}(A) = A$.) In fact, if $A \subset X$ were f-invariant and $0 < \mu(A) < 1$, then $\mu = \mu(A)\mu_1 + (1-\mu(A))\mu_2$, where the elements μ_1 and μ_2 of $\mathcal{M}(X,f)$ are given by $\mu_1(B) = \mu(A)^{-1} \cdot \mu(A \cap B)$ and $\mu_2(B) = \mu(X \setminus A)^{-1} \cdot \mu(B \setminus A)$ for all Borel sets $B \subset X$. Another condition equivalent to ergodicity is that any f-invariant ($\phi(f(x)) = \phi(x)$ for all $x \in X$) measurable (or, L^2) function $\phi : X \to \mathbb{R}$ is constant μ-a.e.

Also, recall that the Choquet Representation Theorem asserts the existence of unique Borel probability measures τ_μ, $\mu \in \mathcal{M}(X, f)$, on $\mathcal{E}(X, f)$ for which

$$\mu = \int_{\mathcal{E}(X,f)} m d\tau_\mu(m), \tag{3.1.16}$$

that is,

$$\int_X \phi d\mu = \int_{\mathcal{E}(X,f)} \left(\int_X \phi dm\right) d\tau_\mu(m)$$

for any $\phi \in C(X)$.

Let us also recall that a point $x \in X$ is said to be *wandering* when there exists an open neighbourhood U of x such that all the sets $f^{-n}(U)$, $n \geq 0$, are mutually disjoint. Points which are not wandering are called *non-wandering.* All the non-wandering points form the *non-wandering set* $\Omega(f)$ of a transformation f. It is easy to prove that, if X is compact, it is always non-empty and closed, so compact. Also, $\Omega(f)$ is f-invariant in the sense that $f(\Omega(f)) \subset \Omega(f)$. If f is a homeomorphism, $f(\Omega(f)) = \Omega(f)$.

Finally, let us recall that the *support* $\operatorname{supp} \mu$ of a measure μ on X is defined as the complement of the collection of all the points $x \in X$ which admit an open neighbourhood U such that $\mu(U) = 0$.

If μ is an f-invariant probability measure on X, x is wandering and U is a neighbourhood of x such that $f^{-n}(U) \cap f^{-m}(U) = \emptyset$ for all $m \neq n$, then $1 = \mu(X) \geq \mu(\bigcup_n f^{-n}(U)) = \sum_n \mu(f^{-n}(U))$ and $\mu(f^{-n}(U) = \mu(U)$ for all n. Therefore, $\mu(U) = 0$ and we get the following.

Proposition 3.1.6. $\operatorname{supp} \mu \subset \Omega(f)$ *for any* $\mu \in \mathcal{M}(X, f)$. □

3.1.3 Measure-theoretic entropy

Let $f : X \to X$ be a measurable transformation preserving a probability measure μ.

Definition 3.1.7. A *partition* of X is a finite family $\mathcal{A}$ of pairwise disjoint measurable subsets of X such that $\cup\mathcal{A} = X$. Given two partitions $\mathcal{A}$ and $\mathcal{B}$ of X we put

$$\mathcal{A} \vee \mathcal{B} = \{A \cap B; A \in \mathcal{A} \text{ and } B \in \mathcal{B}\}, \tag{3.1.17}$$

$$f^{-1}\mathcal{A} = \{f^{-1}(A); A \in \mathcal{A}\} \tag{3.1.18}$$

and

$$\mathcal{A}^{(n)} = \mathcal{A} \vee f^{-1}\mathcal{A} \vee \cdots \vee f^{-(n-1)}\mathcal{A}. \tag{3.1.19}$$

We say that $\mathcal{A} \prec \mathcal{B}$ when $\mathcal{B}$ is a refinement of $\mathcal{A}$, i.e., when each element of $\mathcal{A}$ coincides with the union of some elements of $\mathcal{B}$. Also, let us put

$$H_\mu(\mathcal{A}) = -\sum_{A\in\mathcal{A}} \mu(A)\log\mu(A), \tag{3.1.20}$$

$$H_\mu(\mathcal{A}|\mathcal{B}) = -\sum_{A\in\mathcal{A}}\sum_{B\in\mathcal{B}} \mu(A\cap B)\log\frac{\mu(A\cap B)}{\mu(B)}, \tag{3.1.21}$$

and

$$H_\mu(f,\mathcal{A}) = \lim_{n\to\infty}\frac{1}{n}H_\mu(\mathcal{A}^{(n)}). \tag{3.1.22}$$

$H_\mu(\mathcal{A})$ is the *entropy* of $\mathcal{A}$, $H_\mu(\mathcal{A}|\mathcal{B})$ — the *conditional entropy* of $\mathcal{A}$ given $\mathcal{B}$ and $H_\mu(f,\mathcal{A})$ — the *entropy* of f with respect to $\mathcal{A}$. Finally,

$$H_\mu(f) = \sup_{\mathcal{A}} H_\mu(f,\mathcal{A}) \tag{3.1.23}$$

is the *measure theoretic entropy* of f with respect to μ.

To deal with H_μ we have to collect some properties of the entropies defined above. The proofs are rather elementary and probably can be done by the reader. If this proves to be difficult, we refer to [**Wa**], pp. 77–94.

Proposition 3.1.8. *For any partitions $\mathcal{A}$, $\mathcal{B}$ and $\mathcal{C}$ of X we have the following.*

(i) $H_\mu(\mathcal{A}\vee\mathcal{B}|\mathcal{C}) = H_\mu(\mathcal{A}|\mathcal{C}) + H_\mu(\mathcal{B}|\mathcal{A}\vee\mathcal{C})$,

(ii) $H_\mu(\mathcal{A}\vee\mathcal{B}) = H_\mu(\mathcal{A}) + H_\mu(\mathcal{B}|\mathcal{A})$,

(iii) $\mathcal{A}\prec\mathcal{B}\Rightarrow H_\mu(\mathcal{A}|\mathcal{C})\le H_\mu(\mathcal{B}|\mathcal{C})$,

(iv) $\mathcal{A}\prec\mathcal{B}\Rightarrow H_\mu(\mathcal{A})\le H_\mu(\mathcal{B})$,

(v) $\mathcal{B}\prec\mathcal{C}\Rightarrow H_\mu(\mathcal{A}|\mathcal{B})\ge H_\mu(\mathcal{A}|\mathcal{C})$,

(vi) $H_\mu(\mathcal{A})\ge H_\mu(\mathcal{A}|\mathcal{B})$,

(vii) $H_\mu(\mathcal{A}\vee\mathcal{B}|\mathcal{C})\le H_\mu(\mathcal{A}|\mathcal{C}) + H_\mu(\mathcal{B}|\mathcal{C})$,

(viii) $H_\mu(\mathcal{A}\vee\mathcal{B})\le H_\mu(\mathcal{A}) + H_\mu(\mathcal{B})$,

(ix) $H_\mu(f^{-1}\mathcal{A}|f^{-1}\mathcal{B}) = H_\mu(\mathcal{A}|\mathcal{B})$, *and*

(x) $H_\mu(f^{-1}\mathcal{A}) = H_\mu(\mathcal{A})$. □

Proposition 3.1.8 has also the following consequences.

Corollary 3.1.9. *If $\mathcal{A}$ and $\mathcal{B}$ are partitions of X, $f: X\to X$ and μ is an invariant probability measure on X, then*

(i) $H_\mu(f,\mathcal{A})\le H_\mu(\mathcal{A})$,

(ii) $H_\mu(f,\mathcal{A}\vee\mathcal{B})\le H_\mu(f,\mathcal{A}) + H_\mu(f,\mathcal{B})$,

(iii) $\mathcal{A}\prec\mathcal{B}\Rightarrow H_\mu(f,\mathcal{A})\le H_\mu(f,\mathcal{B})$,

(iv) $H_\mu(f,\mathcal{A})\le H_\mu(f,\mathcal{B}) + H_\mu(\mathcal{A}|\mathcal{B})$,

(v) $H_\mu(f, f^{-1}\mathcal{A}) = H_\mu(f, \mathcal{A})$,

(vi) $m > 0 \Rightarrow H_\mu(f, \mathcal{A}) = H_\mu(f, \mathcal{A}^{(m)})$, *and*

(vii) $H_\mu(f, \mathcal{A}) = H_\mu(f, \bigvee_{k=-m}^{m} f^k \mathcal{A})$ *whenever f is invertible and $m > 0$.* □

3.1.4 Examples

In this section, we collect some examples.

Example 3.1.10. The normed Lebesgue measure λ on S^1 is invariant with respect to the maps $f_k : S^1 \to S^1$, $f_k(z) = z^k$, $k \in \mathbb{N}$. It is ergodic when $k > 1$ and $h_{\text{top}}(f_k) = H_\lambda(f_k) = \log k$.

In fact, if $A \subset S^1$ is an arc of length a, then $f_k^{-1}(A)$ is the union of k pairwise disjoint arcs of length a/k each and, therefore, $\lambda(f_k^{-1}(A)) = \lambda(A)$. This proves that $\lambda \in \mathcal{M}(S^1, f_k)$. If $\phi \in L^2(\lambda)$ is f_k-invariant and (a_n) are the Fourier coefficients of ϕ, $\phi(z) = \sum_{n=-\infty}^{\infty} a_n z^n$, then $(\phi \circ f_k)(z) = \sum_{n=-\infty}^{\infty} a_n z^{kn}$. Therefore, $a_n = a_{kn} = a_{k^m n}$ for all $m \in \mathbb{N}$ and $a_n = 0$ for all $n > 0$ since the series $\sum_{n=-\infty}^{\infty} |a_n|$ has to converge. Consequently, $\phi =$const. a. e. and this proves that λ is ergodic.

Moreover, if ε is small enough and $n \in \mathbb{N}$, then two points x and y of S^1 are (n, ε) separated if and only if the distance between them exceeds $k^{-n}\varepsilon$. Therefore, $N(n, \varepsilon) \approx 2\pi\varepsilon \cdot k^n$ up to an error not greater than 1. Consequently, $N(\varepsilon) = \log k$ for all $\varepsilon > 0$ and this gives the desired value of the topological entropy of f_k. Also, if $\mathcal{A}$ is the partition of S^1 into the arcs with end points at the k-roots of unity, then $f_k^{-m}\mathcal{A}$ consists of all the arcs with end points at the k^m-roots and, therefore, $\mathcal{A}^n$ consists of k^m arcs of measure k^{-m} each. Consequently,

$$H_\lambda(f_k) \geq H_\lambda(f_k, \mathcal{A}) = -\lim_{m\to\infty} \frac{1}{m} \log k^{-m} = \log k. \tag{3.1.24}$$

The desired equality $H_\lambda(f_k) = \log k$ follows from the above estimate and the variational principle below (Section 3.1.5).

Example 3.1.11. For any measure preserving transformation f of a probability measure space (X, μ) and any $k > 0$ the equality

$$H_\mu(f^k) = kH_\mu(f) \tag{3.1.25}$$

holds. In fact, if $\mathcal{A}$ is a partition, then

$$H_\mu(f^k, \mathcal{A}) \leq H_\mu(f^k, \mathcal{A}^{(k)}) \tag{3.1.26}$$

by Corollary 3.1.9 (iii) and

$$\begin{aligned} H_\mu(f^k, \mathcal{A}^{(k)}) &= \lim_{n\to\infty} \frac{1}{n} H_\mu\Big(\bigvee_{j=0}^{n-1} f^{-kj} \bigvee_{i=0}^{k-1} f^{-i}\mathcal{A}\Big) \\ &= k \cdot \lim_{n\to\infty} \frac{1}{nk} H_\mu\Big(\bigvee_{j=0}^{nk-1} f^{-j}\mathcal{A}\Big) = kH_\mu(f, \mathcal{A}). \end{aligned} \tag{3.1.27}$$

Relations (3.1.26) and (3.1.27) yield (3.1.25) directly.

Similarly, Proposition 3.1.8 (x) shows that

$$H_\mu(\bigvee_{j=0}^{n-1} f^j\mathcal{A}) = H_\mu(f^{-(n-1)}\bigvee_{j=0}^{n-1} f^j\mathcal{A}) = H_\mu(\bigvee_{j=0}^{n-1} f^{-j}\mathcal{A}),$$

and, therefore, $H_\mu(f^{-1},\mathcal{A}) = H_\mu(f,\mathcal{A})$ whenever f is invertible and $\mathcal{A}$ is a finite partition. Consequently,

$$H_\mu(f^{-1}) = H_\mu(f) \tag{3.1.28}$$

and

$$H_\mu(f^k) = |k| \cdot H_\mu(f) \tag{3.1.29}$$

for any invertible transformation f of X, any $\mu \in \mathcal{M}(X,f)$ and $k \in \mathbb{Z}$.

Example 3.1.12. The topological entropy of any homeomorphism f of a circle vanishes. In fact, suppose that our circle has length 1 and choose $\varepsilon > 0$ such that $d(f^{-1}(x), f^{-1}(y)) < \frac{1}{4}$ if only $d(x,y) < \varepsilon$, d being the natural distance function on the circle. We shall show that

$$r(n,\varepsilon) \le n([(1/\varepsilon)] + 1) \tag{3.1.30}$$

for all $n \in \mathbb{N}$.

Clearly, (3.1.30) holds for $n = 1$. Assume that A is an (n,ε)-spanning set of cardinality $\#A = r(n,\varepsilon) \le n([1/\varepsilon] + 1)$. Consider the set $f^n(A)$ and all the arcs determined by its consecutive points. At most $[1/\varepsilon]+1$ of the arcs has length $\ge \varepsilon$. Therefore, there exists a subset B of the circle such that $\#B \le [1/\varepsilon] + 1$ and all the arcs determined by consecutive points of $f^n(A) \cup B$ have length less than ε. Let $A' = A \cup f^{-n}(B)$. For any point x there exists $y \in A$ such that

$$\max_{0\le k\le n-1} d(f^k(x), f^k(y)) \le \varepsilon. \tag{3.1.31}$$

If $d(f^n(x), f^n(y)) \le \varepsilon$, then

$$\max_{0\le k\le n} d(f^k(x), f^k(y)) \le \varepsilon. \tag{3.1.32}$$

If there are no points $y \in A$ satisfying (3.1.32), then we can find $y \in A$ and $z \in A'$ such that one of the arcs, say I, with end points $f^n(x)$ and $f^n(y)$ is mapped by f^{-1} onto an arc I_1 of length $\le \varepsilon$, $f^n(z) \in I$ and $d(f^n(x), f^n(z)) \le \varepsilon$. Then, $f^{n-1}(z) \in I_1$ and therefore $d(f^{n-1}(x), f^{n-1}(z)) \le \varepsilon$. Next, the arc I_1 is mapped by f^{-1} onto an arc I_2 of length $< \frac{1}{4}$. By (3.1.31), the length of I_2 is $\le \varepsilon$. Since, $f^{n-2}(z) \in I_2$, $d(f^{n-2}(x), f^{n-2}(z)) \le \varepsilon$. By induction,

$$\max_{0\le k\le n} d(f^k(x), f^k(z)) \le \varepsilon. \tag{3.1.33}$$

Inequalities (3.1.32) and (3.1.33) show that A' is $(n+1,\varepsilon)$-spanning. Clearly, $\#A' \le (n+1)([1/\varepsilon]+1)$.

Finally, (3.1.30) shows immediately that $h_{\text{top}}(f) = 0$. Moreover, the variational principle below implies that $H_\mu(f) = 0$ for any f-invariant measure μ and any homeomorphism f of a circle. It is not too hard to see that the same results hold for homeomorphisms of closed intervals.

3.1.5 Variational principle

Again, let f be a continuous transformation of a compact metric space X. The topological entropy and measure-theoretic entropies of f are related by the following theorem called the variational principle. The proof presented here is due to Misiurewicz [**Mis1**].

Theorem 3.1.13. $h_{\text{top}}(f) = \sup\{H_\mu(f);\ \mu \in \mathcal{M}(X,f)\}$.

Proof. First, we shall prove that

$$H_\mu(f) \le h_{\text{top}}(f) \tag{3.1.34}$$

for any $\mu \in \mathcal{M}(X,T)$.

To this end, take any finite partition $\mathcal{A} = \{A_1,\dots,A_m\}$ of X and a positive number ε such that $\varepsilon < 1/(m\log m)$. Find compact sets $B_j \subset A_j$, $j = 1,\dots,m$, with $\mu(A_j \setminus B_j) < \varepsilon$. Let $\mathcal{B}$ be the partition of X consisting of all B_j's, $j = 1,\dots,m$, and of $B_0 = X \setminus \bigcup_{j\ge1} B_j$. Since $\mu(B_0) \le m\varepsilon$ and the quotient $\mu(B_i \cap A_j)/\mu(B_i)$, $i > 0$, can be equal either to 0 (when $i \ne j$) or to 1 (otherwise), we have

$$\begin{aligned}
H_\mu(\mathcal{A}|\mathcal{B}) &= -\sum_{i=0}^{m}\sum_{j=1}^{m}\mu(B_i)\phi\left(\frac{\mu(B_i\cap A_j)}{\mu(B_i)}\right)\\
&= -\mu(B_0)\sum_{j=1}^{m}\phi\left(\frac{\mu(B_0\cap A_j)}{\mu(B_0)}\right) \le \mu(B_0)\log m < m\varepsilon\log m < 1,
\end{aligned}$$

where ϕ is the function defined by

$$\phi(0) = 0 \quad \text{and} \quad \phi(t) = t\cdot\log t \quad \text{when} \quad t > 0 \tag{3.1.35}$$

.

The sets $B_0 \cup B_j$, $j \ge 1$, are open and form a finite covering $\mathcal{U}$ of X. From the properties of ϕ it follows easily that, for any $n \ge 1$, $H_\mu(\mathcal{B}^{(n)}) \le \log N(\mathcal{B}^{(n)})$, $N(\mathcal{B}^{(n)})$ being the number of non-empty sets in the partition $\mathcal{B}^{(n)} = \mathcal{B} \vee f^{-1}\mathcal{B} \vee \dots \vee f^{1-n}\mathcal{B}$. Also, $N(\mathcal{B}^{(n)}) \le 2^n \cdot N(\mathcal{U}^{(n)})$ and therefore

$$H_\mu(f,\mathcal{B}) \le h(f,\mathcal{U}) + \log 2 \le h_{\text{top}}(f) + \log 2.$$

Also, Corollary 3.1.9 (iv) implies that

$$H_\mu(f,\mathcal{A}) \le H_\mu(f,\mathcal{B}) + H_\mu(\mathcal{A}|\mathcal{B}) \le h_{\text{top}}(f) + \log 2 + 1.$$

Consequently,

$$H_\mu(f) \le h_{\text{top}}(f) + \log 2 + 1 \tag{3.1.36}$$

for all f and $\mu \in \mathcal{M}(X, f)$. Applying (3.1.36) to the transformations f^m and making use of (3.1.25) and Theorem 3.1.4 (ii) we arrive at the inequalities

$$H_\mu(f) \le h_{\text{top}}(f) + m^{-1}(\log 2 + 1), \quad m = 1, 2, \dots.$$

Passing to the limit as $m \to \infty$ we get (3.1.34).

Now, given $\varepsilon > 0$, we shall find a measure $\mu \in \mathcal{M}(X, f)$ such that

$$H_\mu(f) \ge s(\varepsilon). \tag{3.1.37}$$

Clearly, this will complete the proof.

Let C_n be an (n, ε)-separated set of cardinality $s(n, \varepsilon)$. Let

$$\nu_n = \frac{1}{s(n,\varepsilon)} \sum_{x \in C_n} \delta_x, \tag{3.1.38}$$

where δ_x is the Dirac measure concentrated at x, and

$$\mu_n = \frac{1}{n} \sum_{k=0}^{n-1} f^{k*} \nu_n. \tag{3.1.39}$$

As in Section 3.1.2, we can choose a subsequence (μ_{n_j}) converging to a limit $\mu \in \mathcal{M}(X, f)$. We shall show that μ satisfies (3.1.37).

Choose a partition $\mathcal{A} = \{A_1, \dots, A_m\}$ of X such that $\operatorname{diam} A_i < \varepsilon$ and $\mu(\partial A_i) = 0$ for $i = 1, \dots, k$. Such $\mathcal{A}$ exists. In fact, for any $x \in X$ and $\eta > 0$ there exists $\eta' \in (0, \eta)$ such that $\mu(\partial B(x, \eta')) = 0$. Therefore, we can cover X by open balls B_i, $i = 1, \dots, m$, of radius less than $\varepsilon/2$ and such that $\mu(\partial B_i) = 0$ for all i. The sets $A_1 = \overline{B}_1$ and $A_j = \overline{B}_j \setminus (\overline{B}_1 \cup \dots \cup \overline{B}_{j-1})$, $j = 2, \dots, m$, form a partition as desired.

For this partition $\mathcal{A}$ we have

$$H_{\nu_n}(\mathcal{A}^{(n)}) = \log s(n, \varepsilon). \tag{3.1.40}$$

In fact, since any member of $\mathcal{A}^{(n)}$ contains at most one point of C_n, there are exactly $s(n, \varepsilon)$ members of $\mathcal{A}^{(n)}$ of ν_n-measure $s(n, \varepsilon)^{-1}$ each while all the other members A of $\mathcal{A}^{(n)}$ satisfy $\nu_n(A) = 0$.

Now, fix $q \in \mathbb{N}$ and let $a(k) = [(n-k)/q]$ for $k = 0, 1, \dots, q-1$. Then, for any $k \le q - 1$,

$$\{1, \dots, n\} = S \cup S', \tag{3.1.41}$$

where

$$\begin{aligned} S &= \{0, 1, \dots k-1, k + a(k)q, k + a(k)q + 1, \dots, n\} \quad \text{and} \\ S' &= \{k + rq + j; 0 \le r \le a(k) - 1, 0 \le j \le q - 1\}. \end{aligned} \tag{3.1.42}$$

Clearly, $\#S \le 2q$.

From (3.1.41) and (3.1.42) it follows that

$$\mathcal{A}^{(n)} = \bigvee_{r=0}^{a(k)-1} f^{-(rq+k)} \bigvee_{j=0}^{q-1} f^{-j}(\mathcal{A}) \vee \bigvee_{i\in S} f^{-1}\mathcal{A}$$

and

$$\begin{aligned}\log s(n,\varepsilon) &\le \sum_{r=0}^{a(k)-1} H_{\nu_n}\left(f^{-(rq+k)} \bigvee_{j=0}^{q-1} f^{-j}\mathcal{A}\right) + \sum_{i\in S} H_{\nu_n}(f^{-i}\mathcal{A}) \\ &\le \sum_{r=0}^{a(k)-1} H_{f^{(rq+k)*}\nu_n}(\mathcal{A}^{(q)}) + 2q\log m.\end{aligned} \tag{3.1.43}$$

Since the function ϕ, $\phi(t) = -t\log t$, is convex,

$$H_{\sum_i p_i\sigma_i}(\mathcal{C}) \ge \sum_i p_i H_{\sigma_i}(\mathcal{C}) \tag{3.1.44}$$

for any partition $\mathcal{C}$ of X, arbitrary measures $\sigma_i \in \mathcal{M}(X)$ and non-negative numbers p_i such that $\sum_i p_i = 1$. Therefore, summing inequalities (3.1.43) over k ranging from 0 to $q-1$ we obtain the relation

$$\frac{q}{n}\log s(n,\varepsilon) \le H_{\mu_n}(\mathcal{A}^{(q)}) + \frac{2q^2}{n}\log m. \tag{3.1.45}$$

Finally, since the boundaries of the members of $\mathcal{A}$ have μ-measure 0, $\mu(A) = \lim_{j\to\infty}\mu_{n_j}(A)$ for any $A \in \mathcal{A}^{(q)}$. Therefore,

$$H_\mu(\mathcal{A}^{(q)}) = \lim_{j\to\infty} H_{\mu_{n_j}}(\mathcal{A}^{(q)}) \tag{3.1.46}$$

for any $q \in \mathbb{N}$. Replacing n by n_j in (3.1.45) and letting j go to ∞ we arrive at

$$qs(\varepsilon) \le H_\mu(\mathcal{A}^{(q)}), \quad q = 1, 2, \ldots. \tag{3.1.47}$$

Clearly, (3.1.47) implies (3.1.37). □

From the above variational principle and Proposition 3.1.6 we obtain immediately the following.

Corollary 3.1.14. *For any continuous transformation f of a compact space X the equality*

$$h_{\text{top}}(f) = h_{\text{top}}(f|\Omega(f)), \tag{3.1.48}$$

where $\Omega(f)$ is the set of all the non-wandering points of f, holds. □

3.2 Entropy of pseudogroups

Let $\mathcal{G}$ be a good pseudogroup on a compact metric space (X, d) generated by a good finite symmetric set $\mathcal{G}_1$. As in Section 2.4, let $s(n, \varepsilon, \mathcal{G}_1)$ denote the maximal number of (n, ε)-separated points of X. Let

$$s(\varepsilon, \mathcal{G}_1) = \limsup_{n \to \infty} \frac{1}{n} \log s(n, \varepsilon, \mathcal{G}_1) \tag{3.2.1}$$

and

$$h(\mathcal{G}, \mathcal{G}_1) = \lim_{\varepsilon \to 0} s(\varepsilon, \mathcal{G}_1). \tag{3.2.2}$$

Obviously, the limit (either finite or infinite) in (3.2.2) exists.

Definition 3.2.1. $h(\mathcal{G}, \mathcal{G}_1)$ is called the (topological) *entropy* of $\mathcal{G}$ with respect to $\mathcal{G}_1$.

Remark 3.2.2. Clearly, $h(\mathcal{G}, \mathcal{G}_1)$ can be defined in terms of open covers, analogously to the case of single continuous transformations (Definition 3.1.1). Recently, Blanchard, Host and Maas [**BHM**] considered the topological complexity function $c(f, \mathcal{U}) : \mathbb{N} \to \mathbb{N}$ related to a continuous transformation f of a compact metric space X and a finite open cover $\mathcal{U}$ of X: $c(f, \mathcal{U})(n)$ is just equal to $N(f, \mathcal{U}, n)$ in (3.1.3). This notion can be easily generalized to the case of a finitely generated pseudogroup $\mathcal{G}$ of local transformations of X. The type of growth of a sequence of such complexity functions $c(\mathcal{G}, \mathcal{G}_1, \mathcal{U}_n)$, where $\operatorname{diam}(\mathcal{U}_n) \to 0$ as $n \to \infty$, can be compared to the Egashira's expansion growth (Section 2.4) of $\mathcal{G}$. We leave this to the reader as an exercise.

If G is a finitely generated group of homeomorphisms of X, then the entropy of G with respect to a finite generating set G_1 is defined as the entropy of the pseudogroup $\mathcal{G} = \mathcal{G}(G_1)$ generated by G_1 (with respect to the same G_1, of course). Therefore, we may write

$$h(G, G_1) = h(\mathcal{G}(G_1), G_1). \tag{3.2.3}$$

Clearly, the entropy of $\mathcal{G}$ depends strongly on the choice of $\mathcal{G}_1$. In fact, if $\mathcal{G}_1'$ is another good generating set, then there exists $m \in \mathbb{N}$ such that

$$s(n, \varepsilon, \mathcal{G}_1) \le s(mn, \varepsilon, \mathcal{G}_1') \quad \text{and} \quad s(n, \varepsilon, \mathcal{G}_1') \le s(mn, \varepsilon, \mathcal{G}_1) \tag{3.2.4}$$

for all n and ε (compare (2.4.4) in Section 2.4). Consequently,

$$\frac{1}{m} h(\mathcal{G}, \mathcal{G}_1) \le h(\mathcal{G}, \mathcal{G}_1') \le m h(\mathcal{G}, \mathcal{G}_1). \tag{3.2.5}$$

However, if $h(\mathcal{G}, \mathcal{G}_1) = 0$ for one generating set $\mathcal{G}_1$, then $h(\mathcal{G}, \mathcal{G}_1') = 0$ for any other good generating set $\mathcal{G}_1'$. Therefore, we can distinguish pseudogroups with vanishing entropy from those with non-vanishing entropy. Using the terminology of Sections 2.1 and 2.4 we can say that $\mathcal{G}$ has vanishing (resp., non-vanishing) entropy if and only if $\mathcal{G}$ has subexponential (resp., quasi-exponential) expansion growth. The results of Chapter 2 imply directly the following

Proposition 3.2.3. *If X is either the closure of a relatively compact open subset of $\mathbb{R}^m$ or an m-dimensional Riemannian manifold, $\mathcal{G}$ is a good pseudogroup on X generated by a finite set $\mathcal{G}_1$ of Lipschitz homeomorphisms and a is the maximum of Lipschitz constants of the members of $\mathcal{G}_1$, then*

$$h(\mathcal{G}, \mathcal{G}_1) \leq m \log a. \tag{3.2.6}$$

Proof. Our statement follows directly from inequality (2.4.11) of Chapter 2. In fact, this inequality shows that $s(\varepsilon, \mathcal{G}_1) \leq m \log a$ for any $\varepsilon > 0$. □

Corollary 3.2.4. *The entropy of any good pseudogroup of local C^1-diffeomorphisms of any compact Riemannian manifold M is finite.* □

Let us discuss now some simple examples.

Example 3.2.5. Assume that $\mathcal{G}'$ is a subpseudogroup of a pseudogroup $\mathcal{G}$ generated by a finite set $\mathcal{G}_1$. Choose a finite set $\mathcal{G}'_1$ generating $\mathcal{G}'$. Then, $\mathcal{G}'_1 \subset \mathcal{G}_k$ for some $k \geq 1$, any two points (n, ε)-separated with respect to $\mathcal{G}'_1$ are (kn, ε)-separated w.r.t $\mathcal{G}_1$ and therefore

$$h(\mathcal{G}', \mathcal{G}'_1) \leq \frac{1}{k} h(\mathcal{G}, \mathcal{G}_1).$$

In particular, if $\mathcal{G}'_1 \subset \mathcal{G}_1$, then $h(\mathcal{G}', \mathcal{G}'_1) \leq h(\mathcal{G}, \mathcal{G}_1)$.

Example 3.2.6. Pseudogroups of local isometries have entropy zero ([**GLW2**], Ex. 2.3). If $f : X \to X$ is a homeomorphism and $\mathcal{G}(f)_1 = \{\mathrm{id}, f, f^{-1}\}$, then

$$h(\mathcal{G}(f), \mathcal{G}(f)_1) = 2h_{\mathrm{top}}(f).$$

In fact, a set $A \subset X$ is (n, ε)-separated with respect to $\mathcal{G}(f)$ if and only if the set $f^{-n}(A)$ is $(2n, \varepsilon)$-separated with respect to f. Consequently,

$$s(n, \varepsilon, \mathcal{G}(f)_1) = s(2n, \varepsilon, f)$$

for all n and ε. Also, if $f : S^1 \to S^1$, $f(z) = z^k$ ($k \in \mathbb{N}$), $\mathcal{G} = \mathcal{G}(f)$, $\mathcal{G}_1 = \{\mathrm{id}, f^{\pm 1}|U, U \in \mathcal{U}\}$ and $\mathcal{U}$ is a finite covering of S^1 such that $f|U$ and $f^{-1}|U$, $U \in \mathcal{U}$, are one-to-one, then $h(\mathcal{G}, \mathcal{G}_1) = 2h_{\mathrm{top}}(f) = 2 \log k$. If G is a group generated by a finite set G_1 of homeomorphisms of X, then

$$h(G, G_1) \geq 2 \max\{h_{\mathrm{top}}(f); f \in G_1\}. \tag{3.2.7}$$

The equality in (3.2.7) need not hold. In fact, if $X = T^2$,

$$f = \begin{pmatrix} 1 & 1 \\ 0 & 1 \end{pmatrix} \quad \text{and} \quad h = \begin{pmatrix} 1 & 0 \\ 1 & 1 \end{pmatrix}$$

are Dehn twists, then $h_{\mathrm{top}}(f) = h_{\mathrm{top}}(h) = 0$ while

$$f \circ h = \begin{pmatrix} 2 & 1 \\ 1 & 1 \end{pmatrix}$$

has positive topological entropy. Therefore $h(G(f,h),G_1) > 0$ and $\max\{h_{\text{top}}(g); g \in G_1\} = 0$ when $G_1 = \{\text{id}, f^{\pm 1}, h^{\pm 1}\}$. This shows that the entropy of $\mathcal{G}$ depends not only on the entropies of generators but also on other factors like the growth of $\mathcal{G}$ (or, $\mathcal{G}$-orbits), for instance.

Remark 3.2.7. Friedland [**Fr**] proposed to define the entropy $h(G)$ of a finitely generated group G of transformations of a space X by the formula

$$h(G) = \inf_{G_1} h(G, G_1), \tag{3.2.8}$$

where G_1 ranges over all finite symmetric generating sets. Of course, the same idea could be applied to pseudogroups. If $G = G(f)$ is generated by a single homeomorphism f of X and $G_1 = \{\text{id}, f^{\pm k_1}, \ldots, f^{\pm k_m}\}$ generates G, then (3.2.7) implies the inequality

$$h(G, G_1) \geq 2\max\{k_1, \ldots, k_m\} \cdot h_{\text{top}}(f)$$

which yields

$$h(G(f)) = 2 \cdot h_{\text{top}}(f). \tag{3.2.9}$$

It is of great interest to understand the geometric meaning of $h(G)$ in the case of a finitely generated Kleinian (or, just hyperbolic) group G acting on the boundary ∂G of its Cayley graph (compare Section 1.5). Observe that the metric on ∂G depends on the choice of generators but its topology does not. Therefore, the entropy $h(G)$ is independent of the choice of a generating set defining the Cayley graph. The interest in understanding $h(G)$ comes from the fact that it is defined entirely in terms of the group G .

For a free group F_k of k generators one has the equality

$$h(F_k) = \log(2k-1). \tag{3.2.10}$$

In fact, from relations (1.5.16), (1.5.20) and (1.5.21) in Chapter 1 it follows easily that two sequences (x_n) and (y_n) with $|x_n| = |y_n| = n$, $|\cdot|$ being the distance from the origin induced on F_k by the standard set G_1 of generators, determine points x_∞ and y_∞ of the boundary ∂F_k for which

$$d_1(x_\infty, y_\infty) = \exp(-\max\{j; x_j = y_j\}).$$

Therefore, given any set G_1' generating F_k, one has the inequality

$$s(n, \varepsilon, G_1') \geq \#\partial B(n), \quad n \in \mathbb{N},\ \varepsilon \leq 1, \tag{3.2.11}$$

where $B(r)$ is the ball of centre e and radius r in the distance induced on F_k by G_1'. It is clear that $s(n, \varepsilon, G_1') = \#\partial B(n) = 2k \cdot 3^{2k-1}$ when $G_1' = G_1$. This yields inequality "$\leq$" in (3.2.10). On the other hand, it is well known ([**GLP**], Ex. 5.13, see also [**GrH**], Section 1) that

$$\lim_{n\to\infty} \frac{1}{n} \log \#B(n) \geq \log(2k-1) \tag{3.2.12}$$

for any generating set G_1'. Clearly, (3.2.12) implies the similar inequality for the sets $\partial B(n)$ and hence inequality "$\geq$" in (3.2.10) follows from (3.2.11) and (3.2.12).

For other results in this direction, we refer to [**BW2**], where some relations between $h(G)$ and the growth of G have been established, and to [**Wat**] for the particular case of $G = \pi_1(\Sigma_g)$, the fundamental group of a closed surface Σ_g of genus $g \geq 2$ considered with the standard generating set $G_0 = \{a_1, \dots a_g, b_1, \dots b_g\}$ such that $a_1 b_1 a_1^{-1} b_1^{-1} \cdot \dots \cdot a_g b_g a_g^{-1} b_g^{-1} = e$ (compare Section 3.6.)

3.3 Geometric entropy of foliations

Orbits of a non-singular flow provide a foliation of dimension 1. For such a flow, the topological entropy is defined by formula (3.1.12) and it could be considered as entropy of the underlying foliation. However, this foliation is independent of the parametrization of the flow while the entropy of the flow does depend on it. Therefore, we have to find a suitable "substitute" of time parametrizing the flow. This role will be played here by a Riemannian metric on the foliated manifold under consideration.

The topological entropy of the flow measures the exponential rate of growth (as $R \to \infty$ and $\varepsilon \to 0$) of its orbits which are pairwise separated to distance $\geq \varepsilon$ for some time $t \leq R$. The role of orbits will be played here by curves contained in leaves (and of length $\leq R$). A curve like this is followed by another one, which lies on another leaf and is as close to the first as possible, i.e., is obtained from it by an "orthogonal projection" which has to be defined carefully. Note that the leaves can be even dense, so — while projecting curves — we have to choose at each time a suitable plaque where the image of our projection is to be situated.

So, let $\mathcal{F}$ be a foliation of a compact Riemannian manifold (M, g). Let d and $d_{\mathcal{F}}$ denote, respectively, the Riemannian distance function on M and the distance on the leaves coming from the induced Riemannian structure on the leaves. Also, let $l(\gamma)$ denote the length of a curve γ on M.

Choose for a moment a nice covering $\mathcal{U}$ of $(M, \mathcal{F})$. Since M is compact, its geometry is bounded and there exists a positive number $\varepsilon_0 > 0$ such that any point $x \in U$ ($U \in \mathcal{U}$) can be projected orthogonally (in the unique way) to the plaque $P_y \subset U$ passing through a point y as long as the distance of x and y does not exceed ε_0.

Now, given a leaf curve $\gamma : [a, b] \to L$, its origin $x = \gamma(a)$, a chart $U \in \mathcal{U}$ containing x and another point $y \in U$ within distance $\varepsilon < \varepsilon_0$ from x we can project orthogonally an initial piece of γ to the plaque P_y. Changing the chart, if necessary, and replacing x and y by end points of the projected piece and its image γ', we can continue the process of projecting as long as the distance between $\gamma(t)$ and $\gamma'(t)$ does not exceed ε_0.

In other words, $\gamma'(t)$ can be defined by

$$\gamma'(t) = \exp_{\gamma(t)} X(t), \tag{3.3.1}$$

where X is a continuous vector field along γ such that $X(t) \in T_{\gamma(t)}M$ is orthogonal to $\mathcal{F}$, $\|X\| < \varepsilon_0$, $\exp_{\gamma(a)} X(a) \in P_y$ and $\exp_{\gamma(t)} X(t)$ belongs to L', the leaf through y, for all t.

Since the projection γ' depends on γ and y, it will be denoted by $p_y\gamma$.

To define the entropy of $\mathcal{F}$, let us fix once again a nice covering $\mathcal{U}$, find ε_0 as above and denote by T the complete transversal determined by $\mathcal{U}$. Also, let us take any $\varepsilon \in (0, \varepsilon_0)$ and $R > 0$. Figure 3.1 depicts separation of points of T along the leaves of $\mathcal{F}$. Roughly speaking, points x and y of T are (R, ε)-separated whenever the distance of $\gamma(t)$ and $p_y\gamma(t)$ exceeds ε for some t and some leaf curve γ of origin x and length $\leq R$.

More precise description of this phenomenon is given in the following.

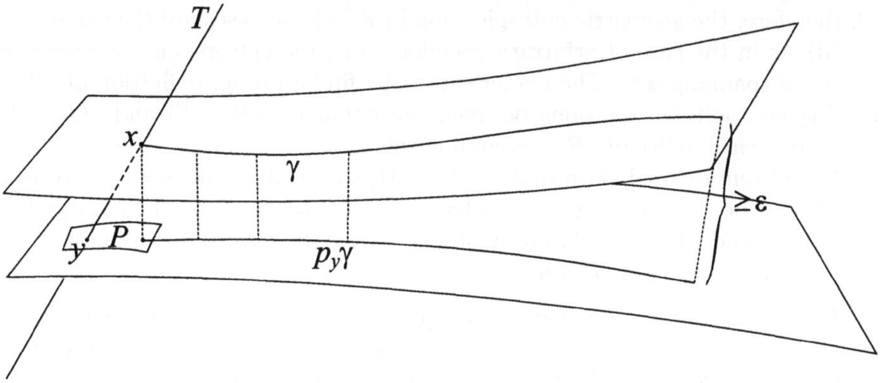

Figure 3.1: Separation of points.

Definition 3.3.1. ([**GLW2**], [**LW2**]) Two points x and y of T are said to be (R, ε)-*separated* (by $\mathcal{F}$) if either $d(x, y) \geq \varepsilon_0$ or there exists a leaf curve $\gamma : [0, 1] \to L_x$ such that $\gamma(0) = x$, $l(\gamma) \leq R$ and $d(\gamma(1), p_y\gamma(1)) \geq \varepsilon$ (or, a leaf curve $\delta : [0, 1] \to L_y$ such that $\delta(0) = y$, $l(\delta) \leq R$ and $d(\delta(1), p_x\delta(1)) \geq \varepsilon$). A subset A of T is (R, ε)-*separated* when any two points x and y of A, $x \neq y$, are (R, ε)-separated. The *geometric entropy* $h(\mathcal{F}, g)$ of $\mathcal{F}$ (with respect to g) is defined by

$$h(\mathcal{F}, g) = \lim_{\varepsilon \to 0^+} s(\varepsilon, \mathcal{F}), \tag{3.3.2}$$

where

$$s(\varepsilon, \mathcal{F}) = \limsup_{R \to \infty} \frac{1}{R} \log s(R, \varepsilon, \mathcal{F}) \tag{3.3.3}$$

and $s(R, \varepsilon, \mathcal{F})$ is the maximal cardinality of an (R, ε)-separated subset of T.

Remark 3.3.2. (i) The nice covering $\mathcal{U}$ plays a minor role here. It has been used only to determine how leaf curves can be projected to other leaves. If $\mathcal{U}'$ and T' are, respectively, another nice covering and the corresponding complete transversal, and $\varepsilon > 0$ is small enough, then, again because the geometry of M is bounded, one can project locally T to T' along the leaves in such a way that any two (R, ε)-separated points x and y of T project to points x' and y' of T' which are $(R+R_0, \varepsilon)$-separated, where R_0 is a constant independent of R and ε. The constant R_0 can be seen as the "distance" between T and T' measured along the leaves. More precisely, R_0 equals the maximum of distances $d_{\mathcal{F}}(x, x')$, where $x \in T \cap U$, $x' \in T' \cap U'$, $U \in \mathcal{U}$, $U' \in \mathcal{U}'$ and the plaques $P_x \subset U$ and $P'_{x'} \subset U'$ overlap. Therefore, the numbers $s(R, \varepsilon, \mathcal{F})$ and $s'(R, \varepsilon, \mathcal{F})$ obtained respectively from $\mathcal{U}$ and $\mathcal{U}'$ satisfy the inequalities

$$s'(R - R_0, \varepsilon, \mathcal{F}) \leq s(R, \varepsilon, \mathcal{F}) \leq s'(R + R_0, \varepsilon, \mathcal{F})$$

and, therefore, the geometric entropies obtained in both cases are the same.

(ii) As in the case of arbitrary pseudogroups, the entropy can be expressed in terms of spanning sets. The reader can easily find a proper definition of (R, ε)-spanning and suitable relations between the numbers $s(R, \varepsilon, \mathcal{F})$ and $r(R, \varepsilon, \mathcal{F})$, the minimal cardinality of (R, ε)-spanning sets.

(iii) Given a closed saturated set $K \subset M$, the relative entropy $h(\mathcal{F}, K, g)$ of $\mathcal{F}$ on K can be defined by the formulae analogous to (3.3.2) and (3.3.3), where $s(R, \varepsilon, \mathcal{F})$ should be just replaced by $s(R, \varepsilon, \mathcal{F}, K)$, the maximal cardinality of an (R, ε)-separated subset of $T \cap K$.

It is clear that the geometric entropy depends *a priori* on the choice of a Riemannian metric g. This is analogous to the situation in classical dynamical systems. If two vector fields X and Y have the same trajectories, and (ϕ_t) and (ψ_t) are their flows, then *a priori* there is no relation between the topological entropies $h_{\mathrm{top}}(\phi_t)$ and $h_{\mathrm{top}}(\psi_t)$. As was mentioned before, a Riemannian metric plays the role of time. We shall see that in fact our geometric entropy depends only on the restriction of the Riemannian structure to $T\mathcal{F}$, the bundle of vectors tangent to $\mathcal{F}$. To this end, let us assume first that we have on M two Riemannian metrics g and g' which, when restricted to $T\mathcal{F}$, are homothetic.

Proposition 3.3.3. *If $c \in \mathbb{R}_+$ is such that $g'(v, v) = c^2 \cdot g(v, v)$ for any vector v tangent to $\mathcal{F}$, then*

$$h(\mathcal{F}, g') = c^{-1} \cdot h(\mathcal{F}, g). \tag{3.3.4}$$

Proof. There exists a constant $a > 1$ such that for any leaf curve γ, the orthogonal projections $p_y\gamma$ and $p'_y\gamma$ determined by g and g' satisfy the following:

$$d(\gamma(t), p'_y\gamma(t)) \leq a \cdot d(\gamma(t), p_y\gamma(t))$$

whenever $d(\gamma(t), p_y\gamma(t)) \leq \varepsilon$ and ε is small enough. Therefore, any two points of T which are (R, ε)-separated with respect to g' are $(cR, \varepsilon/a)$-separated with respect to g'. This yields an inequality in (3.3.4). The converse can be obtained in the same way. □

It is easy to see that arbitrary Riemannian metrics g and g' on a compact manifold M are quasi-isometric in the sense that

$$c^{-2}g(v,v) \leq g'(v,v) \leq c^2 g(v,v) \tag{3.3.5}$$

for some $c \in \mathbb{R}_+$ and all $v \in TM$. Therefore, the argument of the proof of Proposition 3.3.3 implies also the following.

Proposition 3.3.4. *Inequality* (3.3.5) *implies the relation*

$$\frac{1}{c} \cdot h(\mathcal{F}, g) \leq h(\mathcal{F}, g') \leq c \cdot h(\mathcal{F}, g) \tag{3.3.6}$$

for arbitrary foliation $\mathcal{F}$ on M. In the same way, (3.3.6) *holds for a given $\mathcal{F}$ when* (3.3.5) *holds for all v in $T\mathcal{F}$. In particular ($c = 1$), $h(\mathcal{F}, g') = h(\mathcal{F}, g)$ when $g' = g$ on $T\mathcal{F} \otimes T\mathcal{F}$.* □

Therefore, in the same manner as in the case of pseudogroups in Section 3.2, we can distinguish between foliations with vanishing and non-vanishing entropy.

3.4 Relating various entropies

Given a foliation $\mathcal{F}$ of a compact Riemannian manifold (M, g) and a nice covering $\mathcal{U}$ we have the holonomy pseudogroup $\mathcal{H}_\mathcal{U}$ generated by a finite set $\mathcal{H}^1_\mathcal{U}$ of elementary holonomy maps corresponding to overlapping charts. Therefore, we have $h(\mathcal{F}, g)$, the geometric entropy of $\mathcal{F}$, and $h(\mathcal{H}_\mathcal{U}, \mathcal{H}^1_\mathcal{U})$, the entropy of its holonomy pseudogroup. We could expect that these two entropies are related somehow. Indeed, they are and we have the following.

Theorem 3.4.1. *For any nice covering $\mathcal{U}$ of $(M, \mathcal{F})$, denote by* $\operatorname{diam}(\mathcal{U})$ *the maximum of the diameters of the plaques of $\mathcal{U}$ measured with respect to the Riemannian structures induced on the leaves. Then*

$$h(\mathcal{F}, g) = \sup_{\mathcal{U}} \frac{1}{\operatorname{diam}(\mathcal{U})} \cdot h(\mathcal{H}_\mathcal{U}, \mathcal{H}^1_\mathcal{U}). \tag{3.4.1}$$

Proof. First, we shall show that

$$h(\mathcal{H}_\mathcal{U}, \mathcal{H}^1_\mathcal{U}) \leq \operatorname{diam}(\mathcal{U}) \cdot h(\mathcal{F}, g), \tag{3.4.2}$$

i.e., that

$$s(n, \varepsilon, \mathcal{H}^1_\mathcal{U}) \leq s(n \cdot \operatorname{diam}(\mathcal{U}), C^{-1}\varepsilon, \mathcal{F}), \quad n \in \mathbb{N},\ \varepsilon \in (0, \varepsilon_0), \tag{3.4.3}$$

where $C > 0$ is a constant which depends only on $\mathcal{U}$ and the geometry of $(M, \mathcal{F}, g)$.

If two points x and y in T_U, the transversal determined by $U \in \mathcal{U}$, are (n, ε)-separated with respect to $\mathcal{H}_\mathcal{U}$, then there exists a chain $u = (U_1, \ldots, U_n)$ of charts of $\mathcal{U}$ such that suitable chains of plaques $(P_1, \ldots, P_n)$ and $(Q_1, \ldots, Q_n)$

satisfy $x \in P_1$, $y \in Q_1$, $P_i, Q_i \subset U_i$, $P_i \cap P_{i+1} \neq \emptyset$, $Q_i \cap Q_{i+1} \neq \emptyset$ for all i and $d(x_n, y_n) \geq \varepsilon$, where $x_n \in P_n \cap T_{U_n}$ and $y_n \in Q_n \cap T_{U_n}$ are the images of x and y under the holonomy map determined by the chain u. Choose points $x_i \in P_i \cap P_{i+1}$, $i = 1, \ldots, n-1$, let $x_0 = x$ and connect x_i to x_{i+1} by a segment γ_i of a minimizing leaf geodesic. Then, $l(\gamma_i) \leq \operatorname{diam}(\mathcal{U})$ and the union $\gamma : [0,1] \to L_x$ of all these segments is a leaf curve of length $l(\gamma) \leq n \cdot \operatorname{diam}(\mathcal{U})$ joining x to x_n.

Shortening γ, if necessary, we may assume that the distance between $\gamma(t)$ and the orthogonal projection $p_y\gamma(t)$ is all the time less than ε_0, so the whole curve γ can be projected onto L_y. Again, since the transversal $T = \cup T_U$ is compact, there exists a constant $C > 0$ such that

$$d(z, p(z)) \leq C \cdot d(z, w)$$

if only $z, w \in T_U$, $U \in \mathcal{U}$ and $p(z)$ is the orthogonal projection of z to the plaque of U passing through w. With this constant C we have

$$d(\gamma(1), p_y\gamma(1)) \geq C^{-1}\varepsilon$$

and therefore the points x and y are $(n \operatorname{diam}(\mathcal{U}), C^{-1}\varepsilon)$-separated with respect to $\mathcal{F}$. This proves (3.4.3) and, consequently, (3.4.2).

To complete the proof choose an arbitrary positive number η and a number $\Delta_0 > 0$ such that the leafwise exponential map $\exp^{\mathcal{F}}$ maps all the balls $B^{\mathcal{F}}(0_x, 4\Delta_0) \subset T_x\mathcal{F}$, $x \in M$, diffeomorphically onto strictly convex balls $B^{\mathcal{F}}(x, 4\Delta_0) \subset L_x$. Take $\rho > 0$ and let $S_x = \exp B^{\perp}(0_x, 2\rho)$, where exp is the exponential map on M and $B^{\perp}(0_x, r)$ is the ball of radius r, centred at the origin and contained in $T_x^{\perp}\mathcal{F}$, the orthogonal complement of $T_x\mathcal{F}$ in T_xM. Also, let

$$T_x = \exp B^{\perp}(0_x, \rho), \quad V_x = \bigcup_{y \in S_x} B^{\mathcal{F}}(y, 2\Delta) \quad \text{and} \quad U_x = \bigcup_{y \in T_x} B^{\mathcal{F}}(y, \Delta).$$

Clearly, if ρ and Δ are small enough, both U_x and V_x are the domains of distinguished charts for all $x \in M$. They have the shape of round "coins" (Figure 3.2). All the plaques have diameter 2Δ.

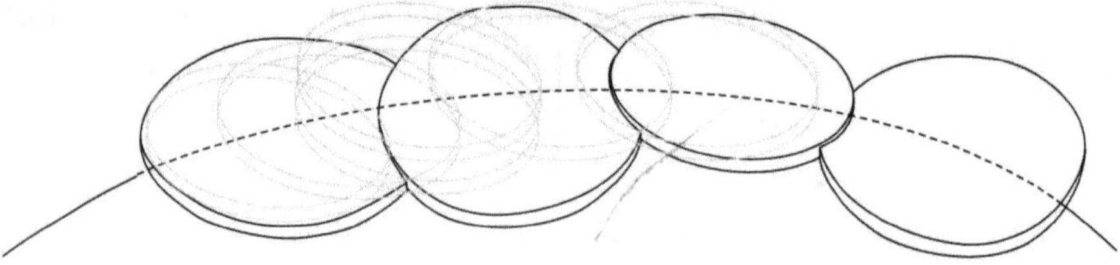

Figure 3.2: A chain of coins.

Also, it is easy to see that (again if ρ, Δ and η are small enough) there exists $\delta_1 > 0$ such that for any two points x_1 and x_2 in the same leaf L and any points $y_i \in T_{x_i}$ $(i = 1, 2)$ which lie in the same leaf ball of radius $4\Delta_0$, the inequalities

$$d_{\mathcal{F}}(x_1, x_2) < 2\Delta - \eta, \quad d(x_1, y_1) \leq \delta_1 \text{ and } d(x_2, y_2) \leq \delta_1 \tag{3.4.4}$$

imply that $d_{\mathcal{F}}(y_1, y_2) \leq 2\Delta - \eta/2$.

Next, let Z be a finite δ-dense subset of M with $\delta < \delta_1$ being small enough. If x_1 and x_2 lie in a leaf $L \in \mathcal{F}$ and satisfy the first inequality in (3.4.4), $z_1, z_2 \in Z$, $d(z_i, x_i) \leq \delta$, $w_i \in T_{z_i}$ and $d(w_i, z_i) \leq \delta$ for $i = 1, 2$, and both w_1 and w_2 belong to the same leaf ball of radius $4\Delta_0$, then the minimal leaf geodesic joining w_1 to w_2 is entirely contained in the union $B^{\mathcal{F}}(w_1, \Delta) \cup B^{\mathcal{F}}(w_2, \Delta)$.

Let now $\mathcal{U}_\Delta = \{U_z; z \in Z\}$ and $\mathcal{V}_\Delta = \{V_z; z \in Z\}$. It is easy to see that modifying Z slightly, if necessary, we may assume that the closures $\bar{U}_z$ and $\bar{U}_{z'}$, z and $z' \in Z$, overlap if and only if the open sets U_z and $U_{z'}$ do. In this case, $\mathcal{U}_\Delta$ becomes a nice covering of $(M, \mathcal{F})$. Obviously,

$$\operatorname{diam}(\mathcal{U}_\Delta) = 2\Delta.$$

Finally, take $\varepsilon > 0$ small enough and two points x and y such that $d(x, y) < \varepsilon$ but x and y are (R, ε)-separated with respect to $\mathcal{F}$. Then, there exists a leaf curve $\gamma : [0, R] \to L$ of length $l(\gamma) \leq R$, originated at, say, x and such that its projection $p_y\gamma$ is well defined on an interval $[0, r]$, $r \leq R$, and satisfies the inequality

$$d(\gamma(r), p_y\gamma(r)) \geq \varepsilon.$$

Assume that $R = 2(1 - \eta)n\Delta$ and let $x_k = \gamma(kr/n)$ for $k = 0, 1, \dots, n$. For each k find a point $z_k \in Z$ δ-close to x_k. From the previous observations it follows that the charts U_{z_k}, $k \leq n$, form a chain along $\gamma|[0, r]$, the corresponding holonomy map $h \in \mathcal{H}_{\mathcal{U}_\Delta}$ is defined at the plaques P and $Q \subset U_{z_0}$ containing, respectively, x and y and

$$d(h(P), h(Q)) \geq C \cdot \varepsilon,$$

where (as in the first part of the proof) $C > 0$ is a universal constant. Therefore,

$$s(2(1 - \eta)n\Delta, C\varepsilon, \mathcal{F}) \leq N(\varepsilon) \cdot s(n, \varepsilon, \mathcal{H}_{\mathcal{U}_\Delta}),$$

where $N(\varepsilon)$ is the minimal cardinality of a covering of M by balls of radius ε. Consequently,

$$s(C\varepsilon, \mathcal{F}) \leq \frac{1}{2(1 - \eta)\Delta} s(\varepsilon, \mathcal{H}_{\mathcal{U}_\Delta}) \text{ and } h(\mathcal{F}, g) \leq \frac{1}{2(1 - \eta)\Delta} h(\mathcal{H}_{\mathcal{U}_\Delta}). \tag{3.4.5}$$

This ends the proof. □

Now, let us consider some particular cases, where holonomy pseudogroups are generated by global maps.

Example 3.4.2. If $(M,\mathcal{F})$ is a foliated bundle (see Example 1.3.8), B is the base, F is a fibre, G_1 is a finite set generating the fundamental group $G=\pi_1(B)$ and g is a Riemannian metric on M such that the projection $\pi: M \to B$ is isometric on the tangent spaces $T_x\mathcal{F}$, $x \in M$, then the inequalities

$$ah(\mathcal{F},g) \le h(G,G_1) \le bh(\mathcal{F},g), \tag{3.4.6}$$

where a (resp., b) is the minimum (resp, the maximum) of lengths $l(z)$ of the homotopy classes $z \in G_1$, hold. (Clearly, $l(z)=\min\{l(\gamma);\gamma\in z\}$.) In fact, without losing generality we may assume that the fibres are orthogonal to the leaves. Then the orthogonal projection onto a leaf L' of a lift to $L \in \mathcal{F}$ of any curve $\gamma:[0,1]\to B$ coincides with a suitable lift of γ to L'. Therefore, any two points of F which are (n,ε)-separated with respect to G_1 become (bn,ε)-separated with respect to $\mathcal{F}$ and any two points which are (R,ε)-separated with respect to $\mathcal{F}$ become (n,ε)-separated with respect to G_1 with $n=[R/a]+1$.

The second example is that of an oriented foliation $\mathcal{F}$ of dimension 1. Such an $\mathcal{F}$ consists of orbits of a non-singular vector field X and it should be possible to compare its entropy with the topological entropy of the flow of X. Since the entropy of $\mathcal{F}$ depends on the Riemannian structure, one can guess the most promising case is that of a vector field of norm 1. As we shall see, the result is very simple and easy to predict. However, the proof is pretty complicated, so we formulate it in the form of a theorem.

Theorem 3.4.3. *If $\mathcal{F}$ is the one-dimensional foliation built of the orbits of a unit vector field X, then*

$$h(\mathcal{F},g) = 2h_{\text{top}}(\phi), \tag{3.4.7}$$

where $\phi=(\phi_t)$ is the flow of X.

Proof. If T is a complete transversal and $A \subset T$ is (R,ε)-separated by $\mathcal{F}$, then (since $d(\phi_t x, p_y\phi_t x) \le d(\phi_t x,\phi_t y)$ for all x and y of T and all $t\in\mathbb{R}$) $\phi_{-R}(A)$ is $(2R,\varepsilon)$-separated by ϕ. Therefore, $s(R,\varepsilon,\mathcal{F}) \le s(2R,\varepsilon,\phi)$ for all R and ε, and this yields "$\le$" in (3.4.7).

The proof of "$\ge$" in (3.4.7) is much more difficult because one has a "gliding effect" along the orbits. To describe this effect, take two positive numbers ε and η and consider the fibre bundle $\pi:\bar{M}_\eta \to M$ built of the balls $B^\perp(0_x,\eta)\subset T_x^\perp\mathcal{F}$, $x\in M$. Given $x\in M$ let $\text{Tub}_\eta(x)=\phi_x^{-1}\bar{M}_\eta$ be the bundle over $\mathbb{R}$ induced by the map $\phi_x: t\mapsto \phi_t(x)$. If η is small enough, the exponential map on M yields a natural immersion $\iota_x:\text{Tub}_\eta(x)\to M$. Equip $\text{Tub}_\eta(x)$ with the induced Riemannian metric and with the induced vector field $\bar{X}$. $\bar{X}$ generates a local flow $(\bar{\phi}_t)$. The "gliding effect" is related to the fact that the family $\pi^{-1}(s)$, $s\in\mathbb{R}$, of fibres of $\text{Tub}_\eta(x)$ is not invariant under $(\bar{\phi}_t)$. However, since M is compact, this family is "almost invariant" in the following sense: for any $\tau\in(0,1)$ there exists $\eta>0$ such that the inequality $t\ge\tau$ (resp., $t\le-\tau$) holds whenever $x\in M$, $y\in\text{Tub}_\eta(x)\cap\pi^{-1}(0)$, $\bar{\phi}_t(y)$ is defined and $\pi\bar{\phi}_t(y)=1$ (resp., -1). If τ and η satisfy the above, then $t \geq n\tau$ ($t \leq -n\tau$) whenever $\bar{\phi}_t(y)$ is defined and $\pi\bar{\phi}_t(y) = n$ (resp., $-n$), $n \in \mathbb{N}$.

Now, decompose $\mathrm{Tub}_\eta(x)$ into the cylinders $C_n(x) = \pi^{-1}([(2n-1)\varepsilon, (2n+1)\varepsilon])$, $n = 1, 2, \ldots$. If η is small enough again and $x \in M$, then the sets $\bar{\phi}_1(C_0(x))$ and $\bar{\phi}_{-1}(C_0(x))$ intersect at most three cylinders of the form $C_n(x)$. For any $y \in C_0(x)$ one can consider the sequences (n_k) of integers such that $\bar{\phi}_k(y) \in C_{n_k}(x)$ (with the natural convention $n_k = \infty$ when $\bar{\phi}_k(y)$ is undefined). If η satisfies the above, the number of all such sequences of length $2n-1$ is not greater than 3^{2n}. If so, all the cylinders $C_0(x)$, $x \in M$, can be decomposed into unions

$$C_0(x) = K_1(x) \cup \cdots \cup K_{m(x)}(x),$$

of subsets $K_j(x)$ such that $m(x) \leq 3^{2n+2}$ and the following condition is satisfied:

(*) If y and z lie in $K_j(x)$, $|k| \leq n$, $\bar{\phi}_k(y)$ and $\bar{\phi}_k(z)$ are defined, then $\bar{\phi}_k(y)$ and $\bar{\phi}_k(z)$ belong to the same cylinder $C_{n(k)}(x)$.

Next, take a set $A \subset T$ which is $(n, \eta/3)$-separated by $\mathcal{F}$ and such that $\#A = s(n, \eta/3, \mathcal{F})$. The sets

$$A(x) = \{y \in T;\ \sup_{|t| \leq n} d(\phi_t(x), p_y\phi_t(x)) \leq \eta/3\}, \quad x \in A,$$

cover T and have diameter less than $2\eta/3$. Therefore, $A(x) \subset \iota_x(C_0(x))$, we may decompose $C_0(x)$ as above and choose one point y_j^x in each non-empty piece $A(x) \cap K_j(x)$. All the points y_j^x, $x \in A$, $j \leq m(x)$, form a set $B \subset T$. Clearly,

$$\#B \leq 3^{2n+2} s(n, \eta/3, \mathcal{F}). \tag{3.4.8}$$

Finally, take any point $y \in M$. There exists $R_0 > 0$ independent of y and such that $\phi_t(y) \in T$ for some $t \in (-R_0, R_0)$. There exists $x \in A$ and $j \leq m(x)$ such that $\phi_t(y) \in A(x) \cap \iota_x K_j(x)$. From (*) it follows that $\bar{\phi}_{t+i}(y)$ and $\phi_i y_j^x$ belong to the same cylinder $C_{n(i)}(x)$ and, therefore, $d(\phi_{t+i}(y), \phi_i(y_j^x) \leq 2\eta + 2\varepsilon$ for all $i = 0, \pm 1, \ldots, \pm[\tau n]$. Also, there exists a constant $c > 0$ such that for small η and all the points z and z' of M the inequality $d(z, z') \leq \eta$ implies the relation $d(\phi_t(z), \phi_t(z')) \leq c\eta$ for all $t \in [-R_0, R_0]$. Therefore, the set $\phi_{-[\tau n]}(B)$ is $(2\tau n, 2c(\eta + \varepsilon))$-spanning with respect to ϕ and (3.4.8) implies that

$$r(2\tau n, 2c(\eta + \varepsilon), \phi) \leq 3^{2n+2} s(n, \eta/3, \mathcal{F}). \tag{3.4.9}$$

Consequently,

$$r(2c(\eta + \varepsilon), \phi) \leq \frac{1}{\tau} \log 3 + \frac{1}{2\tau} s(\eta/3, \mathcal{F}). \tag{3.4.10}$$

When ε and η tend to 0, one can choose τ arbitrarily close to 1, and therefore (3.4.10) implies

$$h_{\mathrm{top}}(\phi) \leq \log 3 + \frac{1}{2} h(\mathcal{F}, g). \tag{3.4.11}$$

To complete the proof observe that (3.4.11) holds for all Riemannian metrics g, so we can change g replacing it by $\tilde{g}$ such that $\tilde{g} = \lambda^2 g$ ($\lambda \in \mathbb{R}_+$) along the leaves. Then, we have to replace X by $\tilde{X} = \lambda^{-1}X$ and ϕ by $\tilde{\phi}$ with $\tilde{\phi}_t = \phi_{t/\lambda}$. Since $h_{\text{top}}(\tilde{\phi}) = \lambda^{-1}h_{\text{top}}(\phi)$ and $h(\mathcal{F},\tilde{g}) = \lambda^{-1}h(\mathcal{F},g)$ by Proposition 3.3.3, inequality (3.4.11) implies

$$h_{\text{top}}(\phi) \leq \lambda \log 3 + \frac{1}{2}h(\mathcal{F},g).$$

Letting $\lambda \to 0$ we get finally the required inequality. □

3.5 Examples and constructions

From the considerations of Section 2.4 it follows immediately that the geometric entropy of any Riemannian foliation vanishes while the entropy of the Hirsch foliation (Example 2.4.6 (ii)) is strictly positive. Below we show how the geometric entropy behaves under some operations of constructing foliations.

3.5.1 Pullback

One way of building new foliations is that of pullback (Section 1.3). If $\mathcal{F}' = f^*\mathcal{F}$, where $f : M' \to M$ is a differentiable map transverse to a foliation $\mathcal{F}$ of $\mathcal{M}$, (M,g) and (M',g') are compact Riemannian manifolds and f maps isometrically the leaves of $\mathcal{F}'$ into the leaves of $\mathcal{F}$ (i.e., $\|f_*v\| = \|v\|$ for any $v \in T\mathcal{F}'$), then for any nice covering $\mathcal{U}'$ of $(M',\mathcal{F}')$ one can construct a nice covering $\mathcal{U}$ of $(M,\mathcal{F})$ such that the images $f(U')$, $U' \in \mathcal{U}'$, are contained in some members of $\mathcal{U}$ and the quotient $\operatorname{diam}\mathcal{U}' / \operatorname{diam}\mathcal{U}$ is arbitrarily close to 1. From Proposition 1.3.9, Example 3.2.5 and Theorem 3.4.1 the following is immediate.

Proposition 3.5.1. *$h(\mathcal{F}',g') \leq h(\mathcal{F},g)$ when $\mathcal{F}'$ is the pullback of $\mathcal{F}$ via a map f which is isometric along the leaves of $\mathcal{F}'$. In particular, $h(\mathcal{F}',g') = h(\mathcal{F},g)$ when $(M,\mathcal{F})$ and $(M',\mathcal{F}')$ are diffeomorphic via a diffeomorphism which is isometric along the leaves.* □

In the case of foliated bundles over the same base B, the above proposition can be expressed in terms of suspended homomorphisms. In fact, for arbitrary group homomorphisms $h_i : G \to \text{Homeo}(X_i)$, $i = 1,2$, we have the following, analogous to part (v) of Theorem 3.1.4, result.

Proposition 3.5.2. *If homomorphisms $h_i : G \to \text{Homeo}(X_i)$, $i = 1,2$, are semiconjugated via a continuous surjective map $F : X_1 \to X_2$ (i.e., $F \circ h_1(g) = h_2(g) \circ F$ for any $g \in G$), then the entropies of groups $G_i = h_i(G)$ (with respect to generating sets $G_{i,0} = h_i(G_0)$, G_0 being a finite symmetric set generating G) satisfy the inequality $h(G_1,G_{1,0}) \geq h(G_2,G_{2,0})$. In particular, these entropies are equal when F is a homeomorphism.* □

3.5.2 Gluing

Assume that $(M_i, \mathcal{F}_i)$, $i = 1, 2$, are compact foliated manifolds with boundaries ∂M_i, $\dim M_1 = \dim M_2$ and $\dim \mathcal{F}_1 = \dim \mathcal{F}_2$.

First, assume that $S_i \subset \partial^{\top} M_i$ is a union of connected components of the tangential boundary and $\Phi : S_1 \to S_2$ is a differentiable map which maps the foliation $\mathcal{F}_1|S_1$ onto $\mathcal{F}_2|S_2$. Then, identifying any point $x \in S_1$ with its image $\Phi(x) \in S_2$ we obtain a manifold $M = M_1 \cup_{\Phi} M_2$ equipped with a foliation $\mathcal{F}$ such that $\mathcal{F}|M_i = \mathcal{F}_i$ for $i = 1, 2$. We shall say that $(M, \mathcal{F})$ is obtained by *tangential gluing* of $(M_i, \mathcal{F}_i)$ along Φ. *A priori*, the class of differentiability of $\mathcal{F}$ can be lower than these of $\mathcal{F}_1$ and $\mathcal{F}_2$. However, if both $\mathcal{F}_1$ and $\mathcal{F}_2$ are C^r-smooth and the holonomy maps h_i of $\mathcal{F}_i$, along arbitrary closed chains of plaques contained in ∂M_i, $i = 1, 2$, are C^r-tangent to identity (that is, their differentials $h_{i,*}(x)$, $x \in \partial M_i$, are equal to identity maps while all their partial derivatives at x of order k, $2 \leq k \leq r$, calculated with the aid of arbitrary charts, are equal to 0), then $\mathcal{F}$ is also C^r-differentiable.

Assume also that g_i is a Riemannian metric on M_i and that $\Phi : S_1 \to S_2$ is isometric with respect to the induced metric. Then, after some modification in a direction transverse to $\mathcal{F}_i$ in a neighbourhood of S_i we can get new Riemannian structures, denoted by g_1 and g_2 again, which match to produce a smooth Riemannian structure g on M such that $g|M_i = g_i$. Since we modified g_i in the transverse direction only, we did not influence the entropies of $\mathcal{F}_i$. So, if T is a complete transversal for $\mathcal{F}$, then $T_i = T \cap M_i$ is a complete transversal for $\mathcal{F}_i$ and if $A \subset T$ is an (R, ε)-separated set for $\mathcal{F}$, then $A \cap T_i$ is (R, ε)-separated for $\mathcal{F}_i$. Therefore,

$$s(R, \varepsilon, \mathcal{F}) \leq s(R, \varepsilon, \mathcal{F}_1) + s(R, \varepsilon, \mathcal{F}_2)$$

for all R and ε. This shows that

$$h(\mathcal{F}, g) \leq \max\{h(\mathcal{F}_1, g_1),\ h(\mathcal{F}_2, g_2)\}. \tag{3.5.1}$$

Also, any set $A \subset T_i$ which is (R, ε)-separated for $\mathcal{F}_i$ becomes (R, ε)-separated for $\mathcal{F}$ and thus

$$h(\mathcal{F}, g) \geq \max\{h(\mathcal{F}_1, g_1),\ h(\mathcal{F}_2, g_2)\}. \tag{3.5.2}$$

Comparing (3.5.1) and (3.5.2) we arrive at the following.

Proposition 3.5.3. *The geometric entropy of a foliation obtained by tangential gluing of two foliated components is equal to the maximum of geometric entropies of these components.* □

A similar construction can be performed when $S_i \subset \partial^{\pitchfork} M_i$ is a union of connected components of the transverse boundary. In this case, we have to suppose that Φ maps leaves of $\mathcal{F}_1^{\pitchfork}$, the foliation induced by $\mathcal{F}_1$ on $\partial^{\pitchfork} M_1$, onto leaves of $\mathcal{F}_2^{\pitchfork}$, the foliation induced by $\mathcal{F}_2$ on $\partial^{\pitchfork} M_2$. Then, we have to choose tubular neighbourhoods $N(S_i)$ of S_i, some vector fields X_i on $N(S_i)$ tangent to $\mathcal{F}_i$ and pointing inwards along the boundary, and glue M_1 to M_2 along the flows (ϕ_t^i) of the fields X_i. More precisely, any point $\phi_t^1(x)$ ($x \in S_1$ and $t \in [0, \delta)$) should be identified with $(\phi_{\delta-t}^2(\Phi(x))$, where $\delta > 0$ is a fixed small constant. The foliations $\mathcal{F}_i$ match and yield a foliation $\mathcal{F}$ of the manifold M obtained in this way. The topology and the dynamics of $(M, \mathcal{F})$ do not depend on the choice of δ and X_i, so we may write $M = M_1 \cup_\Phi M_2$ and $\mathcal{F} = \mathcal{F}_1 \cup_\Phi \mathcal{F}_2$ and say that $(M, \mathcal{F})$ is obtained by *transverse gluing* of $(M_i, \mathcal{F}_i)$ along Φ.

If the foliations $\mathcal{F}_i$ are differentiable of the class C^{r+1}, then their tangent bundles $T\mathcal{F}_i$ are C^r-manifolds, so we can choose X_i to be differentiable of the class C^r. Therefore, in general $\mathcal{F} = \mathcal{F}_1 \cup_\Phi \mathcal{F}_2$ is only C^r-differentiable. Clearly, if $\mathcal{F}_i$ are C^∞-differentiable, then $\mathcal{F} = \mathcal{F}_1 \cup_\Phi \mathcal{F}_2$ is C^∞-smooth as well.

Again, if g_i are arbitrary Riemannian metrics on M_i, then we can modify them to make the identification of $N(S_1)$ with $N(S_2)$ isometric. In this way we can produce a new smooth Riemannian metric g on M which coincides with g_i on $M \setminus N(S_i)$. The modification could be performed arbitrarily in the direction transverse to $\mathcal{F}_i$ and in such a way that $g(v, v) \leq g_i(v, v)$ for all $v \in T_x\mathcal{F}$, $x \in M_i$.

This time, if $A \subset M_i$ is (R, ε)-separated on $(M_i, \mathcal{F}_i, g_i)$, then A becomes $(R, C\varepsilon)$-separated on $(M, \mathcal{F}, g)$ for some constant $C > 0$ which depends only on the modification of the metrics g_i in the transverse direction. Therefore,

$$s(R, \varepsilon, \mathcal{F}) \geq \max\{s(R, C\varepsilon,\ \mathcal{F}_1), s(R, C\varepsilon, \mathcal{F}_2)\}$$

for all R and ε. As before, we immediately arrive at inequality (3.5.2). In other words, we have the following.

Proposition 3.5.4. *The geometric entropy of a foliation obtained by transverse gluing of two foliated components is not less than the maximum of geometric entropies of these components.* □

One can predict that this time the equality of geometric entropies need not hold. Indeed, as we explain more formally below, one can glue foliations without entropy (for example, product ones) using a map which itself has positive topological entropy. This map generates holonomy providing positive entropy of the holonomy pseudogroup of the resulting foliation.

Example 3.5.5. If $M_i = N \times [0, 1]$, $i = 1, 2$, $\mathcal{F}_i$ is the trivial foliation by segments $\{x\} \times [0, 1]$, $x \in N$, then $h(\mathcal{F}_i, g_i) = 0$ for any Riemannian metrics g_i. If $S_i = \partial M_i = N \times \{0\} \cup N \times \{1\}$, $\Phi(x, 0) = (x, 0)$ and $\Phi(x, 1) = (f(x), 1)$, where $f : N \to N$ is a diffeomorphism of positive topological entropy, then $(M_1 \cup_\Phi M_2, \mathcal{F}_1 \cup_\Phi \mathcal{F}_2)$ coincides with the suspension of f and, therefore, $h(\mathcal{F}_1 \cup_\Phi \mathcal{F}_2, g) > 0$ for any Riemannian metric g on $M_1 \cup_\Phi M_2$.

3.5.3 Turbulization

First, let us study a classical example.

Example 3.5.6. One of the oldest nontrivial examples in the theory of foliations is the *Reeb foliation* $\mathcal{F}$ of S^3. It is built of two *Reeb components* (Figure 3.3) $D^2 \times S^1$

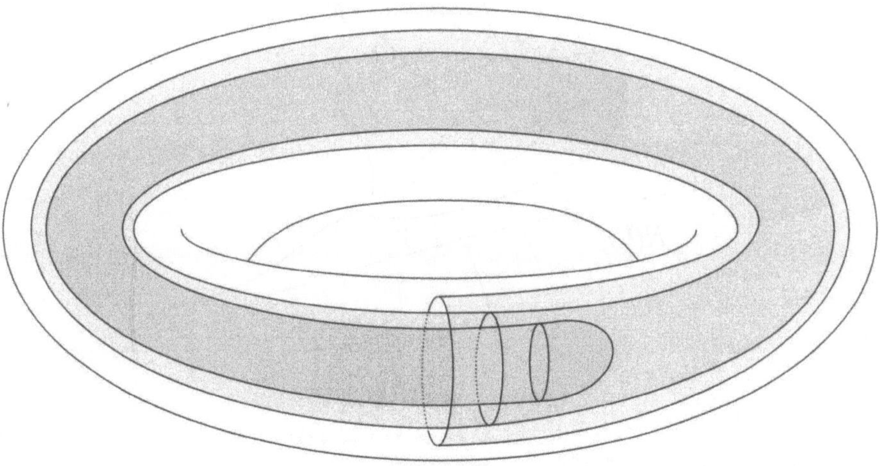

Figure 3.3: A leaf of a Reeb component.

obtained from a graph Γ of a function $f : (-1,1) \to \mathbb{R}$ which is smooth, symmetric, satisfies $f(0) = 0$, $f(-t) = f(t)$ and $\lim_{t\to\pm 1} f(t) = \infty$. To obtain $\mathcal{F}$ one has to shift Γ up and down to fill the rectangle $[-1,1] \times [0,1]$, to rotate this rectangle around the segment $\{0\} \times [0,1]$, to glue together the discs $D^2 \times \{0\}$ and $D^2 \times \{1\}$ to get foliated solid tori $D^2 \times S^1$, to take two such Reeb components and to glue them together along the common boundary $T^2 = S^1 \times S^1$.

To calculate the geometric entropy of the Reeb foliation $\mathcal{F}$ note first that the holonomy pseudogroup of a three-dimensional Reeb component is the same as the holonomy pseudogroup of a two-dimensional Reeb component obtained just by gluing together the sides $[-1,1]\times\{0\}$ and $[-1,1]\times\{1\}$ of the rectangle $[-1,1]\times[0,1]$ foliated as in the construction above. For such a foliated annulus A (equipped with the standard flat metric) it is easy to see that two points are separated if and only if they are separated along the leaves of the foliation $\tilde{\mathcal{F}}$ of A obtained in the same way from the graph of the function $\tilde{f}$ given by $\tilde{f}(t) = \operatorname{sgn}(t) \cdot f(t)$. The foliation $\tilde{\mathcal{F}}$ can be obtained as the suspension of a homeomorphism ϕ of the interval $[-1,1]$. As observed in Example 3.1.12, $h_{\text{top}}(\phi) = 0$. Therefore, the geometric entropy of $\tilde{\mathcal{F}}$ vanishes by either Theorem 3.4.3 or Example 3.4.2. Consequently, the classical Reeb foliation $\mathcal{F}$ of S^3 has entropy zero.

The above argument together with the well-known old classification of foliations of T^2 ([**Kn**], see also [**Go1**] or [**HH**], Part A) shows that any one-dimensional foliation of T^2 has vanishing geometric entropy.

The foliation $\mathcal{F}$ within the Reeb component is given by the equation $\omega = 0$, where

$$\omega(r,t) = \cos\lambda(r)dr + \sin\lambda(r)dt, \tag{3.5.3}$$

$\lambda(r) = \arctan f'(r)$, (r,θ) are the polar coordinates in D^2 and t is a parameter

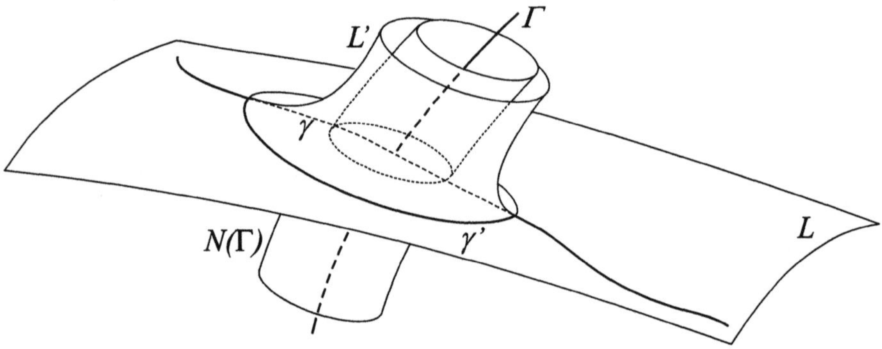

Figure 3.4: Turbulization

along S^1. It is clear that the same equation can be written in $D^p \times S^1$ providing $(p+1)$-dimensional Reeb components.

Now, let $\mathcal{F}$ be a transversely orientable foliation of codimension 1 on a manifold M and Γ be a closed transverse loop. Take a tubular neighbourhood $N(\Gamma) \approx D^{n-1} \times S^1$ built from small discs on the leaves passing through the points of Γ and define a foliation $\mathcal{F}_0$ of $N(\Gamma)$ by the equation $\omega = 0$ with ω given by (3.5.3), where $\lambda : [0,1] \to [-\frac{\pi}{2}, \frac{\pi}{2}]$ is a differentiable function such that $\lambda(0) = -\frac{\pi}{2}$, $\lambda(t) = \frac{\pi}{2}$ for $t \in [1-\varepsilon, 1]$, $\lambda'(t) > 0$ for $t \in (0, 1-\varepsilon)$ and $\lambda^{(k)}(0) = 0$ for $k = 1, 2, \ldots$. Since the foliations $\mathcal{F}$ and $\mathcal{F}_0$ coincide in a neighbourhood of the boundary of $N(\Gamma)$, they can be glued together to produce a new foliation $\mathcal{F}'$ of M. Of course, $\mathcal{F}'$ is as smooth as $\mathcal{F}$.

Definition 3.5.7. The foliation $\mathcal{F}'$ is said to be obtained from $\mathcal{F}$ by the process of *turbulization.*

The discussion in Example 3.5.6 can be applied to get the following.

Proposition 3.5.8. *If $\mathcal{F}'$ is obtained from $\mathcal{F}$ by turbulization, then*

$$h(\mathcal{F}', g) \leq h(\mathcal{F}, g) \leq C \cdot h(\mathcal{F}', g), \tag{3.5.4}$$

where $C \geq 1$ is a constant which depends on the geometry of M and R, the Reeb component introduced during the process of turbulization. Consequently, the entropy of $\mathcal{F}'$ vanishes if and only if the entropy of $\mathcal{F}$ does.

Proof. By Proposition 3.5.3 and Example 3.5.6, $h(\mathcal{F}') = h(\mathcal{F}'')$, where $\mathcal{F}'' = \mathcal{F}'|M \setminus \operatorname{int} R$.

Let T be a complete transversal for $\mathcal{F}$ such that $T \cap R = \emptyset$. There exists a constant $c > 0$ such that if $\varepsilon > 0$ is small enough, then any two points of T which are separated to the distance ε along a curve γ which lies on a leaf L of $\mathcal{F}$ can be separated to the distance $c\varepsilon$ along a curve γ' which lies on a leaf L' of $\mathcal{F}'$ and has length $l(\gamma') \leq Cl(\gamma)$ for a universal constant C (Figure 3.4). (The constant C can be estimated in the following way: any leaf curve δ contained in R can be replaced by a leaf curve δ' contained in $M \setminus R$ and of length $\leq c_0 \cdot l(\delta)$, where c_0 depends on M and R only. Arbitrary points of two different discs of $N(\Gamma)$ have leaf distance at least d_0, therefore the number of such discs met by a leaf curve γ is at most $k = 1 + l(\gamma)/d_0$ and any constant C bigger than $1 + \Delta \cdot c_0/d_0$, where Δ is the maximal leaf diameter of leaf discs contained in $N(\Gamma)$, satisfies our requirements.) Consequently, $s(r, \varepsilon, \mathcal{F}) \leq s(Cr, \varepsilon, \mathcal{F}')$ for all r and small positive ε. This yields the second inequality in (3.5.4).

To get the first inequality, choose a short segment T_0 transverse to $\partial N(\Gamma)$ and put $T' = T \cup T_0$. T' turns out to be a complete transversal for $\mathcal{F}''$. Since the distance between the leaves of $\mathcal{F}''$ decreases along curves entering $N(\Gamma)$, points of T which are separated with respect to $\mathcal{F}''$ are separated (to the same distance) also along curves staying outside $N(\Gamma)$, therefore are separated with respect to $\mathcal{F}$. Moreover, an argument similar to that of Example 3.5.6 shows that the maximal number $s_0(r, \varepsilon, \mathcal{F}'')$ of points of T_0 which are separated to the distance ε along curves γ ($l(\gamma) \leq r$) inside $N(\Gamma)$ grows subexponentially with r. Also, if $V(\varepsilon)$ is the ε-neighbourhood of ∂R and r_ε denotes the Hausdorff distance (measured along the leaves) of $T_0 \cap V(\varepsilon)$ and T, then points of T_0 which are (r, ε)-separated along a curve γ which eventually quits $N(\Gamma)$ can be replaced by points of T which are $(r + r_\varepsilon, \varepsilon)$-separated along a curve in $M \setminus N(\Gamma)$. Therefore,

$$s(r, \varepsilon, \mathcal{F}'') \leq s(r + r_\varepsilon, \varepsilon, \mathcal{F}) + s_0(r, \varepsilon, \varepsilon, \mathcal{F}''), \quad s(\varepsilon, \mathcal{F}'') \leq s(\varepsilon, \mathcal{F}), \quad h(\mathcal{F}'') \leq h(\mathcal{F})$$

and this yields the required inequality in (3.5.4). □

Note that closed transverse loops often exist. Indeed, if $\mathcal{F}$ possesses a non-compact leaf L, then L intersects a connected transversal $T \subset U$, U being a distinguished chart infinitely many times. If p and q are two distinct close points in $L \cap T$, then one can find a curve $\gamma : [0,1] \to L$ joining p to q. Modifying it slightly one can obtain a curve $\tilde{\gamma} : [0,1] \to M$ which is transverse to $\mathcal{F}$ and connects points p' and $q' \in T$ being arbitrarily close to p and q, respectively. Also q' can be connected to p' by a curve $\delta : [1,2] \to T$. The curve $\tilde{\gamma} \cup \delta$ is closed and smooth except at, maybe, p' and q'. It can be modified easily to provide a differentiable loop transverse to $\mathcal{F}$. Since the process of turbulization can be repeated several times, it is possible to construct, for example, pretty complicated foliations with vanishing entropy.

3.6 Entropy and resiliency

In this Section, we concentrate on C^2-foliations of codimension 1. We prove the relation between entropy and existence/non-existence of resilient leaves (recall Definition 2.3.9 and Figure 2.1). More precisely, the main result here reads as follows.

Theorem 3.6.1. *If $\mathcal{F}$ is a codimension-one* C^2*-foliation transverse to the fibres of a bundle $\pi : M \to B$, M being compact and connected manifold equipped with a Riemannian structure g, then the following conditions are equivalent:*

(i) $h(\mathcal{F}, g) > 0$,

(ii) *$\mathcal{F}$ has some resilient leaves.*

In [**GLW2**], this result was established for arbitrary codimension-1 C^2-foliations. The proof in general needs a delicate discussion of the structure of open saturated sets without holonomy. Since recently, the same result was obtained by a different method for C^1-foliations [**Hu4**], we restrict our attention to the simpler case of foliated bundles here and we shall prove the result in full generality in Section 4.6. (Note that *Step 7* is the only point in the proof where this restriction is in force.)

The implication "(ii) $\Rightarrow$ (i)" is satisfied for arbitrary foliations and follows directly from Proposition 2.4.10. To prove the converse we have to recall several known facts about the structure of codimension-one C^2-foliations (see [**CC1**], [**CC3**], [**He**], [**CaC1**] and [**HH**]).

Let $U \subset M$ be a non-empty open subset of M saturated by the leaves of such a foliation $\mathcal{F}$. A subset of U is said to be *local minimal* when it is closed in U, saturated and contains no non-empty proper subsets enjoying these properties. Any local minimal set either consists of a single proper leaf, or coincides with an open connected component of U saturated by dense leaves, or is *exceptional*. The union $\mathcal{Z}(U)$ of all local minimal subsets of U is called the *centre* of U. Since $\mathcal{F}$ is differentiable of class C^2, any set of the form $\bar{L} \cap U$, L being a leaf contained in U, contains a local minimal set ([**CC1**], Thm. 3.0) and, therefore, $\mathcal{Z}(U)$ is always non-empty. The set $\mathcal{Z}(\mathcal{F}) = \mathcal{Z}(M)$ is called the *centre* of $\mathcal{F}$. Since M is compact and connected, either $\mathcal{F}$ is minimal and $\mathcal{Z}(\mathcal{F}) = M$, or $\mathcal{Z}(\mathcal{F}) = \mathcal{C}(\mathcal{F}) \cup E_1 \cup \cdots \cup E_m$, where $\mathcal{C}(\mathcal{F})$ is the union of all the compact leaves and E_i, $i = 1, \ldots, m$, are all the exceptional minimal sets. The sets $\mathcal{C}(\mathcal{F})$, E_i and $\mathcal{Z}(\mathcal{F})$ are compact and all the leaves accumulate on $\mathcal{Z}(\mathcal{F})$.

Let $M(0) = \mathcal{Z}(\mathcal{F})$ and $M(k) \setminus M(k-1) = \mathcal{Z}(M \setminus M(k-1))$ for $k = 1, 2, \ldots$. The leaves of $M(k) \setminus M(k-1)$ are said to be at *level* k. The union $\bigcup_{k \geq 0} M(k)$ is dense in M. The leaves of $M \setminus \bigcup_{k \geq 0} M(k)$ are said to be at *infinite level.*

If $\mathcal{F}$ has no resilient leaves, then (by Sacksteder Theorem, [**Sa**]) M contains no exceptional minimal sets and every open local minimal set U has trivial holonomy. Moreover, such a U admits a flow (ϕ_t) such that $\phi_t = \mathrm{id}$ on $M \setminus U$ and ϕ_t, $t \in \mathbb{R}$, maps leaves to the leaves ([**CC3**], Prop. 1.6). The set P_U of all the reals t such that $\phi_t(L) = L$ for any leaf $L \subset U$ has the structure of an additive subgroup of $\mathbb{R}$ and is called the *group of periods* of U. In general, P_U is not finitely generated. Also, $\partial U \subset M(k-1)$ when $U \subset M(k)$. All the proper leaves are at finite level and have polynomial growth (see [**CaC1**], pp. 324–325 for a sketch of the proof).

Passing to the proof of the implication "(i) ⇒ (ii)" in Theorem 3.6.1 assume that $\mathcal{F}$ has no resilient leaves, hence M contains no exceptional minimal sets. Fix a nice covering $\mathcal{U}$ of M, denote by T the corresponding complete transversal and by $\mathcal{H}$ the holonomy pseudogroup on T generated by the set $\mathcal{H}_1$ of all elementary holonomy maps corresponding to overlapping charts of $\mathcal{U}$. We have to show that $\mathcal{H}$ has vanishing entropy. We shall split the proof into a number of steps.

Step 1. Let $P \subset T$ be the union of all the orbits of $\mathcal{H}$ of polynomial growth. Let K be the closure of P and Ω a connected component of $T \setminus K$. Ω is a segment open in T, therefore the boundary $\partial\Omega$ is either empty or consists of at most two points. Ω corresponds to an open saturated subset of M with trivial holonomy. In fact, since the union $\bigcup_k M(k)$ is dense in M, there exists a point $x \in \Omega \cap M(k)$, $k \in \mathbb{N} \cup \{0\}$. Since the leaf L_x through x has non-polynomial growth, it is non-proper and, therefore, x belongs to a local minimal set U which is open and without holonomy. The connected component U_0 of U containing x coincides with the saturation of Ω. Moreover, the leaves passing through the points of $\partial\Omega$ are proper, hence of polynomial growth, i.e., $\partial\Omega \subset P$.

Step 2. Take a maximal (n,ε)-separated (with respect to $\mathcal{H}$) set $A \subset T$. Consider all the connected components I of $T \setminus A$ with $\partial I \subset A$. The number of such components I differs from $s(n,\varepsilon)$ by at most the number n_0 of connected components of T. Let us say that such a component I is of type (1) when $I \cap K \neq \emptyset$, of type (2), otherwise. Denote by $N_j(n,\varepsilon)$ the number of components I of type (j), $j = 1, 2$. Then

$$\#A \leq 2N_1(n,\varepsilon) + 2N_2(n,\varepsilon) + n_0. \tag{3.6.1}$$

We shall show that if ε is small enough, then the quantities $N_j(n,\varepsilon)$, $j = 1, 2$, grow polynomially with n. This will imply vanishing of the entropy of $\mathcal{H}$ and complete the proof of Theorem 3.6.1.

Step 3. First, let $j = 1$. Find a finite subset Z of P which is $(\varepsilon/3)$-dense in K and such that any connected component of $T \setminus K$ of length $\geq \frac{\varepsilon}{3}$ has its endpoints in Z. Such Z exists by Step 1 of the proof. Let $Z_n = \mathcal{H}_n(Z)$. If I is a component of type (1), then its endpoints are (n,ε)-separated, that is there exists $h \in \mathcal{H}_n$ such that $I \subset D_h$ and $h(I)$ is of length $\geq \varepsilon$. Since $I \cap K \neq \emptyset$, $h(I) \cap K \neq \emptyset$. Also, $h(I) \cap Z \neq \emptyset$, $I \cap h^{-1}(Z) \neq \emptyset$ and $I \cap Z_n \neq \emptyset$. It follows that $N_1(n,\varepsilon) \leq \#Z_n$. Since $Z \subset P$, $\#Z_n$ has polynomial growth and the same holds for $N_1(n,\varepsilon)$.

Step 4. To estimate the growth of $N_2(n,\varepsilon)$ define $N_3(n,\varepsilon)$ and $N_2(n,\varepsilon,\Omega)$, respectively, as the number of these connected components Ω of $T \setminus K$ which contain some intervals of type (2) and, for each such Ω, the number of intervals of type (2) contained in Ω.

If Ω is such a component, then $h(\Omega) \cap \Omega' \neq \emptyset$ for some $h \in \mathcal{H}_n$ and some component Ω' of $T \setminus K$ of length $\geq \varepsilon$. Clearly, the number of such Ω' is finite. Also, by Step 1 of our proof, the boundary $\partial\tilde{\Omega}'$, $\tilde{\Omega}'$ being the $\mathcal{H}$-saturation of Ω', consists of finitely many proper orbits of polynomial growth. It follows that the numbers $N_3(n,\varepsilon)$ grow polynomially with n. Say,

$$N_3(n,\varepsilon) \leq an^p \tag{3.6.2}$$

for all $n \in \mathbb{N}$ and some $a > 0$, $p \geq 1$.

Step 5. Let Ω_0 be any component of $T \setminus K$ and $h \in \mathcal{H}_n$ satisfy $\Omega_0 \cap h(\Omega_0) \neq \emptyset$. Then, $h = \phi_{\tau(h)}$ for some $\tau(h) \in \mathbb{R}$. Here, (ϕ_t) is the flow mentioned before permuting the leaves. We claim that there exists a universal constant $c > 0$ such that

$$|\tau(h)| \leq cn. \tag{3.6.3}$$

To prove this claim let us observe that any such h can be written as a product of at most n basic loops, i.e., elements $g \in \mathcal{H}$ which can be written in the form $g = h_m \circ \cdots \circ h_1$, $h_j \in \mathcal{H}_1$, in such a way that the components Ω_k of $T \setminus K$ containing $h_k \circ \cdots \circ h_1(\Omega_0)$ form a sequence

$$\Omega_0, \Omega_{a_1}, \ldots, \Omega_{a_l}, \Omega_{b_1}, \ldots, \Omega_{b_m}, \Omega_{a_l}, \ldots, \Omega_{a_1}, \Omega_0$$

with $\Omega_0, \Omega_{a_1}, \ldots, \Omega_{a_l}, \Omega_{b_1}, \ldots, \Omega_{b_m}$ different. If $x \in \Omega_0$, then by Theorem 2.4 of [**CC1**] there exists a compact segment $J \subset \Omega_0$ such that $g(x) \in J$ for any basic loop g. Therefore, there exists $c > 0$ such that $|\tau(g)| \leq c$ for any such g. This implies (3.6.3) for all $h \in \mathcal{H}_n$ as claimed.

Step 6. Again, let Ω_0 be a component of $T \setminus K$. This time, we claim that there exists a constant $c' = c'(\varepsilon)$ satisfying the following: If $x_1 < \cdots < x_k$ are points of Ω_0, $h_1, \ldots h_{k-1} \in \mathcal{H}_n$, $h_i([x_i, x_{i+1}]) \subset \Omega_0$ and the length of each segment $h_i([x_i, x_{i+1}])$ is $\geq \varepsilon$, then

$$k \leq c'(\varepsilon) \cdot n. \tag{3.6.4}$$

To prove the claim observe first that since the maps ϕ_t, $-1 \leq t \leq 1$, are equicontinuous, there exists $\delta = \delta(\varepsilon)$ such that any segment I of length $< \delta$ is mapped by ϕ_t's, $|t| \leq 1$, onto segments $\phi_t(I)$ of length $< \varepsilon$. Next, let Z be a minimal δ-dense subset of Ω_0 and let $m(\varepsilon) = \#Z$. If $x_1, \ldots, x_k$ and $h_1, \ldots, h_{k-1}$ satisfy the conditions above, $\tau_i = [\tau(h_i)]$ when $\tau(h_i) \geq 0$ and $\tau_i = [\tau(h_i)] + 1$ otherwise, then all the maps ϕ_{τ_j}, $j = 1, \ldots, k-1$, are defined on the intervals $[x_j, x_{j+1}]$ and the segments $\phi_{\tau_j}([x_j, x_{j+1}])$ have length $\geq \delta$. Therefore, any interval $[x_j, x_{j+1}]$ contains a point of $\phi_{-\tau_j}(Z)$ and

$$k \leq \# \bigcup_{j=-[cn]}^{[cn]} \phi_j(Z) \leq (2c+1)n\#Z,$$

c being the constant found in Step 5. In this way, we obtained (3.6.4) with $c' = (2c+1)m(\varepsilon)$.

Step 7. Now, let Ω be any component of $T \setminus K$ and $x_1 < \cdots < x_k$ be all the points of $A \cap \Omega$. Let $\Omega_1, \ldots, \Omega_m$ be all the components of $T \setminus K$ which are contained in the $\mathcal{H}$-orbit of Ω and have length $\geq \varepsilon$. For each $i \leq k-1$ there exists $h_i \in \mathcal{H}_n$ such that $[x_i, x_{i+1}] \subset D_{h_i}$ and $h_i([x_i, x_{i+1}])$ has length $\geq \varepsilon$.

Since $(M, \mathcal{F})$ is the foliated bundle, T can be identified with S^1, the typical fibre of the bundle, and the pseudogroup $\mathcal{H}$ can be generated by a finite set $\mathcal{H}_1$ of global diffeomorphisms of S^1. Therefore, there exist global diffeomorphisms

$f_2, \dots, f_m \in \mathcal{H}$ such that $f_i(\Omega_i) = \Omega_1$. Moreover, there exists another global diffeomorphism $g \in \mathcal{H}$ of S^1 which maps Ω onto Ω_1. If n is large enough, then f_i and $g \in \mathcal{H}_n$. For such n the maps $h'_i = f_{a_i} \circ h_i \circ g^{-1}$ belong to $\mathcal{H}_{3n}$ and map the intervals $[y_i, y_{i+1}]$, $y_i = g(x_i)$, onto the intervals $h'_i([y_i, y_{i+1}]) \subset \Omega_1$ of length $\geq \varepsilon' = \varepsilon/\lambda$, where λ is the maximum of Lipschitz constants for $f_2, \dots, f_m$. By *Step 6*, $m \leq 3c'(\varepsilon') \cdot n$. Therefore,

$$N_2(n, \varepsilon, \Omega) \leq c''(\varepsilon)n \tag{3.6.5}$$

for a suitable constant $c''(\varepsilon)$.

Step 8. Comparing the results of Step 4 and Step 7 we see that the numbers $N_2(n, \varepsilon)$ grow polynomially with n. In fact, inequalities (3.6.2) and (3.6.5) yield

$$N_2(n, \varepsilon) \leq an^p \cdot c''(\varepsilon)n = ac''(\varepsilon)n^{p+1} \tag{3.6.6}$$

for all $n \in \mathbb{N}$. This observation together with the result of Step 3 completes the proof. □

Recall now that if a codimension-one foliation $\mathcal{F}$ on M is transversely oriented, then it is given by the equation $\omega = 0$, ω being a 1-form on M. By the Frobenius Theorem, $d\omega = \omega \wedge \eta$ for another 1-form η. It is easy to check that the 3-form $\theta = \eta \wedge d\eta$ is closed and therefore represents a cohomology class in $\mathrm{H}^3(M, \mathbb{R})$. Certainly, the forms ω and η are not determined uniquely by $\mathcal{F}$, but — as can be verified easily — the cohomology class of θ does not depend on the choices of ω and η. This cohomology class was discovered by Godbillon and Vey [**GV**], is called the *Godbillon–Vey class* of $\mathcal{F}$ and denoted by $\mathrm{gv}(\mathcal{F})$. Theorem 3.6.1 together with Duminy's result on the Godbillon–Vey class ([**Du2**], see also [**CC2**]) yields immediately the following.

Corollary 3.6.2. *If $\mathcal{F}$ is a codimension-one C^2-foliation of a compact manifold M and $h(\mathcal{F}, g) = 0$ for some (equivalently, any) Riemannian structure g on M, then* $\mathrm{gv}(\mathcal{F}) = 0$. □

The most classical example of a foliation with non-trivial Godbillon–Vey class is that provided by Roussarie (compare [**GV**] again) of the weak stable foliation of an Anosov flow (of the geodesic flow of a compact hyperbolic manifold, in particular). By Corollary 3.6.2 (compare also Remark 2.4.15), this foliation has also positive entropy. However, the converse of Corollary 3.6.2 does not hold. One can suspend over Σ_2, the closed surface of genus 2, the group of orientation preserving diffeomorphisms of S^1 which has positive entropy. The resulting foliation has (by inequalities (3.4.6)) positive entropy too but its Godbillon–Vey class is trivial. Intuitively, for a non-trivial Godbillon–Vey class one needs enough separation among the leaves together with "wobbling" of the leaves, the behaviour similar to this which can be seen on the surface of a helicoid. Positive entropy needs separation of the leaves only.

Watanabe [**Wat**] applied Theorem 3.6.1 to get the following results.

Corollary 3.6.3. (i) *If $(M, \mathcal{F})$ is an oriented foliated bundle over a closed oriented surface Σ_g of genus $g \geq 2$ and the Euler number $\chi(M)$ of M does not vanish, then the geometric entropy of $\mathcal{F}$ is positive.*

(ii) *If $\chi(M) = \chi(\Sigma_g)$, then the entropy of Γ_g, the fundamental group of Σ_g acting on the fibres of M, coincides with the exponential rate of growth of Γ_g with respect to the standard generating set $\Gamma_0 = \{a_1^{\pm 1}, \ldots, a_{2g}^{\pm 1}, b_1^{\pm 1}, \ldots, b_{2g}^{\pm 1}\}$ of elements satisfying the relation*

$$a_1 b_1 a_1^{-1} b_1^{-1} \cdot \ldots \cdot a_{2g} b_{2g} a_{2g}^{-1} b_{2g}^{-1} = e.$$

Part (i) follows directly from the mentioned theorem and the following result due to Matsumoto [**Mat2**]. Indeed, this result and assumptions of (i) imply the existence of a resilient leaf, therefore non-vanishing entropy of $(M, \mathcal{F})$.

Proposition 3.6.4. *If $\phi : \Gamma \to \mathrm{Homeo}^+(S^1)$ is a homomorphism and $(M, \mathcal{F})$ its suspension, then the pullback $\phi^*(e_{\mathbb{R}})$ of the bounded Euler class $e_{\mathbb{R}}$ does not vanish if and only if $(M, \mathcal{F})$ contains a minimal set M_0 such that $\phi(\gamma)|M_0 \neq \mathrm{id}$ and $\phi(\gamma)(x) = x$ for some $\gamma \in \Gamma$ and $x \in M_0$.* □

The proof of part (ii) is a bit more complex. First, let us observe that the arguments similar to those in Example 3.1.12 show that the entropy $h(G, G_1)$ of any group G of orientation preserving homeomorphisms of the circle (with respect to any finite symmetric generating set G_1) does not exceed the rate

$$\mathrm{expgr}(G, G_1) = \lim_{n \to \infty} \frac{1}{n} \log \# G_n = \lim_{n \to \infty} \frac{1}{n} \log \# S_n$$

of exponential growth of G (with respect to G_1). Here, $S_n = G_n \setminus G_{n-1}$ is the sphere of radius n and centre e in the Cayley graph $\mathcal{C}(G, G_1)$.

Next, recall another result by Matsumoto ([**Mat3**]). There, a map $f : S^1 \to S^1$ is said to be *monotonic* if $f(e^{it}) = \tilde{f}(t)$ for all $t \in \mathbb{R}$ and some monotonic function $\tilde{f} : \mathbb{R} \to \mathbb{R}$. *A priori*, neither f nor $\tilde{f}$ has to be continuous.

Proposition 3.6.5. *If two actions of the fundamental group $\pi_1(\Sigma_g)$ of a closed oriented surface Σ_g of genus $g \geq 2$ provide suspensions whose Euler numbers coincide (up to the absolute value) with that of Σ_g, then these actions are semiconjugate via a monotonic map $f : S^1 \to S^1$.* □

Therefore, the homomorphism $\phi : \Gamma_g = \pi_1(\Sigma_g) \to \mathrm{Homeo}^+(S^1)$ providing our foliated bundle $(M, \mathcal{F})$ is semiconjugate to the Fuchsian action $\psi : \Gamma_g \to S^1$ by maps induced by covering transformations of Σ_g, S^1 being considered as the ideal boundary of the hyperbolic plane H^2. Since the action ψ is minimal, the semiconjugation is realised by a continuous surjective map $f : S^1 \to S^1$.

Now, choose a fundamental domain $D \subset H^2$ of the action of Γ_g on H^2 *via* covering transformations; D is a $(4g)$-gon bounded by consecutive geodesic arcs $\gamma_1, \ldots, \gamma_{4g}$. To fix ideas, assume that the boundary of D is oriented in the anticlockwise way. Extend arcs γ_i to maximal geodesics (denoted still by γ_i) and let $p_i = \gamma_i(-\infty)$ be one of the end points of γ_i. For any i, denote by g_i the member of our set Γ_0 generating Γ_g for which $\|dg_i|\gamma_i\| = 1$ and define a map $h : S^1 \to S^1$ by

$$h(x) = g_i(x),$$

whenever $x \in [p_i, p_{i+1})$. Given $x \in S^1$, define x_h as the unique sequence $(g_{i_1}, g_{i_2}, \dots)$ of generators for which $h^n(x) \in [p_{i_n}, p_{i_n+1})$, $n \in \mathbb{N}$. Let $S = \{x_h, x \in S^1\}$ and FS be the set of all finite subsequences of members of S. In [**Se**], it has been shown that any element $g \in \Gamma_g$ has a unique shortest representative s_g in FS. Let $\omega(g)$ be the end element of s_g. Given $\alpha \in SF$, let $Z(\alpha) = \{x \in S^1;\ x_h = (\alpha, \dots)\}$. Certainly, $Z(\alpha)$ is a non-empty subarc of S^1.

Finally, let $\varepsilon = (1/3) \cdot \min_i d_{S^1}(p_i, p_{i+1})$, denote by a_i the centre of $[p_i, p_{i+1}]$ and set

$$A_n = \{g^{-1}(a_{i_n});\ g \in S_n \quad \text{and} \quad \omega(g) = g_{i_n}\}.$$

(As before, S_n is the sphere of radius n in the Cayley graph, this time in $\mathcal{C}(\Gamma_g, \Gamma_0)$.) Since each $g \in S_n$ maps $Z(\alpha)$ into one of the arcs $[p_i, p_{i+1})$, the set A_n is (n, ε)-separated with respect to Γ_0. Therefore, the entropy under consideration is not less than $\operatorname{expgr}(\pi_1(\Sigma_g), \Gamma_0)$. Applying Proposition 3.5.2 ends the proof of Corollary 3.6.3. □

The Godbillon–Vey class is one of so-called secondary characteristic classes [**Bo**] for foliations. Its geometric meaning is relatively well understood and this allows us to link it with the geometric entropy as above. One of the questions formulated in [**GLW2**], still open, is whether all the secondary classes of foliations with zero entropy must vanish. Some hope for the positive answer comes from the following example: Riemannian foliations have vanishing entropy (Example 3.2.6) and all their secondary characteristic classes are trivial ([**KT**], Theorem 4.52).

Chapter 4

Invariant measures

In this chapter, we consider invariant measures for groups and pseudogroups of transformations, and for foliations, in the last case defined as measures on transversals invariant under holonomy maps. The study of such measures (called transverse invariant ones) was inaugurated by J. Plante [**Pl**] who has shown that the existence of such measures has some influence on the topology of a foliated manifold and is related to the growth types of leaves. Some of his results are discussed here, in Sections 4.2 and 4.3. Since non-trivial invariant measures need not exist, we consider also two wider classes of measures: harmonic measures on foliated Riemannian manifolds and quasi-invariant measures for Kleinian groups. In both cases, the considered measures have several interesting, very geometric, properties and reflect some dynamics of the systems, foliations and Kleinian groups. Also, in Section 4.6, we show how measures invariant under the geodesic flow of a foliation can be applied to the proof of the implication "if no resilient leaves, then vanishing entropy" for arbitrary codimension-one C^1-foliations. This interesting idea is due to Hurder, was originated at the very beginning of the theory of geometric entropy in [**Hu1**] and developed to a final form recently in [**Hu4**].

4.1 Basic definitions and facts

As we could observe in Section 3.1.2, for any continuous transformation f of a compact metric space X there exist f-invariant Borel probability measures on X. Given two such transformations f and g, it might happen that the corresponding sets $\mathcal{M}(X, f)$ and $\mathcal{M}(X, g)$ are disjoint. However, if our transformations f and g of X commute, $f \circ g = g \circ f$, these sets may be different but have a non-empty intersection. In fact, if $\mu \in \mathcal{M}(X, g)$ is g-invariant and (ν_{n_j}) is a convergent sequence of Cesàro means (ν_n) defined by formula (3.1.15) in Section 3.1.2, where $\mu_k = \mu$ for $k = 1, 2, \ldots$, then the limit $\nu = \lim_{j\to\infty} \nu_{n_j}$ is f-invariant and, at the same time,

$$g^*\nu = \lim_{j\to\infty} \frac{1}{n_j} \sum_{k=0}^{n_j} g^* f^{k*} \mu = \lim_{j\to\infty} \frac{1}{n_j} \sum_{k=0}^{n_j} f^{k*} g^* \mu = \lim_{j\to\infty} \frac{1}{n_j} \sum_{k=0}^{n_j} f^{k*} \mu = \nu,$$

i.e., ν is g-invariant. Therefore, $\nu \in \mathcal{M}(X, f) \cap \mathcal{M}(X, g)$.

In the same way, if $f_1, \ldots, f_{m+1}$ are pairwise commuting transformations of X, $\mu \in \mathcal{M}(X)$ is f_k-invariant for $k = 1, \ldots, m$, $\nu_n = \frac{1}{n} \sum_{j=0}^{n-1} f_{m+1}^{j*} \mu$ and ν is the limit of a convergent subsequence of (ν_n), then ν is f_k-invariant for all $k \le m+1$. Therefore, for any finitely generated abelian group G of transformations of X there exist G-invariant Borel probability measures on X. Clearly, a measure μ is said to be *G-invariant* when $g^*\mu = \mu$ for any $g \in G$. As in Section 3.1.2, all such measures form a compact convex subset $\mathcal{M}(X, G)$ of the space $\mathcal{M}(X)$ of all Borel probability measures on X.

For general non-abelian groups, it might happen that $\mathcal{M}(X, G) = \emptyset$.

Example 4.1.1. Let $X = S^1$ and G be the group generated by two maps f_i, $i = 0, 1$, such that each of them has exactly two fixed points: a source p_i and a sink q_i which split S^1 into two arcs A_i and B_i each of them being contracted to q_i. By Proposition 3.1.6, any measure μ which is f_i-invariant has to be supported in the set $\{p_i, q_i\}$. Therefore, $\mathcal{M}(X, G) = \emptyset$ if only $\{p_0, q_0\} \cap \{p_1, q_1\} = \emptyset$.

Clearly, the same may occur in the case of measures invariant under the action of a pseudogroup. First, introduce the following.

Definition 4.1.2. A measure $\mu \in \mathcal{M}(X)$ is *$\mathcal{G}$-invariant*, $\mathcal{G}$ being a pseudogroup of local homeomorphisms of X, whenever

$$\mu(g(A)) = \mu(A) \tag{4.1.1}$$

for all $g \in \mathcal{G}$ and all Borel subsets A of D_g.

The set of all $\mathcal{G}$-invariant Borel probability measures on X will be denoted by $\mathcal{M}(X, \mathcal{G})$.

Now, observe that the set of non-trivial $\mathcal{G}$-invariant measures can be empty even if $\mathcal{G}$ is generated by a single local homeomorphism f (in spite of the fact that — compare Section 3.1.2 of Chapter 3 — the set $\mathcal{M}(X, f)$ of f-invariant probability measures is non-empty).

Example 4.1.3. Again, let $X = S^1$, $f(z) = z^2$ and $\mathcal{G} = \mathcal{G}(f)$ as in Section 1.2. For any open arc A containing number 1, one has $f^n(A) = S^1$ for n large enough. If $\mu \in \mathcal{M}(S^1, \mathcal{G}(f))$, then $\mu(f^n(A)) = \mu(A)$ as long as f^n maps A homeomorphically onto its image. Thus, $\mu = \delta_1$, the Dirac measure at the point $1 \in S^1$. On the other hand, $\delta_1(\{-1\}) = 0$ while $f(-1) = 1$ and $\delta_1(\{1\}) = 1$. This contradiction shows that non-trivial $\mathcal{G}(f)$ invariant measures do not exist.

Analogously to the classical theory (Section 3.1) one can consider ergodicity of $\mathcal{G}$-invariant measures with $\mathcal{G}$ being an arbitrary pseudogroup. A good part of the theory is almost the same as in the case of a single transformation. However, one has to be concerned with domains of maps $g \in \mathcal{G}$ and this is why we provide the proofs below.

First, let us say that a subset A of X is $\mathcal{G}$-*invariant* whenever $g(A \cap D_g) \subset A$ for any $g \in \mathcal{G}$, equivalently, whenever $g(A \cap D_g) = A \cap R_g$ for all $g \in \mathcal{G}$. Then, let us introduce the following.

Definition 4.1.4. A $\mathcal{G}$-invariant Borel probability measure μ on X is called *ergodic* when the following condition is satisfied: if $A \subset X$ is $\mathcal{G}$-invariant, then either $\mu(A) = 0$ or $\mu(A) = 1$. We shall denote by $\mathcal{E}(X, \mathcal{G})$ the set of all ergodic members of $\mathcal{M}(X, \mathcal{G})$.

Ergodicity can be expressed also in the terms of $\mathcal{G}$-invariant functions:

Lemma 4.1.5. *If μ is ergodic, $\phi : X \to \mathbb{R}$ is measurable, bounded and $\phi \circ g = \phi$ μ-a.e. for all $g \in \mathcal{G}$, then ϕ is constant μ-a.e.*

Proof. Fix an almost everywhere $\mathcal{G}$-invariant function ϕ and put

$$X(k,n) = \phi^{-1}\left(\left[\frac{k}{2^n}, \frac{k+1}{2^n}\right)\right)$$

for $k \in \mathbb{Z}$ and $n \in \mathbb{N}$. Then the symmetric difference

$$X(k,n) \triangle \bigcup_{g \in \mathcal{G}} g(X(k,n))$$

is contained in

$$\{x \in X; (\exists g \in \mathcal{G})\ x \in D_g \text{ and } \phi(g(x)) \neq \phi(x)\}.$$

Therefore,

$$\mu(X(k,n) \triangle \bigcup_g g(X(k,n)) = 0, \quad \mu(X(k,n)) = \mu(\bigcup_g g(X(k,n))$$

and either $\mu(X(k,n)) = 0$ or $\mu(X(k,n)) = 1$ since the set $\bigcup_g g(X(k,n))$ is $\mathcal{G}$-invariant and the measure μ ergodic. It follows that for any $n \in N$ there exists a unique k, denoted by k_n, such that $\mu(X(k_n,n)) = 1$. The set $Y = \bigcap_n X(k_n,n)$ has measure 1 and ϕ is constant on Y. □

If $\mathcal{M}(X, \mathcal{G})$ is non-empty, then it is closed and convex, hence one can consider the set of its extreme points.

Proposition 4.1.6. *If $\mathcal{G}$ admits non-trivial invariant measures, then $\mathcal{E}(X, \mathcal{G})$ coincides with the set of all extreme points of $\mathcal{M}(X, \mathcal{G})$.*

Proof. If $\mu \notin \mathcal{E}(X,\mathcal{G})$, then there exists a $\mathcal{G}$-invariant set $Z \subset X$ such that $0 < \mu(Z) < 1$. The measures μ_0 and μ_1 defined by

$$\mu_0(A) = \frac{\mu(A \cap Z)}{\mu(Z)} \quad \text{and} \quad \mu_1(A) = \frac{\mu(A \setminus Z)}{\mu(X \setminus Z)}$$

for all Borel sets $A \subset X$ are $\mathcal{G}$-invariant, $\mu_0(X) = \mu_1(X) = 1$ and $\mu = \mu(Z)\mu_0 + \mu(X \setminus Z)\mu_1$, so μ is not extreme in $\mathcal{M}(X,\mathcal{G})$.

Conversely, if μ is ergodic, $0 < t < 1$ and $\mu = t\mu_0 + (1-t)\mu_1$ for some μ_0 and μ_1 of $\mathcal{M}(X,\mathcal{G})$, then μ_0 is absolutely continuous with respect to μ and the Radon–Nikodym derivative $\phi = d\mu_0/d\mu$ is μ-almost everywhere $\mathcal{G}$-invariant. In fact, if $g \in \mathcal{G}$, then

$$\int_A (\phi \circ g - \phi) d\mu = \mu_0(g(A)) - \mu_0(A) = 0$$

for any Borel set $A \subset D_g$ and, consequently, $\phi \circ g = \phi$ μ-a.e. on D_g. By Lemma 4.1.5, ϕ is μ-almost everywhere constant. Since $\int_X \phi d\mu = 1$, $\phi \equiv 1$ μ-a.e. and $\mu_0 = \mu_1 = \mu$, i.e., μ is extreme in $\mathcal{M}(X,\mathcal{G})$. □

Now, we shall concentrate on the case of holonomy pseudogroups of foliations.

Definition 4.1.7. Given a foliation $\mathcal{F}$ of a manifold M, a *transverse invariant measure* is a Borel probability measure on a complete transversal T which is invariant under the holonomy pseudogroup $\mathcal{H}$ of $\mathcal{F}$ acting on T.

From the discussion of Section 1.3 it follows that the existence of a transverse invariant measure on one complete transversal implies the existence of such a measure on any other transversal. Therefore, one can say that $\mathcal{F}$ has *measure preserving holonomy* if a transverse invariant measure exists for some complete transversal T.

Example 4.1.8. (i) Riemannian foliations of compact manifolds have measure preserving holonomy. In fact, the holonomy maps of a Riemannian foliation preserve a Riemannian volume form on a complete transversal. (ii) From Examples 4.1.3 and 2.4.6 it follows that the Hirsch foliation admits no transverse invariant measures. (iii) Any foliation containing a compact leaf admits transverse invariant measures. In fact, if L is a compact leaf and T is a complete transversal, then the set $L \cap T$ is finite and the combination of Dirac measures

$$\mu = \frac{1}{\#(L \cap T)} \cdot \sum_{x \in L \cap T} \delta_x \tag{4.1.2}$$

is invariant under the holonomy pseudogroup of $\mathcal{F}$ acting on T.

Finally, recall that in general the $\mathcal{F}$-saturation of a closed set need not be closed while the saturation of an open set is always open. However, if μ is a transverse invariant measure defined on a complete transversal T, then the saturation

$\operatorname{supp}_{\mathcal{F}}(\mu)$ of the support $\operatorname{supp}(\mu)$ of μ (compare Section 3.1.2) is closed. In fact, the saturation S_μ of $T \setminus \operatorname{supp}(\mu)$ is open and

$$M \setminus \operatorname{supp}_{\mathcal{F}}(\mu) = S_\mu. \tag{4.1.3}$$

The relation "$\subset$" in (4.1.3) is obvious. To show the other one, "$\supset$", take any point $x \in S_\mu$. Then there exists $y \in T \cap L_x$, L_x being the leaf of $\mathcal{F}$ through x, and a neighbourhood $U \subset T$ of y such that $\mu(U) = 0$. Let h be a holonomy map defined in a neighbourhood of x and mapping x to y. Then $\mu(h^{-1}(U)) = \mu(U \cap R_h) = 0$ and $x \notin \operatorname{supp}_{\mathcal{F}}(\mu)$.

4.2 Transverse invariant measures and homology

The existence of transverse invariant measures has some influence on the topology of a foliated manifold $(M, \mathcal{F})$. For example, if $\mathcal{F}$ is transversely oriented and T, a closed and transverse to $\mathcal{F}$ submanifold of dimension $q = \operatorname{codim} \mathcal{F}$, intersects the support of a transverse invariant measure μ, then T represents a non-zero element in $H_q(M, \mathbb{R})$.

The proof of this fact consists in a construction of a suitable $(n-q)$-*current* (i.e., a real continuous linear functional on the space of $(n-q)$-forms on M) which is closed (that is, vanishes on the set of all exact forms) and positive on a form which represents the volume form of T (in the sense described below). To get such a current, it is convenient to proceed as follows

Consider $N(T)$, a tubular neighbourhood of T built of small disks contained in the leaves of $\mathcal{F}$, and find a reduction of the structure group of the bundle $\pi : N(T) \to T$ to the orthogonal group $\mathrm{O}(n-q)$, $n = \dim M$. Then, choose an $\mathrm{O}(n-q)$-invariant C^∞-function $f : \mathbb{R}^{n-q} \to [0,1]$ such that $f(0) = 1$, $\operatorname{supp} f \subset B(0,r)$ and $\int_{\mathbb{R}^{n-q}} f = 1$. Here, $r > 0$ is a small constant. For any open set $U \subset T$ such that the bundle $\pi : N(T) \to T$ is trivial over U, choose a suitable trivialization $\phi : \pi^{-1}(U) \to U \times \mathbb{R}^{n-q}$ and define a closed $(n-q)$-form ω on $\pi^{-1}(U)$ by

$$\omega = \phi^*(f dx_1 \wedge \cdots \wedge dx_{n-q}).$$

Since the structure group is orthogonal, ω is well defined on $N(T)$ and extends to a global form ω on M by $\omega = 0$ outside $N(T)$.

Now, let $\mathcal{U} = \{U_1, \dots, U_k\}$ be a nice covering of $(M, \mathcal{F})$ and (α_i) a smooth partition of unity subordinated to $\mathcal{U}$. For any $(n-q)$-form η put

$$\Phi_\mu(\eta) = \sum_{i=1}^{k} \int_{D_i'} \left(\int_{\{x\} \times D_i} \psi_i^*(\alpha_i \eta) \right) d\psi_i^* \mu, \tag{4.2.1}$$

where $\psi_i : U_i \to D_i \times D_i' \subset \mathbb{R}^{n-q} \times \mathbb{R}^q$ is a distinguished chart on U_i which induces on D_i' the standard orientation. Then — as one can easily see from Stokes Theorem — the equality $\Phi_\mu(d\xi) = 0$ holds for any $(n-q-1)$-form ξ and therefore

Φ_μ induces a linear map $H^{n-q}(M) \to \mathbb{R}$. The form $\omega|T$ is a volume form on T and $\Phi_\mu(\omega) = \mu(T) > 0$. Therefore, $T \neq 0$ in $H_q(M, \mathbb{R})$.

In this way we have proved the following.

Theorem 4.2.1. [**Pl**] *If μ is a transverse invariant measure for an $(n-q)$-dimensional foliation $\mathcal{F}$ of an n-dimensional manifold M and T is a closed transverse q-dimensional submanifold of M which intersects* $\mathrm{supp}_{\mathcal{F}}(\mu)$, *then T represents a non-trivial element of $H_q(M, \mathbb{R})$.* □

If $\mathcal{F}$ is of codimension 1, then — as was explained in Section 3.5.3 — any noncompact leaf of $\mathcal{F}$ meets a closed transverse loop. Therefore, if $\mathcal{C}(\mathcal{F})$ is the union of all the compact leaves of $\mathcal{F}$ and μ is a transverse invariant measure with $\mathrm{supp}(\mu) \cap \mathcal{C}(\mathcal{F}) = \emptyset$, then $\mathrm{supp}\,\mu$ meets a closed transverse loop and, consequently, Φ_μ represents a non-trivial cohomology class.

Corollary 4.2.2. *If μ is a transverse invariant measure for a transversely oriented codimension-one foliation $\mathcal{F}$ of a compact oriented manifold M and* $\mathrm{supp}(\mu)$ *meets no closed leaves of $\mathcal{F}$, then Φ_μ represents a non-trivial element of $H^1(M, \mathbb{R})$.* □

4.3 Measures and orbit growth

From the observations of the last section it follows that we should search for some conditions which are sufficient (or, necessary) for the existence of invariant measures. A very simple example is that of a pseudogroup $\mathcal{G}$ generated by maps which fix a given point $x \in X$. Then, $\delta_x \in \mathcal{M}(X, \mathcal{G})$. Also, as we observed already in the previous section, if the $\mathcal{G}$-orbit of a point x is finite, $\mathcal{G}(x) = \{x_1, \dots, x_m\}$, then the formula

$$\mu = \frac{1}{m} \sum_{i=1}^{m} \delta_{x_i} \tag{4.3.1}$$

defines a $\mathcal{G}$-invariant probability measure.

These observations have been generalized by Plante [**Pl**] who formulated a pretty nice condition providing the existence of measures invariant under a pseudogroup action.

Theorem 4.3.1. *If the orbit $\mathcal{G}(x)$ of a point $x \in X$ has non-exponential growth with respect to a given finite set $\mathcal{G}_1$ generating a pseudogroup $\mathcal{G}$, then there exists a $\mathcal{G}$-invariant Borel probability measure on X supported in the closure of $\mathcal{G}(x)$.*

Proof. From the Riesz Representation Theorem it follows directly that we have to construct a normalized non-negative $\mathcal{G}$-invariant functional $\Lambda : C(X) \to \mathbb{R}$, $C(X)$ being, as before, the space of all continuous real functions on X. In this context, $\mathcal{G}$-invariant means that $\Lambda(\phi \circ g) = \Lambda(\phi)$ for any $g \in \mathcal{G}$ and any function $\phi \in C(X)$ supported in R_g. Hereafter, $\phi \circ g$ is extended all over X by $\phi \circ g = 0$ on $X \setminus D_g$.

First, let us observe that, since $\mathcal{G}(x)$ has non-exponential growth, there exists an increasing sequence (n_k) of natural numbers such that $n_k \to \infty$ as $k \to \infty$ and

$$\lim_{k\to\infty} \frac{\#(\mathcal{G}_{n_k+1}(x) \setminus \mathcal{G}_{n_k-1}(x))}{\#\mathcal{G}_{n_k}(x)} = 0. \tag{4.3.2}$$

In fact, if $t_n = \#\mathcal{G}_n(x)$ and $t = \liminf_{n\to\infty}(t_{n+1} - t_{n-1})/t_n > 0$, then $t_{n+1} \geq \frac{1}{2}tt_n + t_{n-1} \geq (\frac{1}{2}t + 1)t_{n-1}$ for n large enough. Therefore, $t_{n+2k-1} \geq (\frac{1}{2}t+1)^k t_{n-1}$ and

$$\begin{aligned}\frac{1}{n+2k-1}\log t_{n+2k-1} &\geq \frac{k}{n+2k-1}\log(1+\frac{t}{2}) + \frac{1}{n+2k-1}\log t_{n-1}\\ &\geq \frac{1}{4n}\log(1+\frac{t}{2}) > 0\end{aligned}$$

for $k = 1, 2, \ldots$ and n large enough. In such a case, (t_n) would have exponential growth.

Next, choose a sequence (n_k) as above and define functionals Λ_k, $k = 1, 2, \ldots$, by

$$\Lambda_k(\phi) = \frac{1}{\#\mathcal{G}_{n_k}(x)} \sum_{z\in\mathcal{G}_{n_k}(x)} \phi(z). \tag{4.3.3}$$

Of course, each of Λ_k is non-negative (that is, $\Lambda_k(\phi) \geq 0$ when $\phi \geq 0$) and normalized ($\Lambda_k(1) = 1$). Since the space $\mathcal{M}(X)$ is compact, we may assume, passing to a subsequence if necessary, that $\Lambda_k \to \Lambda$ as $k \to \infty$. Again, Λ is non-negative and normalized. It remains to show that Λ is $\mathcal{G}$-invariant.

To this end, take any $g \in \mathcal{G}_1$ and $\phi \in C(X)$ with $\operatorname{supp}(\phi) \subset R_g$. Then

$$\Lambda_k(\phi \circ g) - \Lambda_k(\phi) = \frac{1}{\#\mathcal{G}_{n_k}(x)} \cdot \sum_{z\in A_k(x)} \pm\phi(z),$$

where $A_k(x) = \mathcal{G}_{n_k}(x)\triangle g(\mathcal{G}_{n_k}(x))$ and $\triangle$ denotes, as before, the symmetric difference. Therefore,

$$\begin{aligned}|\Lambda_k(\phi\circ g) - \Lambda_k(\phi)| &\leq \max|\phi| \cdot \frac{\#(\mathcal{G}_{n_k+1}(x)\triangle\mathcal{G}_{n_k-1}(x))}{\#\mathcal{G}_{n_k}(x)}\\ &\leq \max|\phi| \cdot \frac{\#(\mathcal{G}_{n_k+1}(x)\setminus\mathcal{G}_{n_k-1}(x))}{\#\mathcal{G}_{n_k}(x)}.\end{aligned}$$

Passing to the limit with $k \to \infty$ and using (4.3.2) we arrive at $\Lambda(\phi \circ g) = \Lambda(\phi)$. Simple induction shows that the last equality holds for all $g \in \mathcal{G}$, not only for those in $\mathcal{G}_1$. □

Applying Theorem 4.3.1 to a holonomy pseudogroup of a foliation we arrive at

Corollary 4.3.2. *Any foliation $\mathcal{F}$ which possesses a leaf L of non-exponential growth admits a transverse invariant measure supported in the closure $\bar{L}$ of L.* □

4.4 Transverse invariant measures in codimension 1

In this section, all the foliations are transversely oriented, of codimension 1 and class C^2. Foliated manifolds are oriented and compact.

Let us recall once again some results concerning structure of codimension-one foliations (see **[HH]**, **[CaC1]**, etc.).

The union $\mathcal{C}(\mathcal{F})$ of all the compact leaves of a codimension-one foliation $\mathcal{F}$ of a compact manifold M is compact. Also, M contains at most finitely many exceptional minimal sets $E_1, \ldots, E_m$ and the centre $\mathcal{Z}(\mathcal{F})$ of $\mathcal{F}$ defined as the union of all the minimal sets of $\mathcal{F}$ is also compact. *A priori* one has two cases: either M is minimal (and then $\mathcal{Z}(\mathcal{F}) = M$) or

$$\mathcal{Z}(\mathcal{F}) = \mathcal{C}(\mathcal{F}) \cup E_1 \cup \cdots \cup E_m. \tag{4.4.1}$$

In both cases, the closure of any leaf L intersects $\mathcal{Z}(\mathcal{F})$.

Theorem 4.4.1. *The support* $\operatorname{supp}(\mu)$ *of any transverse invariant measure* μ *on a compact manifold* M *equipped with a foliation* $\mathcal{F}$ *of codimension* 1 *is contained in* $\mathcal{Z}(\mathcal{F})$ *and contains no resilient leaves.*

Proof. First, assume that g is an element of a pseudogroup $\mathcal{G} \subset \mathrm{Homeo}_+(\mathbb{R})$, μ is a finite $\mathcal{G}$-invariant measure, $x \in \operatorname{supp}(\mu)$ and $g(x) = x$. If $y \in D_g \cap \operatorname{supp}(\mu)$ and $g(y) \neq y$, say $g(y) < y$, then $\mu(U) > 0$ for a segment U such that $y \in U$ and $g(U) \cap U = \emptyset$. If so, all the images $g^n(U)$, $n = 1, 2, \ldots$, are pairwise disjoint and $\mu(g^n(U)) = \mu(U)$. Therefore, $\mu(\bigcup_n g^n(U)) = \infty$, a contradiction. It follows that

$$g(y) = y \quad \text{for any} \quad y \in D_g \cap \operatorname{supp}(\mu). \tag{4.4.2}$$

In particular, the orbit through x cannot be resilient.

This observation proves the second part of the statement and completes the proof of the theorem in the case when $\mathcal{Z}(\mathcal{F}) = M$.

If $\mathcal{Z}(\mathcal{F}) = \mathcal{C}(\mathcal{F}) \cup E_1 \cup \cdots \cup E_m$, then — by an argument similar to that above — we can find an open saturated set V such that $V \cap \operatorname{supp}(\mu) = \mathcal{C}(\mathcal{F}) \cap \operatorname{supp}(\mu)$. Indeed, since $\mathcal{F}$ is transversely oriented and — as was mentioned in Section 3.6 — all the leaves of $\mathcal{F}$ accumulate on $\mathcal{Z}(\mathcal{F})$, there exists an open saturated set V such that $C(\mathcal{F}) \subset V$ and $\bar{L} \cap C(\mathcal{F}) \neq \emptyset$ for any leaf $L \subset V$. (Otherwise, one would be able to find a sequence of leaves L_n contained in arbitrarily small neighbourhoods V_n of $\mathcal{C}(\mathcal{F})$ and such that their closures would intersect one of the exceptional minimal sets, say E_1; this would allow us to conclude that $E_1 \cap \mathcal{C}(\mathcal{F}) \neq \emptyset$, a contradiction.) If $x \in V \cap T$, T being as always a complete transversal, then there exist holonomy maps h_n such that $x_n = h_n(x) \to x_0 \in \mathcal{C}(\mathcal{F})$ as $n \to \infty$. If $x \notin \mathcal{C}(\mathcal{F})$, then using the transverse orientation of $\mathcal{F}$ and passing to a subsequence if necessary we may assume that (x_n) is strictly monotonic, say decreasing to (x_0). Choose a segment $I \subset T$ of the form $I = [y, x]$ where $x_1 < y < x$. The segments $I_n = h_n(I)$ are pairwise disjoint. If x were contained in $\operatorname{supp}(\mu)$, then we would have as before $\mu(\cup_n I_n) = \infty$, a contradiction which implies that $V \cap \operatorname{supp}(\mu) \subset \mathcal{C}(\mathcal{F})$.

In the same way, using the Sacksteder Theorem (see either [**Sa**] or [**CaC1**], Thm. 8.2.1), we can find an open saturated set W such that $W \cap \operatorname{supp}(\mu) = (E_1 \cup \cdots \cup E_m) \cap \operatorname{supp}(\mu)$.

Since the set $\operatorname{supp}(\mu)$ is holonomy invariant, it follows that it is contained in $\mathcal{Z}(\mathcal{F})$. In fact, if $x \in M \setminus (V \cup W)$ were a point of $\operatorname{supp}(\mu)$, then the leaf L_x through x would accumulate at some points of $\mathcal{Z}(\mathcal{F})$ and, therefore, there would exist a holonomy map g such that $x \in D_g$ and $g(x) \in (V \cup W) \setminus \mathcal{Z}(\mathcal{F})$. Thus, $g(x)$ would not belong to $\operatorname{supp}(\mu)$, a contradiction. □

If $\mathcal{F}$ has trivial holonomy and is C^2-differentiable, then $\mathcal{F}$ is conjugate to a foliation $\mathcal{F}'$ (of the same compact manifold M) defined by a nowhere-vanishing closed 1-form ω ([**HH**], Thm. VIII.3.2.7). If $f : M \to M$ is a conjugating map, then the measure μ defined by

$$\mu(B) = \int_B f^*\omega \tag{4.4.3}$$

for any Borel set B contained in a transversal T is holonomy invariant (by Stokes Theorem), non-trivial and finite. Since $\mathcal{F}$ is transversely oriented, the integrals in (4.4.3) are of constant sign, and, replacing ω by $-\omega$ if necessary, we may assume that they are non-negative. This observation together with Theorem 4.4.1 (condition (4.4.2) particularly) yields the following.

Corollary 4.4.2. *A transversely oriented codimension-one minimal C^2-foliation of a compact manifold M admits a transverse invariant measure if and only if all the leaves of $\mathcal{F}$ have trivial holonomy.* □

The assumption of C^2-differentiability could be dropped but then the argument would become more delicate. We should use the fact that the holonomy pseudogroup of $\mathcal{F}$ is semi-conjugate to a subgroup G of $\mathbb{R}$, construct a G-invariant measure on $\mathbb{R}$ and use the conjugating map to bring it to a complete transversal T.

We shall not go deeper into the study of the structure of codimension-one foliations which admit transverse invariant measures. An interested reader can consult either [**HH**] or [**Pl**], for instance. Let us just mention here that any leaf L of $\mathcal{F}$ contained in $\operatorname{supp}_{\mathcal{F}}(\mu)$, μ being a transverse invariant measure, has polynomial growth of degree $d \leq \max\{0, b_1(M) - 1\}$, where $b_1(M) = \operatorname{rank} H_1(M, \mathbb{R})$ is the first Betti number of M ([**Pl**], Theorem 6.3). From this result and Corollary 4.3.2 above, it follows immediately that any codimension-one foliation with a leaf of non-exponential growth has some leaves of polynomial growth of degree $d \leq \max\{0, b_1(M) - 1\}$. In particular, if $b_1(M) = 1$, then such a foliation $\mathcal{F}$ has a compact leaf, i.e., $\mathcal{C}(\mathcal{F}) \neq \emptyset$ in this case.

4.5 Vanishing entropy and invariant measures

The method of "averaging sequences" applied in the proof of Theorem 4.3.1 allows us to construct transverse invariant measures for foliations of vanishing entropy. The construction is somewhat more technical than that of Theorem 4.3.1. Essentially, it consists in averaging sums of Dirac measures over (n, ε)-separated sets and passing to suitable limits as $n \to \infty$ and $\varepsilon \to 0$.

Let therefore $\mathcal{G}$ be again a finitely generated pseudogroup of local homeomorphisms of a compact metric space (X, d). Assume that $\mathcal{G}_1$ is a good finite symmetric set generating $\mathcal{G}$.

Theorem 4.5.1. *If $h(\mathcal{G}, \mathcal{G}_1) = 0$, then X admits a $\mathcal{G}$-invariant Borel probability measure.*

Proof. Denote by $C_+(X)$ the space of continuous non-negative functions on X. For any $n \in \mathbb{N}$ and $\varepsilon > 0$ denote by $S(n, \varepsilon)$ the family of all (n, ε)-separated subsets of X. Given $f \in C_+(X)$ put

$$\Lambda_{n,\varepsilon}(f) = \frac{1}{s(n, \varepsilon, \mathcal{G}_1)} \sup \left\{ \sum_{x \in A} f(x);\ A \in S(n, \varepsilon) \right\}. \tag{4.5.1}$$

It is easy to see that the functionals $\Lambda_{n,\varepsilon}$ are non-negatively homogeneous (that is, $\Lambda_{n,\varepsilon}(af) = a\Lambda_{n,\varepsilon}(f)$ for all f and $a \in [0, \infty)$), subadditive (i.e., $\Lambda_{n,\varepsilon}(f_1 + f_2) \leq \Lambda_{n,\varepsilon}(f_1) + \Lambda_{n,\varepsilon}(f_2)$ for all f_1 and f_2 of $C_+(X)$), monotonic (i.e., $\Lambda_{n,\varepsilon}(f_1) \leq \Lambda_{n,\varepsilon}(f_2)$ whenever $f_1 \leq f_2$), normalized (i.e., $\Lambda_{n,\varepsilon}(1) = 1$) and bounded in the following sense:

$$\Lambda_{n,\varepsilon}(f) \leq \sup f, \quad f \in C_+(X). \tag{4.5.2}$$

Moreover, if the distance between the supports of f_1 and f_2 exceeds ε, then

$$\Lambda_{n,\varepsilon}(f_1 + f_2) = \Lambda_{n,\varepsilon}(f_1) + \Lambda_{n,\varepsilon}(f_2). \tag{4.5.3}$$

This follows directly from the following observation: if $d(\operatorname{supp} f_1, \operatorname{supp} f_2) \geq \varepsilon$ and sets $A_1 \subset \operatorname{supp} f_1$ and $A_2 \subset \operatorname{supp} f_2$ are (n, ε)-separated, then the union $A_1 \cup A_2$ is (n, ε)-separated as well.

Now, observe that the assumption $h(\mathcal{G}, \mathcal{G}_1) = 0$ implies the equality

$$\lim_{p \to \infty} \liminf_{n \to \infty} \frac{s(n + 1, 1/p, \mathcal{G}_1)}{s(n, 1/p, \mathcal{G}_1)} = 1 \tag{4.5.4}$$

which allows us to choose numbers $n(k, p) \in \mathbb{N}$ in such a way that $\lim_{k \to \infty} n(k, p) = \infty$ for any $p \in \mathbb{N}$, the limits

$$a_p = \lim_{k \to \infty} \frac{s(n(k, p) + 1, 1/p, \mathcal{G}_1)}{s(n(k, p), 1/p, \mathcal{G}_1}$$

exist and satisfy

$$\lim_{p \to \infty} a_p = 1.$$

The proof of (4.5.4) follows Plante's ideas and is similar to that of (4.3.2) in Section 4.3.

Next, note that the numbers $n(k,p)$ can be chosen (passing to a subsequence several times, if necessary) in such a way that the limits

$$\begin{aligned} \Lambda_{1/p}(f) &= \lim_{k\to\infty} \Lambda_{n(k,p),1/p}(f) \quad \text{and} \\ \bar{\Lambda}_{1/p}(f) &= \lim_{k\to\infty} \Lambda_{n(k,p)+1,1/p}(f) \end{aligned} \tag{4.5.5}$$

exist for all $f \in C_+(X)$. In fact, the space $C_+(X)$ contains a countable dense subset C_0 and the above-listed properties of the functionals $\Lambda_{n,\varepsilon}$ imply that the existence of limits (4.5.5) for all $f \in C_0$ yields the existence of these limits for all $f \in C_+(X)$.

In the same way, we can choose a subsequence (p_m) converging to infinity and such that the limits

$$\begin{aligned} \Lambda(f) &= \lim_{m\to\infty} \Lambda_{1/p_m}(f) \quad \text{and} \\ \bar{\Lambda}(f) &= \lim_{m\to\infty} \bar{\Lambda}_{1/p_m}(f) \end{aligned} \tag{4.5.6}$$

exist for all $f \in C_+(X)$.

From the previous observations it follows easily that the functionals Λ and $\bar{\Lambda}$ are non-negatively homogeneous, subadditive, monotonic and normalized. Moreover,

$$\Lambda(f_1 + f_2) = \Lambda(f_1) + \Lambda(f_2) \quad \text{and} \quad \bar{\Lambda}(f_1 + f_2) = \bar{\Lambda}(f_1) + \bar{\Lambda}(f_2) \tag{4.5.7}$$

whenever the supports of f_1 and f_2 are disjoint.

For any compact set $K \subset X$ put

$$\begin{aligned} \mu(K) &= \inf\{\Lambda(f); f \in C_+(X), f \leq 1 \text{ and } f|K \equiv 1\} \quad \text{and} \\ \bar{\mu}(K) &= \inf\{\bar{\Lambda}(f); f \in C_+(X), f \leq 1 \text{ and } f|K \equiv 1\}. \end{aligned} \tag{4.5.8}$$

Following the proof of the Riesz Representation Theorem in [**Rud**] one can prove that μ and $\bar{\mu}$ can be extended to Borel probability measures (denoted by μ and $\bar{\mu}$ again) on X. In fact, the only difficulty comes from the fact that the functionals Λ and $\bar{\Lambda}$ are not linear: equalities (4.5.7) hold for functions with disjoint supports only. However, Steps IV and X are the only points in the proof we want to follow, where the additivity of the functional is used. In Step IV we need it for functions with disjoint supports only, while in Step X it is used to show that "μ represents Λ", the property we are not interested in.

Since any (n,ε)-separated set A is also $(n+1,\varepsilon)$-separated, we have

$$s(n,\varepsilon,\mathcal{G}_1)\Lambda_{n,\varepsilon}(f) \leq s(n+1,\varepsilon,\mathcal{G}_1)\Lambda_{n+1,\varepsilon}(f)$$

for all $n \in \mathbb{N}$, $\varepsilon > 0$ and $f \in C_+(X)$. This implies directly the inequality $\mu(K) \leq \bar{\mu}(K)$ for all compact sets K. Consequently, $\mu(B) \leq \bar{\mu}(B)$ for any Borel set $B \subset X$,

$\mu(B) = 1 - \mu(X \setminus B) \geq 1 - \bar{\mu}(X \setminus B) = \bar{\mu}(B)$ and $\mu = \bar{\mu}$, i.e., the measures μ and $\bar{\mu}$ coincide.

The last step is to prove that the measure μ is $\mathcal{G}$-invariant. To this end, take any $g \in \mathcal{G}_1$ and a function $f \in C_+(X)$ such that $\operatorname{supp} f \subset R_g$. Extend $f \circ g$ all over X putting $f \circ g \equiv 0$ outside D_g. If a set $A \subset D_g$ is (n, ε)-separated, then $g(A)$ is $(n+1, \varepsilon)$-separated and, consequently,

$$\begin{aligned} &\Lambda_{n,\varepsilon}(f \circ g) \leq \frac{s(n+1, \varepsilon, \mathcal{G}_1)}{s(n, \varepsilon, \mathcal{G}_1)} \cdot \Lambda_{n+1,\varepsilon}(f), \\ &\Lambda_{1/p}(f \circ g) \leq \bar{\Lambda}_{1/p}(f) \quad \text{and} \\ &\Lambda(f \circ g) \leq \bar{\Lambda}(f). \end{aligned} \tag{4.5.9}$$

Using (4.5.8) and the equality $\mu = \bar{\mu}$ we arrive at $\mu(g^{-1}(K)) \leq \mu(K)$ for any compact set $K \subset R_g$ and then $\mu(g^{-1}(B)) \leq \mu(B)$ for any Borel subset B of R_g. Applying the same argument to g^{-1} and the set $g^{-1}(B) \subset D_g$ we conclude by

$$\mu(g^{-1}(B)) = \mu(B). \tag{4.5.10}$$

This shows that μ is $\mathcal{G}$-invariant. □

As a direct consequence of Theorems 3.4.1 and 4.5.1 we obtain

Corollary 4.5.2. *Any foliation with vanishing geometric entropy admits a non-trivial transverse invariant measure.* □

4.6 Entropy, geodesic flow and invariant measures

Throughout this Section, $\mathcal{F}$ is a codimension-one transversely oriented C^1-foliation of a compact Riemannian manifold (M, g). It is well known (see [**CaC1**], Sections 1.2 and 5.1) that $(M, \mathcal{F})$ admits a nice foliated atlas $\mathcal{A}$ of class $C^{\infty,1}$, i.e., is such that for arbitrary charts ϕ and $\psi \in \mathcal{A}$ both of the maps α and γ in equation (1.3.1) of Chapter 1 are C^1-differentiable in all variables but α is also C^∞-differentiable with respect to $x \in \mathbb{R}^p$. In particular, the leaves of $\mathcal{F}$ have structures of C^∞-manifolds and we may assume that the Riemannian structures induced by g on the leaves are also C^∞-smooth. Moreover, there exists a smooth 1-dimensional foliation $\mathcal{F}^{\pitchfork}$ transverse to $\mathcal{F}$, so we fix it and make g-orthogonal to $\mathcal{F}$.

Our main goal here is to show the following.

Theorem 4.6.1. *If $\mathcal{F}$ is a transversely oriented codimension-one C^1-foliation of a compact manifold M and the geometric entropy of $\mathcal{F}$ is positive, then $\mathcal{F}$ has a resilient leaf.*

To express the fact that the geometric entropy of $\mathcal{F}$ is positive we shall use a nice covering $\mathcal{U}_\Delta$ by distinguished "coins" similar to those in the proof of Theorem 3.4.1 with the following difference: the "cores" of charts of $\mathcal{U}_\Delta$ are short segments of leaves of $\mathcal{F}^{\pitchfork}$ centred at centres of our coins. We shall keep the notation of that proof and assume that all the constants (coin radius Δ, coin thickness ρ, δ, a constant measuring density of the covering, and the others) involved in the construction of $\mathcal{U}_\Delta$ are sufficiently small. In particular, we assume that the covering $\mathcal{U}_\Delta$ is sufficiently dense, i.e., that any point $x \in M$ is close to a centre $z \in Z$ of one of the charts $U_z \in \mathcal{U}_\Delta$. Therefore (by Theorem 3.4.1 and inequalities (3.2.5)), our assumption of positive geometric entropy can be expressed in the form

$$E := h(\mathcal{H}, \mathcal{H}_1) > 0, \tag{4.6.1}$$

where $\mathcal{H}$ is the holonomy pseudogroup of $\mathcal{F}$ which acts on T, the complete transversal built of the "axes" T_z of "coins" $U_z \in \mathcal{U}_\Delta$, and is generated by the set $\mathcal{H}_1$ of elementary holonomy maps corresponding to all the pairs of overlapping "coins".

The general idea of the proof consists in the following. Given sufficiently large $n \in \mathbb{N}$, small $\varepsilon > 0$ and a maximal (n, ε)-separated set $A \subset T$, find a geodesic ray $\gamma_0 : [0, \infty) \to L$, L being a leaf of $\mathcal{F}$, which separates eventually two neighbouring points of A, show that the closure of γ_0 contains a closed leaf geodesic γ such that the holonomy map $h = h_\gamma$ along γ contracts its domain D_h hyperbolically towards the origin x of γ and D_h contains a point y, $y \neq x$, of the leaf through x. In order to show existence of such leaf geodesics, we shall use the foliated geodesic flow ϕ and consider some ergodic ϕ-invariant measures on $T^1\mathcal{F}$, the unit tangent bundle of $\mathcal{F}$.

To begin with, let us recall that the geodesic flow on any complete Riemannian manifold assigns to any unit tangent vector v and any $t \in \mathbb{R}$ the vector tangent at time t to the geodesic tangent to v at time $t_0 = 0$ (compare Example 2.4.12 (ii)). Since our foliated manifold M is compact, any leaf L of $\mathcal{F}$ is complete with respect to the induced Riemannian metric and has its own geodesic flow. All such flows corresponding to all the leaves constitute the *geodesic flow* $\phi = (\phi_t)$ of $\mathcal{F}$ which acts in the unit tangent bundle $V = T^1\mathcal{F}$ of $\mathcal{F}$: if $x \in L$, $L \in \mathcal{F}$, $v \in T_xL$ and $t \in \mathbb{R}$, then $\phi_t(v) = \dot{\gamma}_v(t)$, where $\gamma_v : \mathbb{R} \to L$ is the unique geodesic on L which satisfies the conditions $\gamma_v(0) = x$ and $\dot{\gamma}_v(0) = v$. Since $\mathcal{F}$ is C^1-smooth, the leaves are differentiable of class C^∞ and differential equations describing leaf geodesics contain only differentiation in the leaf direction, (ϕ_t) is differentiable of class C^1.

Let us consider also the foliation $\hat{\mathcal{F}}$ of V with leaves $\hat{L}$, $L \in \mathcal{F}$, which are just unit tangent bundles T^1L of the leaves of $\mathcal{F}$. In other words, $\hat{\mathcal{F}}$ is the pullback (see Section 3.5.1) of $\mathcal{F}$ *via* the canonical projection $\pi : V \to M$. Clearly, the transformations ϕ_t map the leaves of $\hat{\mathcal{F}}$ into themselves.

Next, consider the normal bundle $Q \to V$ of the foliation $\hat{\mathcal{F}}$: $Q = TV/T\hat{\mathcal{F}}$. Since the differentials $d\phi_t$, $t \in \mathbb{R}$, preserve $T\hat{\mathcal{F}}$, they induce maps, denoted by $d\phi_t$ again, of Q into itself. V carries the natural Riemannian structure (induced from the Sasaki metric on TM — compare Example 2.4.12(ii) again). Therefore, Q can be identified with $T^\perp\hat{\mathcal{F}}$, the bundle of vectors $\xi \in TV$ perpendicular to the leaves of $\hat{\mathcal{F}}$.

Recall now (compare Section 1.3) that for any leaf curve $\gamma : [0, s] \to M$ with end points in T one has the holonomy map $h_\gamma : W_\gamma \to T$ defined on a neighbourhood $W_\gamma \subset T$ of $x = \gamma(0)$ as the composition of elementary holonomy maps corresponding to a chain $u = (U_0, \dots, U_k)$ of pairwise overlapping charts of $\mathcal{U}_\Delta$ for which $x \in U_0$, $y = \gamma(s) \in U_k$ and $\gamma([t_i, t_{i+1}]) \subset U_i$ for some t_i's such that $0 = t_0 < t_1 < \cdots < t_{k+1} = s$. The germ of h_γ is independent of the choice of u, therefore the differential $dh_\gamma(x) : T_xT \to T_yT$ is well defined.

Using canonical isomorphisms $T_v^\perp \hat{\mathcal{F}} \approx T_xT$ and $T_w^\perp \hat{\mathcal{F}} \approx T_yT$ determined by the differential $d\pi$ of the projection $\pi : TM \to M$ one can write

$$dh_\gamma(x) = d\phi_s(v) \tag{4.6.2}$$

whenever $\gamma : [0, s] \to M$ is a leaf geodesic with $\dot\gamma(0) = v$ and $\dot\gamma(s) = w$. Since $\operatorname{codim}\mathcal{F} = 1$ and the connected components of T are built of pieces of geodesics on M perpendicular to $\mathcal{F}$, they have natural parametrizations and can be identified with intervals of real numbers. Therefore, we can consider a real-valued derivative $h'_\gamma(t)$ for any $t \in W_\gamma$, the domain of h_γ. Obviously, it is defined by $dh_\gamma(t) = h'_\gamma(t) \cdot \mathrm{id}_{\mathbb{R}}$. Also, $d\phi_s : Q \to Q$ can be considered as a real-valued function and one can define a function $\nu : \mathbb{R} \times V \to \mathbb{R}$ by

$$\nu(t, v) = \log d\phi_t(v).$$

Clearly, ν satisfies the cocycle property

$$\nu(t + s, v) = \nu(t, \phi_s(v)) + \nu(s, v)$$

for all $s, t \in \mathbb{R}$ and $v \in V$. Moreover, we can define the *infinitesimal transverse expansion* of the flow (ϕ_t) as the function $\psi : V \to \mathbb{R}$ given by

$$\psi(v) = \frac{d}{dt}\nu(t, v)(0). \tag{4.6.3}$$

Since $\mathcal{F}$ is transversally C^1, the function ψ is continuous. Moreover, since V is compact, the norm

$$\|\psi\|_\infty = \sup\{|\psi(v)|; v \in V\}$$

is finite. Finally, since $\phi_t(-v) = -\phi_{-t}(v)$, $\psi(-v) = -\psi(v)$ for all $v \in V$.

Now, let $\gamma : [a, b] \to M$ be a *piecewise leaf geodesic.* This means that γ maps $[a, b]$ into a leaf of $\mathcal{F}$ and that the segment $[a, b]$ can be split into the union of segments $[t_i, t_{i+1}]$ such that $a = t_0 < t_1 < \cdots < t_k = b$ and $\gamma|[t_i, t_{i+1}]$ is a smooth leaf geodesic for all $i = 0, 1, \dots, k-1$. Each leaf geodesic segment can be lifted to a segment of an orbit of the flow (ϕ_t). Therefore, γ also can be lifted to a (perhaps non-continuous at the corners t_i !) map $\hat\gamma : [a, b] \to V$ given by the formula

$$\hat\gamma(t) = \dot\gamma_j(t),$$

where $\gamma_j = \gamma|[t_j, t_{j+1}]$ and $t \in [t_j, t_{j+1})$. In other words,

$$\hat\gamma(t) = \phi_{t - t_j}(v_j),$$

when $t_j \leq t < t_{j+1}$ and $v_j = \dot{\widehat{(\gamma|[t_j, t_{j+1}])}}(t_j)$.

We shall also deal with *piecewise leaf geodesic rays*, that is piecewise leaf geodesics of the form $\gamma : [a, \infty) \to M$, where we admit an infinite number of "corners" $a = t_0 < t_1 < \cdots$ with $t_n \to \infty$ as $n \to \infty$. The same notation $\hat{\gamma}$ will be used in this situation.

Note that here (as well as in the sequel) we use term "corners" for t_j's, but we do not assume that our piecewise geodesic is not smooth at t_j. So, some of these corners may be "fake". Anyway, the set of all the corners contains the set of all real ones, i.e., all the points of nondifferentiability of γ.

Definition 4.6.2. The *transverse rate of expansion* ($\pitchfork$-*expansion rate*, for short) $\lambda(\gamma)$ along a piecewise leaf geodesic $\gamma : [a, b] \to M$ is defined by

$$\lambda(\gamma) = \frac{1}{b-a} \int_a^b \psi(\hat{\gamma}(t))dt. \tag{4.6.4}$$

A piecewise leaf geodesic ray $\gamma : [a, \infty) \to M$ is said to be $\pitchfork$-*hyperbolic with exponent* $\lambda(\gamma)$ if the limit

$$\lambda(\gamma) = \lim_{s \to \infty} \frac{1}{s-a} \int_a^s \psi(\hat{\gamma}(t)dt \tag{4.6.5}$$

exists and is different from zero; γ is $\pitchfork$-*stable* (resp., $\pitchfork$-*unstable*) when $\lambda(\gamma) < 0$ (resp., $\lambda(\gamma) > 0$). Finally, if $\gamma : [a, \infty) \to M$ is $\pitchfork$-stable and $\varepsilon > 0$, then a number $s_0 \geq a$ is said to be ε-*regular* when

$$\int_{s_0}^s (\psi(\hat{\gamma}(t)) + \varepsilon)dt < 0 \tag{4.6.6}$$

for all $s > s_0$.

Intuitively, s_0 is regular if our curve γ (with the origin moved to $\gamma(s_0)$) and following it piecewise geodesics on neighbouring leaves do not "pulse" in average too much (not more than with the ε-rate).

Let us begin by verification of existence of regular values.

Lemma 4.6.3. *If $\gamma : [a, \infty) \to M$ is a $\pitchfork$-stable piecewise leaf geodesic ray, then for any $\varepsilon \in (0, -\lambda(\gamma))$ there exists an increasing sequence (s_n) of ε-regular values for which $\lim_{n \to \infty} s_n = \infty$.*

Proof. Since the function

$$s \mapsto \frac{1}{s-a} \int_a^s (\psi(\hat{\gamma}(t)) + \varepsilon)dt \tag{4.6.7}$$

is continuous and $\lambda(\gamma) + \varepsilon < 0$, there exists the greatest value s_1 of s for which the integral in (4.6.7) equals zero. Clearly, s_1 is ε-regular. Given a sequence $s_1 < s_2 < \cdots < s_k$ of ε-regular values, consider the function (4.6.7) with a replaced by $s_k + 1$ and define s_{k+1} as the largest s for which this new integral equals 0. Again, s_{k+1} is regular. □

Next, we shall investigate the distribution of ε-regular values (in fact, of their discrete analogue called ε-good values here) along a periodic $\pitchfork$-stable piecewise leaf geodesic ray $\gamma : [a, \infty) \to M$. Obviously, periodicity of γ means that there exists $t_\gamma > 0$ such that $\gamma(t + t_\gamma) = \gamma(t)$ for all $t \geq a$. So, let γ be as above and denote by $a = t_0 < t_1 < \cdots < t_N = a + t_\gamma$ all the corners of $\gamma|[a, a + t_\gamma]$. Set

$$a_j = \int_{t_{j-1}}^{t_j} (\psi(\hat{\gamma}(t)) + \varepsilon)dt.$$

Then $a_1 + \cdots + a_N = (\lambda(\gamma) + \varepsilon) \cdot t_\gamma < 0$ and $|a_j| \leq (\|\psi\|_\infty + \varepsilon)(t_j - t_{j-1})$. Extend the finite sequence (a_j) to an infinite one setting $a_{j+kN} = a_j$ for all $j \leq N$ and $k \in \mathbb{N}$. Let

$$S_k(n) = \sum_{j=k}^{n} a_j$$

when $1 \leq k \leq N$ and $n \geq k$. Also, extend the sequence $(t_j; 0 \leq j \leq N\}$ to an infinite one, where $t_{j+kN} = t_j + kt_\gamma$.

Definition 4.6.4. In the situation just described, a value t_k $(1 \leq k \leq N)$ is said to be ε-*good* when $S_k(n) < 0$ for all $n \geq k$.

Lemma 4.6.5. *If γ is $\pitchfork$-stable, $0 < \varepsilon < -\lambda(\gamma)$ and*

$$c = \frac{(-\lambda(\gamma) - \varepsilon) \cdot \min\{t_j - t_{j-1}\}}{\|\psi\|_\infty \cdot \max\{t_j - t_{j-1}\}},$$

then the set $\{t_1, \ldots, t_N\}$ of all corners contains at least $[cN]$, the integer part of N, ε-good values.

Proof. Since $0 < \varepsilon < -\lambda(\gamma)$, there exists at least one ε-good number t_k. Indeed, if $s > 0$ is ε-regular and $t_k > s$ is such that the integral $\int_s^{t_k} (\psi(\hat{\gamma}(t) + \varepsilon)dt$ is maximal among all the integrals $\int_s^{t_j} (\psi(\hat{\gamma}(t) + \varepsilon)dt$, $j \in \{l, l+1, \ldots, l+N\}$, where $t_l > s$, then t_k is ε-good. Reparametrizing γ if necessary, we may assume that $t_0 = a$ is ε-good.

Now, if $k \in \{1, \ldots, N\}$ is the largest value of j for which $a_j \geq 0$, then $t_{k+1}, \ldots, t_N$ are ε-good. Also, if $i \leq k$ is the smallest such that $a_{i+1} + \cdots + a_{k+1} \geq 0$, then $a_i < 0$ and t_{i-1} is ε-good. All the good points can be obtained in this way by induction, after removing step by step segments containing only points t_j which are not good. Therefore, if $\{k_1, \ldots, k_l\}$ is the subset of $\{1, \ldots, N\}$ consisting of all indexes j which correspond to all ε-good values t_j, then

$$a_{k_1} + \cdots + a_{k_l} \leq (\lambda(\gamma) + \varepsilon) \cdot t_\gamma < 0.$$

On the other hand,

$$|a_{k_1} + \cdots + a_{k_l}| \leq l \cdot \|\psi\|_\infty \cdot \max\{t_{j+1} - t_j\}$$

and

$$t_\gamma \geq N \cdot \min\{t_{j+1} - t_j\}.$$

All these inequalities yield $l \geq cN$. □

Regular values can be used to get the property of uniform contraction for holonomy maps along $\pitchfork$-stable piecewise leaf geodesic rays. Moreover, it occurs that the size of the domain of any such holonomy map is bounded from below by a positive constant which depends only on the limit $\lambda(\gamma)$ in (4.6.5). More precisely, we have the following.

Proposition 4.6.6. *Let $\gamma : [a, \infty) \to M$ be a $\pitchfork$-stable piecewise leaf geodesic ray with all the corners contained in the transversal T. There exist constants $C, \varepsilon_1 > 0$ and $s_0 > a$, and a segment $I \subset T$ of length $\geq 2\varepsilon_1$ such that for any sufficiently large $s_* > s_0$ the holonomy map h_{s_*} along $\gamma_{s_*} = \gamma|[s_0, s_*]$ is a contraction which is defined on I and satisfies the condition*

$$0 < h'_{s_*}(y) < C \cdot \exp\left((s_* - s_0) \cdot \lambda(\gamma)/2\right), \quad y \in I. \tag{4.6.8}$$

Moreover, ε_1 depends only on $\lambda(\gamma)$.

Proof. Let ε_0 be a Lebesgue number of our nice covering $\mathcal{U}_\Delta$. Fix $\delta_0 > 0$ and choose an increasing function $\varepsilon_2 : (0, \delta_0) \to (0, \varepsilon_0)$ such that $|\psi(v) - \psi(w)| < \delta$ whenever $v, w \in V$ and $d_V(v, w) < \varepsilon_2(\delta)$, where d_V is the Riemannian distance function on V determined by the Sasaki metric. Choose another increasing function $\varepsilon_3 : (0, \delta_0) \to (0, \varepsilon_0)$ such that $d_V(\phi_t v, \phi_t w) < \varepsilon_2(\delta)$ whenever $d_V(v, w) < \varepsilon_3(\delta)$ and $|t| < 10\Delta$. For any $U \in \mathcal{U}_\Delta$ and $x \in U$ denote by $g_U(x)$ the matrix at x of g, the Riemannian metric on M, in the canonical frame determined by the distinguished chart defined on U. Let $\|g_U(x)\|$, $x \in U$, be the normalized (the normalized norm of the identity matrix equals 1) Euclidean norm of $g_U(x)$,

$$\|g\| = \sup\{\|g_U(x)\|; x \in U, U \in \mathcal{U}\};$$

put also $\varepsilon_1(\delta) = \varepsilon_3(\delta)/\|g\|^2$ and

$$\varepsilon_1 = \varepsilon_1(-\frac{1}{10}\lambda(\gamma)).$$

Let s_0 be ε-regular with $\varepsilon = -\frac{9}{10}\lambda(\gamma)$. Take any $s_* > s_0$ and a chain of plaques $P_0, P_1, \ldots, P_k$ along γ_{s_*} such that $P_i \subset U_i = U_{z_i}$ (remember that we keep the notation of the proof of Theorem 3.4.1, so z is the centre of the coin U_z of our nice covering $\mathcal{U}_\Delta$) where $\gamma(s_i) \in P_i$ for some sequence $s_0 < s_1 < \cdots < s_k = s_*$ and $d(\gamma(s_i), z_i) \leq \delta$. Let y_i, $i = 0, 1, \ldots, k$ be points of intersection of P_i with V_{z_i}, the "core" of U_{z_i}. Let also I be the segment of length $2\varepsilon_1$, contained in V_{z_0} and centred at y_0. From the construction of $\mathcal{U}_\Delta$ it follows that the holonomy map $h_0 = h_{U_1 U_0}$ is defined on I. Moreover, our choice of ε_1 and the Mean Value Theorem imply that the distance between h_0-images of the endpoints of I does not exceed

$$2\varepsilon_1 \cdot \exp \int_{s_0}^{s_*} (\psi(\hat{\gamma}(s)) - \frac{\lambda(\gamma)}{10}) ds \leq 2\varepsilon_1 \exp(\frac{8}{10}\lambda(\gamma)) < 2\varepsilon_1,$$

hence the segment $I_1 = h_0(I)$ is centred at y_1 and has length at most $2\varepsilon_1$. By induction, we can get a sequence $I_0 = I, I_1, \dots, I_{k-1}$ of segments such that $h_i(I_i) = I_{i+1}$, where $h_i = h_{U_{i+1}U_i}$. Therefore, the holonomy map $h_{s_*} = h_{k-1} \circ \cdots \circ h_1 \circ h_0$ is defined on I.

Finally, take an arbitrary point $y \in I$ and denote by γ_y the piecewise geodesic lift ($\gamma_y(0) = y$) of γ to L_y, the leaf through y. The lift γ_y is obtained by projecting the corners orthogonally and joining them by short geodesic segments. Let

$$v_i = (\dot{\widehat{\gamma_y|[s_i, s_{i+1}]}})(s_i).$$

Then — by the same reason as before —

$$|\psi(\phi_t(v_i)) - \psi(\hat{\gamma}(t))| \leq -\lambda(\gamma)/10$$

and consequently

$$\begin{aligned} h'_{s_*}(y) &= \prod_{i=0}^{k-1} d\phi_{s_{i+1}-s_i}(v_i) \\ &= \exp\left(\sum_i \log d\phi_{s_{i+1}-s_i}(v_i)\right) = \exp\left(\sum_i \int_{s_i}^{s_{i+1}} \psi(\phi_t(v_i))dt\right) \\ &\leq \exp\left(\int_{s_0}^{s_*} (\psi(\hat{\gamma}(t)) + \frac{\lambda(\gamma)}{10})dt\right) \leq \exp\left((s_* - s_0)\frac{8}{10}\lambda(\gamma)\right). \end{aligned}$$

The above estimate implies (4.6.8) for large s_*. □

Now, let us switch our attention to Borel probability measures invariant under the geodesic flow $\phi = (\phi_t)$. Such measures exist (this can be proved by several means, either by an averaging process as in Section 3.1.2 or by applying Proposition 4.7.3 of Section 4.7 below to the subspace W of continuous functions on V generated by all the functions of the form $f - f \circ \phi_t$, where $t \in \mathbb{R}$ and $f : V \to \mathbb{R}$ is continuous), form a convex subset $\mathcal{M}(V, \phi)$ in the space of all the Borel probability measures on V. The extreme points of this set coincide with ergodic measures (i.e., those for which any ϕ-invariant set has measure either 0 or 1) and, again as in Section 3.1.2, any ϕ-invariant measure μ has the unique ergodic decomposition of the form

$$\mu = \int_{\mathcal{E}(V,\phi)} m d\tau_\mu(m), \tag{4.6.9}$$

where $\mathcal{E}(V, \phi)$ is the set of all such ergodic measures and τ_μ is a probability measure on $\mathcal{E}(V, \phi)$.

Definition 4.6.7. For any $\mu \in \mathcal{M}(\mathcal{V}, \phi)$ the *transverse support* ($\pitchfork$-*support*, for short) of μ is defined as the intersection of the transversal T with the union of all the plaques of $\mathcal{U}_\Delta$ which have non-empty intersections with $\pi(\mathrm{supp}(\mu))$, $\pi : V \to M$ being, as usual, the canonical projection. Such a measure μ is said to be $\pitchfork$-*discrete* or $\pitchfork$-*finite* (resp., $\pitchfork$-*uncountable*, resp., $\pitchfork$-*non-atomic*) whenever its transverse support is finite (resp., uncountable, resp., consists of points z for which the corresponding plaques P_z ($z \in P_z$) satisfy $\mu(\pi^{-1}(P_z)) = 0$). A ϕ-invariant ergodic measure μ is said to be $\pitchfork$-*hyperbolic* whenever the quantity $\Lambda(\mu)$ defined by

$$\Lambda(\mu) = \int_V \psi d\mu \tag{4.6.10}$$

is different from 0. Given an arbitrary measure $\mu \in \mathcal{M}(\mathcal{V}, \phi)$, we shall say that it is $\pitchfork$-*hyperbolic* when $\Lambda(m) \neq 0$ for τ_μ-almost all $m \in \mathcal{E}(\mathcal{V}, \phi)$ in (4.6.9). Moreover, if there exists $c > 0$ such that $|\Lambda(m)| > c$ for τ_μ-almost all measures m, then μ is said to be *uniformly* $\pitchfork$-*hyperbolic* .

One can observe that $\pitchfork$-hyperbolic measures do not exist if $\mathcal{F}$ is Riemannian (i.e., leaves stay a constant distance apart) or is obtained by a slight deformation of such foliations (in such a way that the distance of two leaves along leaf geodesics varies but is bounded away from zero). Later in this section we shall see that $\pitchfork$-hyperbolic measures are present only when the dynamics of a foliation $\mathcal{F}$ is rich. Roughly speaking, such measures exist whenever $\mathcal{F}$ has resilient leaves.

The geodesic flow ϕ is conjugated to its inverse *via* the symmetry $S : V \to V$ given by $S(v) = -v$ (that is, $\phi_{-t} = S \circ \phi_t \circ S$ for any t). Therefore, any ϕ-invariant measure μ induces another one, $S^*\mu$, which is also ϕ-invariant and satisfies $\Lambda(S^*\mu) = -\Lambda(\mu)$. So, without loss of generality, we may assume that $\pitchfork$-hyperbolic measures μ under consideration satisfy $\Lambda(\mu) < 0$.

Proposition 4.6.8. *If μ is ergodic and $\pitchfork$-hyperbolic, and $\varepsilon_1 = \varepsilon_1(\frac{1}{10}|\lambda(\mu)|)$, then there exists a point z of the transverse support of μ, a constant $c \in (0,1)$ and a holonomy transformation h of $I = (z - \varepsilon_1, z + \varepsilon_1) \subset T$ into itself for which $h(z) = z$ and $0 < h'(y) < c$ for all $y \in I$.*

Roughly speaking, the proof consists in establishing existence of $\pitchfork$-stable leaf geodesics and applying Proposition 4.6.6.

Proof. Assume that $\Lambda(\mu) < 0$. Since μ is ergodic, the classical Birkhoff Ergodic Theorem implies that the limit $\lambda(\gamma)$ in (4.6.5) exists and equals $\Lambda(\mu)$ for μ-almost all leaf geodesic rays γ with $\dot\gamma(0) \in V_0$, $V_0 \subset \mathrm{supp}(\mu)$ being a subset of μ-measure 1. Therefore, there exists a $\pitchfork$-hyperbolic piecewise leaf geodesic ray γ with $x = \gamma(0) \in T$ and $\lambda(\gamma) = \Lambda(\mu) < 0$. Apply Proposition 4.6.6 to this γ. For some s_* large enough, the holonomy transformation h_{s_*} is defined on $I_x = (x - \varepsilon_1, x + \varepsilon_1)$ and contracts I_x into itself. Therefore, h_{s_*} has a fixed point $z \in I_x$. It is easy to see that z lies in the $\pitchfork$-support of μ. Applying Proposition 4.6.6 once again, this time to the piecewise geodesic lift $\tilde\gamma$ of γ to the leaf L_z through z, we get our holonomy transformation h: $h = \tilde h_{s_*}$ where s_* is large enough and $\tilde h_{s_*}$ denotes the holonomy map along $\tilde\gamma|[0, s_*]$. □

Now, we are in a position to prove existence of resilient leaves (Theorem 4.6.1) in a particular case. More precisely, we have the following.

Theorem 4.6.9. *If V admits a $\pitchfork$-hyperbolic ergodic $\pitchfork$-non-discrete ϕ-invariant measure μ, then $\mathcal{F}$ has a resilient leaf.*

Proof. From the definition of resilient leaves (Definition 2.3.9) it follows directly that it suffices to show the existence of a segment $I \subset T$ and elements h_1, h_2 of the holonomy pseudogroup such that $I \subset D_{h_1} \cap D_{h_2}$, $h_1(I) \cup h_2(I) \subset I$ and $h_1(I) \cap h_2(I) = \emptyset$. In this situation, I contains a point fixed by one of these maps and the leaf passing through this point is resilient.

To this end, take a measure μ which satisfies the assumptions of the theorem with $\Lambda(\mu) < 0$. By previous propositions, there exist $v \in \operatorname{supp} \mu$, $\varepsilon_1 > 0$ and $s_0 > 0$ such that $\pi(v) \in T$ and the holonomy transformation h_{s_*} along a leaf geodesic $\gamma_{s_*} : [s_0, s_*] \ni t \mapsto \pi(\phi_t(v))$ is defined on the interval $I = (x_0 - \varepsilon_1, x_0 + \varepsilon_1) \subset T$ and satisfies there the inequality

$$0 < h'_{s_*} < \exp\left(\frac{1}{2}(s_* - s_0)\Lambda(\mu)\right). \tag{4.6.11}$$

(Here, x_0 is the centre of a plaque which contains $\gamma(s_0)$.)

Since μ is ergodic and $\pitchfork$-non-discrete, we can choose v is such a way that the orbit $\sigma = \{\phi_t(v); t \geq 0\}$ intersects infinitely many plaques of $\hat{\mathcal{F}}$. Then, the intersection of T with the union of all the plaques which intersect the projection $\pi(\bar{\sigma})$ of the closure $\bar{\sigma}$ is perfect (in T). Therefore, any open segment $J \subset T$ which intersects a plaque P for which $\pi^{-1}(P) \cap \bar{\sigma} \neq \emptyset$ contains disjoint open intervals I_1 and $I_2 \subset J$ such that each of them intersects such a plaque.

Clearly, we may apply the above to $J = I$. Therefore, our geodesic "visits" each of the segments I_1 and I_2 and one can find inductively sequences $s_0 < s_1 < s_2 < \cdots \to \infty$ and $s_0 < t_1 < t_2 < \cdots \to \infty$ such that the plaques P'_k through $\gamma(s_k)$ meet T at points of I_1 while the plaques P''_k through $\gamma(t_k)$ meet T at points of I_2. For k large enough, the contraction property (4.6.11) of the holonomy along $\gamma = \pi \circ \sigma$ implies that the holonomy maps $h_1 = h_{s_k}$ and $h_2 = h_{t_k}$ satisfy the relations $h_1(I) \subset I_1$ and $h_2(I) \subset I_2$, hence $h_1(I) \cap h_2(I) = \emptyset$. □

Now, we shall prove the existence of ergodic $\pitchfork$-hyperbolic ϕ-invariant measures provided that $\mathcal{F}$ has positive geometric entropy. If one of these measures is $\pitchfork$-non-discrete, the existence of resilient leaves follows from the last theorem. Later on, we shall consider the case when all these measures are $\pitchfork$-discrete. As we shall see, this part of the proof is the most technical.

Theorem 4.6.10. *If the geometric entropy of $\mathcal{F}$ is positive, then there exist Borel probability measures on V which are invariant under the geodesic flow ϕ, ergodic and $\pitchfork$-hyperbolic.*

Proof. If the geometric entropy of $\mathcal{F}$ is positive, then — as was observed at the beginning of this section — the entropy $E = h(\mathcal{H}, \mathcal{H}_1)$ of the holonomy pseudogroup of $\mathcal{F}$ generated by the elementary holonomy maps corresponding to overlapping charts of $\mathcal{U}_\Delta$ is also positive. Therefore, there exists $\varepsilon_0 > 0$ such that $s(\varepsilon, \mathcal{H}_1) > \frac{3}{4}E$ for all $\varepsilon \in (0, \varepsilon_0)$. Fix such an ε and find a sequence $n_1 < n_2 < \cdots \to \infty$ of natural numbers for which the maximal numbers of pairwise (n_k, ε)-separated points of T satisfy the inequalities

$$e_k = s(n_k, \varepsilon, \mathcal{H}_1) > \exp(\frac{1}{2}n_k E), \quad k = 1, 2, \ldots . \tag{4.6.12}$$

Therefore, for any $k \in \mathbb{N}$ there exists a set $A_k = \{x_k^j; j = 1, \ldots, e_k\} \subset T$ which is (n_k, ε)-separated by the maps of $\mathcal{H}_1$. Since the covering $\mathcal{U}_\Delta$ is finite, we may assume that all the sets A_k are contained in a segment T_U of T corresponding to a chart $U \in \mathcal{U}_\Delta$. Renumbering points of A_k if necessary, we may also assume that $x_k^1 < x_k^2 < \cdots < x_k^{e_k}$ (with respect to a given transverse orientation of $\mathcal{F}$). Thus, for any $k \in \mathbb{N}$ and any $j \in \{1, \ldots, e_k - 1\}$ there exists a chain $u_{k,j}$ of charts of $\mathcal{U}_\Delta$ which originates at U, has length $\leq n_k$ and separates points x_k^j and x_k^{j+1}:

$$d_T(h_{k,j}(x_k^j), h_{k,j}(x_k^{j+1})) \geq \varepsilon, \tag{4.6.13}$$

where $h_{k,j}$ is the holonomy map corresponding to $u_{k,j}$. We may also assume that $u_{k,j}$ is the shortest among all the chains separating points x_k^j and x_k^{j+1} to the distance ε.

By the Mean Value Theorem,

$$h'_{k,j}(y_k^j) \geq \varepsilon/(x_k^{j+1} - x_k^j) \tag{4.6.14}$$

for some $y_k^j \in (x_k^j, x_k^{j+1})$. For any segment $I \subset T_U$ let

$$E(I, k) = \max_j \{\sup h'_{k,j}(y); y \in (x_k^j, x_k^{j+1})\}, \tag{4.6.15}$$

where j ranges over the set of all the indexes for which $(x_k^j, x_k^{j+1}) \subset I$, be the maximal infinitesimal expansion rate of the holonomy restricted to I along $u_{k,j}$. Since ρ, the transverse radius of charts of $\mathcal{U}_\Delta$, has been chosen to be small, we may assume that the length of T_U is less than 1. Then, for any k one can find $j \in \{1, \ldots, e_k - 1\}$ for which $x_k^{j+1} - x_k^j < 1/e_k$. Hence,

$$E(T_U, k) \geq \varepsilon \cdot \exp(n_k \frac{E}{2}), \quad k = 1, 2, \ldots . \tag{4.6.16}$$

Let also

$$\tilde{\Lambda}(I) = \liminf_{k \to \infty} \frac{1}{n_k} \log E(I, k). \tag{4.6.17}$$

Then

$$\tilde{\Lambda}(T_U) \geq \frac{E}{2} > 0. \tag{4.6.18}$$

Choose now a small number $\theta \in (0, \frac{1}{2}\tilde{\Lambda}(T_U))$ and set $\lambda = \tilde{\Lambda}(T_U) - \theta$. For any k large enough choose $j_k \in \{1, \dots, e_k - 1\}$ and $y_k \in (x_k^{j_k}, x_k^{j_k+1})$ for which $h_k'(y_k) \geq \varepsilon \exp(n_k \lambda)$, where $h_k = h_{k,j_k}$. The chain $u_k = u_{k,j_k}$ determines a piecewise leaf geodesic τ_k with end points y_k and $h_k(y_k)$. Since M is compact, the leaves of $\mathcal{F}$ are complete (with respect to the induced Riemannian structure) and τ_k is homotopic (relative to its end points) to a leaf geodesic γ_k. Clearly, the length l_k of γ_k does not exceed $2n_k \cdot \Delta$. The holonomy maps along τ_k and γ_k coincide in a neighbourhood of y_k, hence

$$h_{\gamma_k}'(y_k) = h_k'(y_k) \geq \varepsilon \exp(n_k \lambda). \tag{4.6.19}$$

Let $v_k = \dot{\gamma}_k(0)$ and μ_k be the Borel probability measure on V determined — *via* the Riesz Representation Theorem — by the linear functional $\Theta_k : C(V) \to \mathbb{R}$ given by

$$\Theta_k(f) = \frac{1}{l_k} \int_0^{l_k} f(\phi_t(v_k))dt. \tag{4.6.20}$$

As in Section 3.1.2, the sequence (μ_k) contains a subsequence (denoted by (μ_k) again) converging to a ϕ-invariant Borel probability measure μ. Applying (4.6.20) to the infinitesimal ⋔-expansion ψ of (ϕ_t) we obtain the estimate

$$\begin{aligned} \Lambda(\mu) &= \int_V \psi d\mu = \lim_{k\to\infty} \int_V \psi d\mu_k = \frac{1}{l_k} \int_0^{l_k} \psi(\phi_t(v_k))dt \\ &= \frac{1}{l_k} \log h_{\gamma_k}'(y_k) \geq \frac{n_k \lambda}{l_k} \geq \frac{\tilde{\Lambda}(T_U) - \theta}{2\Delta} > 0. \end{aligned}$$

The ergodic decomposition of μ in (4.6.9) implies that there exists a measure $m \in \mathcal{E}(V, \phi)$ for which

$$\Lambda(m) \geq \frac{\tilde{\Lambda}(T_U) - 2\delta}{4\Delta} > 0. \tag{4.6.21}$$

This m is ⋔-hyperbolic, ergodic and ϕ-invariant. □

To complete the proof of Theorem 4.6.1 it remains to show the following.

Theorem 4.6.11. *If the geometric entropy of $\mathcal{F}$ is positive and all the ergodic ⋔-hyperbolic ϕ-invariant measures on V are ⋔-discrete, then $\mathcal{F}$ has a resilient leaf.*

The proof of this theorem is pretty complicated, has to be split into some cases and needs certain technical preparations.

To begin with, let us recall that Proposition 4.6.8 allows us to associate with any ergodic ⋔-hyperbolic ϕ-invariant measure μ a periodic piecewise leaf geodesic γ contained in the ⋔-support of μ and such that the corresponding holonomy map h_γ is defined on the interval $I = (x - \varepsilon_1, x + \varepsilon_1)$ (where x is the origin of γ and ε_1 is a positive constant which depends on $\Lambda(\mu)$ only) and satisfies there the condition $0 < h_\gamma' < c$ for some universal constant $c \in (0, 1)$. We may assume that this γ is built of length minimizing geodesic segments γ_i connecting the centres of plaques P_i and P_{i+1} of a chain $P_0, P_1, \dots, P_n$, where $P_n = P_0$. Such a chain will be called here *irreducible* when $P_i \neq P_j$ for all $i < j$ (except of $i = 0$ and $j = n$). It is clear that any plaque chain $p = (P_0, P_1, \dots, P_n)$ contains irreducible pieces of the form $\tilde{p} = (P_i, P_{i+1}, \dots, P_j)$, where $i < j$ and $P_i = P_j$. One can see that at least one of the corresponding pieces $\tilde{\gamma}$ of γ satisfies the inequality $\lambda(\tilde{\gamma}) \leq \lambda(\gamma)$. The holonomy map $h_{\tilde{\gamma}}$ along such a piece $\tilde{\gamma}$ is defined on a segment $\tilde{I} = (\tilde{x} - \varepsilon_1, \tilde{x} + \varepsilon_1)$ and contracts it uniformly into itself.

Given $a > 0$, denote by $G(a)$ the set of all $\pitchfork$-hyperbolic closed piecewise leaf geodesics γ which correspond to irreducible plaque chains as above and satisfy the condition $\lambda(\gamma) \leq -a$. Let also

$$T(a) = \{x \in T; (\exists \gamma \in G(a))\ (\exists t)\ \gamma(t) = x\}$$

be the set of all the points of intersection of T with the loops of $G(a)$.

Case 1. There exists $a > 0$ such that $T(a)$ is infinite.

In this case, there exist infinitely many irreducible loops of $G(a)$, their lengths diverge to infinity (i.e., for any $m > 0$ only finitely many of them have length $\leq m$) and Lemma 4.6.5 shows that infinitely many of them contain good points. In other words, the set of origins of irreducible periodic piecewise leaf geodesics $\gamma \in G(a)$ is infinite. Therefore, there exists a point $x \in T$ and an infinite sequence (γ_k), $\gamma_k : [0, 1] \to L_k$ (L_k being a leaf of $\mathcal{F}$) of elements of $G(a)$ such that $x_k = \gamma_k(0) = \gamma_k(1) \to x$ as $k \to \infty$. If k and $l \in \mathbb{N}$ are large enough, then the points x, x_k and x_l lie in $I = I_k \cap I_l$, where $I_j = (x_j - \varepsilon_1, x_j + \varepsilon_1)$ and $\varepsilon_1 = \varepsilon_1(\frac{a}{10})$. If $n \in \mathbb{N}$ is sufficiently large, then the maps $h_1 = h^n_{\gamma_k}$ and $h_2 = h^n_{\gamma_l}$ map I into itself and satisfy the condition $h_1(I) \cap h_2(I) = \emptyset$. As in the proof of Theorem 4.6.9, I contains a point z (this time just $z = x_k$) fixed by one of these maps, say h_1, and the leaf L_z through z is resilient. □

Case 2. $T(a)$ is finite for any $a > 0$.

This is the most technical part of the proof of the main result of this Section. Let

$$\Lambda_0 = \min\{\frac{1}{2}, \frac{E}{200\Delta}\}. \tag{4.6.22}$$

The estimate for $\Lambda(m)$ in the proof of Theorem 4.6.10 shows that the set $G(\Lambda_0)$ is non-empty. Since $T(\Lambda_0)$ is finite, there exists a finite collection $\mathcal{P} = \{P_1, \dots, P_N\}$ of plaques of $\mathcal{U}_\Delta$ such that any $\gamma \in G(\Lambda_0)$ is contained in $\bigcup \mathcal{P}$. Therefore, if μ is ergodic, $\pitchfork$-hyperbolic, ϕ-invariant and $\Lambda(\mu) \leq -\Lambda_0$, then the transverse support of μ meets $\bigcup \mathcal{P}$. We may assume that the set $\mathcal{P}$ is minimal, i.e., that each plaque P_j meets the support of such a measure.

Let x_j be the centre of P_j, U_j — the chart of $\mathcal{U}_\Delta$ containing P_j and L_j — the leaf through x_j. Since L_j contains a loop with contracting holonomy, L_j is resilient if only non-proper (compare Section 1.3). Therefore, it remains to consider the case when all the leaves L_j are proper. In this situation, we can choose $\theta_0 > 0$ such that for each $j \leq N$ the intersection $L_j \cap (x_j - \theta_0, x_j + \theta_0)$ consists of a single point x_j.

Given $\theta > 0$, let $\mathcal{P}(\theta)$ be the neighbourhood of $\bigcup \mathcal{P}$ of transverse diameter 2θ. That is, $\mathcal{P}(\theta)$ is the union of all the plaques of $\mathcal{U}_\Delta$ which intersect the transversal T at points of the segments $(x_j - \theta, x_j + \theta)$, $j = 1, \dots, N$.

The following lemma shows that the total expansion along any geodesic segment which is θ-apart from $\cup\mathcal{P}$ is uniformly bounded by quantities which depend only on θ and the length of the segment. Observe that if θ is small (for instance, $\theta \leq \theta_0$), then such segments always exist.

Lemma 4.6.12. *For any small $\theta > 0$ there exists $t(\theta) > 0$ such that any leaf geodesic segment*

$$s \mapsto \pi(\phi_s(v)), \quad s \in [0, u],$$

where $v \in V$ and $u > 0$, satisfies the condition

$$\left| \int_0^u \psi(\phi_s v) ds \right| \leq t(\theta) \|\psi\| + u\Lambda_0, \tag{4.6.23}$$

provided that the segment is disjoint from $\mathcal{P}(\theta)$.

Proof. Define a function $\alpha_\theta : (0, \infty) \to \mathbb{R}$ by

$$\alpha_\theta(t) = \sup\left\{ \frac{1}{t} \left| \int_0^t \psi(\phi_s v) ds \right|; v \in V_{t,\theta} \right\},$$

where $V_{t,\theta}$ is the set of all $v \in V$ such that $\pi\phi_s(v) \notin \mathcal{P}(\theta)$ for all $s \in [0, t]$. Elementary arguments show that α_θ is continuous.

First, we shall prove that

$$\limsup_{t \to \infty} \alpha_\theta(t) < \Lambda_0. \tag{4.6.24}$$

Indeed, otherwise one could find sequences (v_k) and (t_k) such that $v_k \in V_{t_k,\theta}$, $t_k \to \infty$ and (after replacing v_k by $-v_k$ if necessary)

$$\lim_{k \to \infty} \frac{1}{t_k} \int_0^{t_k} \psi(\phi_s v_k) ds \leq -\Lambda_0.$$

Let μ_k denote the measure on V given by

$$\int_V f d\mu_k = \frac{1}{t_k} \int_0^{t_k} f(\phi_s v_k) ds$$

for all $f \in C(V)$. Let μ be the weak-$*$ limit of a converging subsequence of (μ_k). The measure μ is ϕ-invariant and $\Lambda(\mu) \leq -\Lambda_0$. The ergodic decomposition of μ provides us with an ergodic ϕ-invariant measure μ_0 with $\operatorname{supp} \mu_0 \subset \operatorname{supp} \mu$ and $\Lambda(\mu_0) \leq \Lambda(\mu) \leq -\Lambda_0$. However, $\pi(\operatorname{supp}(\mu_0) \cap \mathcal{P}(\theta)) = \emptyset$ while $\pi(\operatorname{supp}(\mu_0)) \subset \overline{\bigcup \mathcal{P}}$, a contradiction.

Now, let $t(\theta) = \sup\{t \in (0,\infty); \alpha_\theta(t) \geq \Lambda_0\}$. We shall show that this quantity satisfies the conditions of our lemma. First, from (4.6.24) it follows that it is finite. Next, take any $t > 0$ and $v \in V_{t,\theta}$. If $0 < t \leq t(\theta)$, then

$$\left|\int_0^t \psi(\phi_s v)ds\right| \leq t\|\psi\|_\infty \leq t(\theta)\|\psi\|_\infty + t\Lambda_0.$$

It $t > t(\theta)$, then $\alpha_\theta(t) < 0$ and

$$\left|\int_0^t \psi(\phi_s v)ds\right| \leq t\Lambda_0 < t(\theta)\|\psi\|_\infty + t\Lambda_0. \qquad \square$$

Let us come back to the main thread of the proof of Theorem 4.6.11. To this end let us keep the notation introduced in the proof of Theorem 4.6.10. That is, we have a sequence $n_k \to \infty$ for which $e_k = s(n_k, \varepsilon, \mathcal{H}_1) > \exp(n_k \frac{E}{2})$, where $E = h(\mathcal{H}, \mathcal{H}_1)$ and $\varepsilon > 0$ is sufficiently small, sets $A_k = \{x_k^j\}$ are maximal (n_k, ε)-separated, etc.

Let $k \in \mathbb{N}$ and $a_k = [\exp(n_k \frac{E}{4})]$, where $[a]$ denotes the largest integer $\leq a$. The transversal T contains a closed interval J_k of length $1/a_k$ and such that $\#(J_k \cap A_k) \geq a_k$. Relabel points of A_k in such a way that $J_k \cap A_k = \{x_k^j; j \leq a_k\}$. For each k and $j < a_k$ there exists a piecewise leaf geodesic $\tau_{k,j}$ separating x_k^j and x_k^{j+1} to the distance ε. That is, the holonomy map $h_{k,j}$ along $\tau_{k,j}$ is defined in a neighbourhood of the set $\{x_k^j, x_k^{j+1}\}$ and satisfies $d_T(h_{k,j}(x_k^j), h_{k,j}(x_k^{j+1}) \geq \varepsilon$. We may assume that $\tau_{k,j}$ is the shortest among all the piecewise leaf geodesics separating these points. Then $l(\tau_{k,j}) \leq 2\Delta n_k$.

Choose $y_k^j \in (x_k^j, x_k^{j+1})$ for which $h'_{k,j}(y_k^j) = \sup\{h'_{k,j}(y); y \in (x_k^j, x_k^{j+1})\}$. Since $d_T(x_k^j, x_k^{j+1}) < 1/a_k$ for some j's, $h'(y_k^j) \geq \varepsilon a_k$ for all such j's. The completeness of the Riemannian metrics induced on the leaves by the Riemannian structure on M implies that $\tilde{\tau}_{k,j}$, the piecewise leaf geodesic shadowing $\tau_{k,j}$ and connecting y_k^j to $h_{k,j}(y_k^j)$, is homotopic (relative to its end points) to a normalised leaf geodesic $\gamma_{k,j} : [0, T_{k,j}] \to M$ obtained by tightening $\tilde{\tau}_{k,j}$. Clearly, $T_{k,j} \leq 2n_k\Delta$ for all k and j. On the other hand,

$$\|\psi\|_\infty \geq \frac{1}{T_{k,j}} \log h'_{k,j}(y_k^j) \geq \frac{\log(\varepsilon a_k)}{T_{k,j}} \geq \frac{\log \varepsilon + n_k E}{8n_k\Delta} \to \frac{E}{8\Delta} \tag{4.6.25}$$

as $k \to \infty$. In particular, $T_{k,j} \to \infty$ uniformly with respect to j.

For each k find an integer j_k such that

$$h'_{k,j_k}(y_k^{j_k}) = \max\{h'_{k,j}(y_k^j)\}.$$

Set $\gamma_k = \gamma_{k,j_k}$, $h_k = h_{k,j_k}$, $y_k = y_k^{j_k}$ etc. Since J_k has length $1/a_k$ and contains a_k points x_k^j, there exists j_0 such that $d_T(x_k^{j_0}, x_k^{j_0+1}) < a_k^{-2}$ and then $h'_{k,j_0}(y_k^{j_0}) \geq \varepsilon a_k^2 \approx \varepsilon \exp(n_k \frac{E}{2})$. Therefore,

$$h'_k(y_k) \geq \varepsilon \exp(n_k \frac{E}{2}).$$

Let also $\lambda_k = \frac{1}{T_k}\log h_k'(y_k)$ and $\lambda_\infty = \limsup_{k\to\infty}\lambda_k$. Passing to a subsequence if necessary, we may assume that

$$\lambda_k > \frac{E}{10\Delta} \tag{4.6.26}$$

for all k.

From Lemma 4.6.12 it follows that if k is large enough, then γ_k intersects $\mathcal{P}(\theta_0)$.

Let $s_k \in [0, T_k]$ be an approximate time of first entry of γ_k into $\mathcal{P}(\theta_0)$. That is, $\gamma_k(s_k) \in \mathcal{P}(\theta_0)$ but $\gamma_k(t) \notin \mathcal{P}(\theta_0)$ for all $t \le s_k - 1$. Let also $\check{\gamma}_k$ denote the piecewise geodesic consisting of $\gamma_k|[0, s_k]$ followed by a geodesic radial segment joining $\gamma(s_k)$ to the centre u_k of the plaque (of one of charts U_j) which contains $\gamma_k(s_k)$. If $\check{f}_k$ is the holonomy along $\check{\gamma}_k$, then — again by Lemma 4.6.12 — the derivative $\check{f}_k'$ satisfies the inequality

$$|\log \check{f}_k'(y_k)| \le \|\psi\|_\infty(t(\varepsilon') + 2\Delta) + \Lambda_0 s_k \le 2\Lambda_0 T_k \tag{4.6.27}$$

for all sufficiently large k. Inequalities (4.6.27) yield

$$\exp(-2\Lambda_0 T_k) \le \check{f}_k'(y_k) \le \exp(2\Lambda_0 T_k). \tag{4.6.28}$$

The following fact can be obtained by a standard use of the "pigeon hole principle".

Lemma 4.6.13. *Let A be the number of charts in $\mathcal{U}_\Delta$. For any $n \in \mathbb{N}$, $\theta > 0$ and any set $B \subset T$ of cardinality $b \ge n[4A\rho/\theta] + 1$ there exist a point $x \in T$ and a set $B' \subset B$ of cardinality n such that $d_T(x, y) \le \frac{\theta}{4}$ for all $y \in B'$. Consequently, $d_T(y, y') \le \frac{\theta}{2}$ for all y and $y' \in B'$.* □

Choose now an integer

$$K_0 \ge 4\Delta\|\psi\|_\infty + 2$$

and set

$$K_1 = K_0 \cdot \left\lceil \frac{4A\rho}{\varepsilon_1} \right\rceil + 1.$$

Denote by γ_k^- the geodesic segment γ_k reparametrized in such a way that $\gamma_k^-(0) = \gamma_k(T_k)$ and $\gamma_k^-(T_k) = \gamma_k(0)$. Our goal now is to construct some ε''-regular points for γ_k^-, k being large enough and ε'' equal to $10\Lambda_0$.

From (4.6.22) and (4.6.26) it follows that

$$\int_0^{T_k} (\psi(\gamma_k^-(t)) + 10\Lambda_0)dt = (-\lambda_k + 10\Lambda_0))T_k < (-\frac{E}{10\Delta} + \frac{E}{20\Delta})T_k < 0.$$

Therefore, there exists the greatest value $s_{k,1}$ of $s \in [0, T_k]$ for which

$$\int_0^{s} (\psi(\gamma_k^-(t)) + 10\Lambda_0)dt = 0.$$

Then, if k is large enough,

$$\int_{s_{k,1}}^{T_k} (\psi(\gamma_k^-(t)) + 10\Lambda_0)dt = (-\lambda_k + 10\Lambda_0)T_k < -(1+\Lambda_0)$$

and there exists $t_{k,1} > s_{k,1}$ for which

$$\int_{s_{k,1}}^{t_{k,1}} (\psi(\gamma_k^-(t)) + 10\Lambda_0)dt = -(1+\Lambda_0).$$

Again, for large k we have

$$\int_{t_{k,1}}^{T_k} (\psi(\gamma_k^-(t)) + 10\Lambda_0)dt = (-\lambda_k + 10\Lambda_0)T_k + 1 + \Lambda_0 < 0$$

and therefore there exists the largest value $s_{k,2}$ of $s \in (t_{k,1}, T_k)$ for which

$$\int_{t_{k,1}}^{s} (\psi(\gamma_k^-(t)) + 10\Lambda_0)dt = 0.$$

Since the constants involved in the above estimates do not depend on k, the above procedure can be repeated K_1 times (if only k is sufficiently large) to get values $s_{k,j}$, $j = 1, \dots, K_1$, such that

$$\int_{s_{k,j}}^{T_k} (\psi(\gamma_k^-(t)) + 10\Lambda_0)dt = (-\lambda_k + 10\Lambda_0)T_k + (j-1)(1+\Lambda_0) < 0.$$

From Lemma 4.6.13 it follows that among the centres $z_{k,j}$ of plaques containing points $\gamma_k^-(s_{k,j})$ there exist K_0 points, say $z_{k,i_1}, \dots, z_{k,i_{K_0}}$ with $1 \le i_1 < \dots < i_{K_0} \le K_1$, such that $d_T(z_{k,i_q}, z_{k,i_r}) < \frac{1}{2}\varepsilon_1$ for all q and $r \in \{1, \dots, K_0\}$.

The holonomy map g_k along $\hat{\gamma}_k = \gamma_k | [s_{k,i_1}, s_{k,i_{K_0}}]$ sends z_{k,i_1} to $z_{k,i_{K_0}}$, is defined (recall Proposition 4.6.6) on the segment $I_k = (z_{k,i_1} - \varepsilon_1, z_{k,i_1} + \varepsilon_1)$ and satisfies there the condition

$$0 < g_k' < \exp\{(1-K_0)(1 + \frac{9}{10}\Lambda_0) + 4\Delta\|\psi\|_\infty\} < e^{-1} < \frac{1}{2}.$$

Therefore, $g_k(I_k) \subset I_k$ and g_k has a fixed point $w_k \in I_k$.

Denote now by τ_k the periodic piecewise leaf geodesic obtained from $\tilde{\gamma}_k$ by lifting to the leaf through w_k. The loop τ_k is $\pitchfork$-hyperbolic and $\lambda(\tau_k) \le -\Lambda_0$. Passing to a suitable subchain of plaques if necessary, we may assume that τ_k is irreducible. Therefore, τ_k is contained in the union of plaques of $\mathcal{P}$ and $w_k \in L_{j_k}$, the leaf containing the plaque P_{j_k}, where $j_k \in \{1, \dots, N\}$; consequently, w_k coincides with one of centres x_i of plaques P_i, $i \le N$.

Let h_k^- be the holonomy map along $\gamma_k^-|[s_{k,i_1}, T_k]$. It satisfies the estimate

$$\begin{aligned}\log(h_k^-)'(y) &\leq \int_{s_{k,i_1}}^{T_k} \psi(\hat{\gamma}^-(t))dt + \frac{1}{10}\Lambda_0 T_k \\ &\leq -\lambda_k T_k + 10\Lambda_0 s_{k,i_1} + (K_1 - 1)(1+\Lambda_0) + \frac{1}{10}\Lambda_0 T_k \\ &\leq (12\Lambda_0 - \lambda_k)T_k,\end{aligned}$$

whenever k is large enough that $T_k \geq K_1(1+\Lambda_0)/\Lambda_0$. Moreover, by (4.6.22) and (4.6.25),

$$12\Lambda_0 T_k \leq \frac{12ET_k}{200\Delta} \leq \frac{3}{5}\log a_k.$$

Therefore,

$$\begin{aligned}\log(h_k^-)'(y) &\leq -\log h_k'(y_k) + \frac{3}{5}\log a_k \leq \log\varepsilon + \frac{1}{2}n_k E + \frac{3}{5}n_k \cdot \frac{E}{4} \\ &\leq -\frac{7}{20}n_k E - \log\varepsilon < 0\end{aligned}$$

for large k. Denote by w_k' the image w_k under the holonomy map h_k^-. For large k, the transverse distance of y_k and w_k' can be estimated by

$$\delta_k' = 2\varepsilon_1 \exp(-\frac{7}{20}n_k - \log\varepsilon).$$

Note that the estimates analogous to (4.6.27) and (4.6.28), and the arguments following them are valid not only for γ_k but also for all $\gamma_{k,j}$'s (where k is large enough), the geodesics homotopic to $\tau_{k,j}$'s, the piecewise leaf geodesics corresponding to the chains $u_{k,j}$ which separate points x_k^j and x_k^{j+1} of A_k. In particular, each holonomy map $h_{k,j}$ along a segment of $\gamma_{k,j}$ connecting some ε''-regular points is defined on a segment $I_{k,j} \subset T$ of length $\geq 2\varepsilon_1$, contracts $I_{k,j}$ uniformly into itself and admits a fixed point $w_{k,j} \in I_{k,j}$.

Let us come back to our geodesics γ_k corresponding to the points $y_k = y_k^{j_k}$, where $h_{k,j}(y_j)$ becomes maximal. Consider the neighbourhood $J_k^{(1)} \subset T$ of $y_k = \gamma_k(0)$ of the form $J_k^{(1)} = (y_k - 2\delta_k, y_k + 2\delta_k)$, where

$$\delta_k = \frac{4\varepsilon_1\sqrt{a_k}}{h_k'(y_k)}.$$

If k is large enough, then — by previous estimates — $w_k' \in J_k^{(1)}$.

If $J_k^{(1)}$ contains at least three points of our (n_k, ε)-separated set A_k, then it contains a pair of the form x_k^j, x_k^{j+1} for which

$$d_T(x_k^j, x_k^{j+1}) \leq \frac{2\delta_k}{m_k - 2}$$

where $m_k = \#(J_k^{(1)} \cap A_k)$. The corresponding holonomy map $h_{k,j}$ has to satisfy the inequality

$$h'_{k,j}(y_k^j) \geq \varepsilon(m_k - 2)/2\delta_k.$$

From definition of δ_k and the estimate $0 < h'_{k,j}(y_k^j) \leq h'_k(y_k)$ it follows that the number b_k of points of A_k outside the interval $J_k^{(1)}$ satisfies the condition $b_k \geq a_k - 2 - c\sqrt{a_k}$, where $c > 0$ is a constant independent on k. Obviously, a similar estimate holds when $m_k \leq 2$. This shows that

$$\lim_{k\to\infty} \frac{b_k}{a_k} = 1. \tag{4.6.29}$$

To complete the proof of existence of resilient leaves in the case under consideration, fix a large integer k and set

$$J_k^{(1)} = J_k, w_k^{(1)} = w_k,\ \tilde{w}_k^{(1)} = w'_k,\ \text{and}\ \tau_k^{(1)} = \tau_k.$$

The complement of $J_k^{(1)}$ contains an interval $J_k^{(2)}$ of length $\geq \varepsilon_1 - \delta_k$. By the same arguments as before (applied now to one of geodesics $\gamma_{k,j}$), we can find a point w_k^2 fixed by a contracting holonomy along a loop $\tau_k^{(2)} \in G(\Lambda_0)$ and such that its image $\tilde{w}_k^{(2)}$ under the holonomy map along the suitable segment of $\gamma_{k,j}$ is located in the δ_k-neighbourhood of $y_{k,j}$. From (4.6.29) it follows that, since our k is large, we can repeat this procedure $N+1$ times to get points $w_k^{(i)}$ fixed by hyperbolic holonomies along loops $\tau_k^{(i)} \in G(\Lambda_0)$ and their images $\tilde{w}_k^{(i)}$ under holonomy maps $h_k^{(i)}$ along segments of geodesics γ_{k,j_i}. We can proceed in such a way that $\tilde{w}_k^{(i+1)}$ is always δ_k-apart from all the points $\tilde{w}_k^{(1)}, \ldots, \tilde{w}_k^{(i)}$. In particular, all the points $\tilde{w}_k^{(i)}$, $i = 1, \ldots, N+1$, are different while among the points $w_k^{(i)}$ there exist at least two, say $w_k^{(1)}$ and $w_k^{(2)}$, which coincide. (This is because, as was mentioned before, all of them belong to the set $\{x_1 \ldots x_N\}$ of all the centres of our plaques P_i.) Denote this common hyperbolic fixed point by x. Since points $\tilde{w}_k^{(1)}$ and $\tilde{w}_k^{(2)}$ do not coincide, one of them, say $y = \tilde{w}_k^{(1)}$, is different from x. The holonomy map $f = h_k^{(1)}$ is defined in a neighbourhood of x and $f(x) = y \neq x$. Moreover, x is fixed by a holonomy map g, for instance the contracting holonomy along the loop $\tau_k^{(1)}$. Therefore, the leaf through x is resilient. (In fact, we ended up with a contradiction which shows that among the leaves L_j there exist some which are not proper.)

This finally ends the proof of Theorem 4.6.1. □

4.7 Harmonic measures

Let $(M, \mathcal{F})$ be a closed (i.e., compact, without boundary) and oriented, foliated manifold. As we observed in Section 4.1, transverse invariant measures on $(M, \mathcal{F})$

need not exist. Therefore, it would be interesting to find a good, always non-empty, class of measures which reflect some dynamics of $\mathcal{F}$. A successful attempt in this direction has been made by L. Garnett [**Gar1**] who has introduced the notion of a harmonic measure.

First, let us equip M with a Riemannian structure g and denote by Δ the Laplace–Beltrami operator on (M, g):

$$\Delta f = \operatorname{div} \nabla f, \tag{4.7.1}$$

where ∇f is the gradient of a C^2-differentiable function $f : M \to \mathbb{R}$ and $\operatorname{div} X = \operatorname{Trace} \nabla X$ is the divergence of a vector field X, ∇ being the Levi–Civita connection on (M, g). Denote this time the Riemannian volume form on M by Ω_M. Stokes' Theorem implies that

$$\int_M \Delta f \cdot \Omega_M = 0. \tag{4.7.2}$$

In the same way, if h is a harmonic function on a Riemannian manifold M and $f : M \to \mathbb{R}$ is a C^2-differentiable function with a compact support, then

$$\Delta(fh) = \Delta f \cdot h + 2\langle \nabla f, \nabla h \rangle,$$

$$\operatorname{div}(f \nabla h) = \langle \nabla f, \nabla h \rangle,$$

and — by Stokes' Theorem again —

$$\int_M \Delta f \cdot h \Omega_M = 0. \tag{4.7.3}$$

Measures μ on M satisfying

$$\int_M \Delta f d\mu = 0 \tag{4.7.4}$$

for all C^2-differentiable functions on M are called *harmonic*. Therefore, by (4.7.2) and (4.7.3), the Riemannian volume Ω_M and all its multiples with harmonic coefficients are harmonic. This notion can be generalized to the case of foliated Riemannian manifolds as follows.

Equip any leaf L of $\mathcal{F}$ with the Riemannian structure g_L induced from M and denote by ∇^L, div_L and Δ_L, respectively, the Levi–Civita connection, the divergence and Laplace–Beltrami operators on (L, g_L). Also, denote by $\Delta_{\mathcal{F}}$ the *foliated Laplace–Beltrami operator* given by

$$\Delta_{\mathcal{F}} f(x) = \Delta_{L_x} f(x), \quad x \in M, \tag{4.7.5}$$

L_x being — as usually — the leaf of $\mathcal{F}$ through x. The operator $\Delta_{\mathcal{F}}$ acts on bounded measurable functions $f : M \to \mathbb{R}$ which are C^2-differentiable along the leaves.

Definition 4.7.1. A Borel probability measure μ on M is *harmonic* if

$$\int_M \Delta_{\mathcal{F}} f d\mu = 0 \tag{4.7.6}$$

for any f.

From (4.7.2) it follows that, if $\operatorname{codim} \mathcal{F} = 0$ (i.e., if $\mathcal{F}$ consists of connected components of M), the standard Lebesgue measure defined by the volume form Ω_M is harmonic.

Our first goal here is to show that harmonic measures always exist.

Theorem 4.7.2. *On any compact foliated Riemannian manifold, harmonic probability measures exist.*

First, consider a compact Hausdorff space X and the linear space $V = C(X)$ of all real continuous functions on X. Assume that W is a linear subspace of V such that any function f of W is non-positive at some points of X. Note that spaces like that often exist. For instance, if $f(x)f(y) < 0$ for some x and y of M, then the 1-dimensional linear space spanned by f satisfies the condition above. Also, if $F : X \to X$ is a homeomorphism, then the family of all the functions of the form $f - f \circ F$ has the structure of a linear space and satisfies this condition.

We shall prove the following.

Proposition 4.7.3. *There exists a Borel probability measure μ on X such that*

$$\int_X f d\mu = 0 \tag{4.7.7}$$

for all f of W.

Proof. Let $l(f) = 0$ for all $f \in W$ and $q(h) = \max(-h)$ for all $h \in V$. The functional $q : V \to \mathbb{R}$ is semi-linear in the following sense:

$$q(h_1 + h_2) \leq q(h_1) + q(h_2)$$

and

$$q(ah) = aq(h)$$

for all $h, h_1, h_2 \in V$ and $a \in \mathbb{R}$, $a \geq 0$. Obviously, $l(f) \leq q(f)$ for all $f \in W$. Therefore, by the Hahn–Banach Theorem (see [**Al**], [**Cn**] etc.), there exists a linear extension $L : V \to \mathbb{R}$ of l such that $L(h) \leq q(h)$ for all $h \in V$. The functional $-L$ is positive (i.e., $-L(h) > 0$ whenever $h > 0$) and therefore, by the Riesz Representation Theorem, corresponds to a Borel measure ν on X. The normalised measure

$$\mu = -\frac{1}{L(1)} \cdot \nu$$

satisfies (4.7.7). □

Now, Theorem 4.7.2 follows directly from Proposition 4.7.3 applied to $X = M$ and $W = \{\Delta_{\mathcal{F}}h; h \in \mathrm{C}^2(M)\}$. Indeed, $\Delta_{\mathcal{F}}h(x_0) \leq 0$ at a point x_0 of M at which h attains its maximum.

In order to get some properties of harmonic measures we need the one-parameter semigroup D_t, $t \geq 0$, of *heat diffusion operators* corresponding to $\Delta_{\mathcal{F}}$. These are characterized by the conditions

$$D_0 = \mathrm{id}, \ D_{t+s} = D_t \circ D_s \quad \text{and} \quad \frac{d}{dt}D_t|_{t=0} = \Delta_{\mathcal{F}}, \tag{4.7.8}$$

and coincide — when restricted to any leaf L — with the heat diffusion operators on L. They are given by

$$D_t f(x) = \int_{L_x} f(y)p(x,y,t)\Omega_{L_x}(y), \tag{4.7.9}$$

where $p(\cdot,\cdot,t)$ is the *foliated heat kernel* on $(M,\mathcal{F})$. The function $p(\cdot,\cdot,t)$ is non-negative and satisfies

$$\int_{L_x} p(x,y,t)\Omega_{L_x}(y) = 1 \tag{4.7.10}$$

for any $t > 0$ and $x \in M$. From (4.7.9) it follows that the operators D_t are linear, preserve constant functions, map non-negative functions to non-negative ones (consequently, are *monotonic* in the sense that $f_1 \leq f_2$ implies $D_t f_1 \leq D_t f_2$ for all t) and satisfy the following condition: if $f \geq 0$ everywhere on a leaf L and $f > 0$ on a set $A \subset L$ of positive leaf volume, then $D_t f$ $(t > 0)$ is strictly positive everywhere on L.

The reader not familiar with differential operators on Riemannian manifolds is referred to [**McK**], [**IW**], [**War**], etc.

We shall begin by the following lemmas.

Lemma 4.7.4. *If f is bounded and measurable, then so is $D_t f$.*

Lemma 4.7.5. *If f is continuous, so is $D_t f$.*

To prove these lemmas, we shall consider first the case of a foliation $\mathcal{F}$ without holonomy (this means that the holonomy groups (defined in Section 1.3) of all the leaves are trivial).

The boundness of $D_t f$ follows directly from (4.7.9) and (4.7.10). Since measurable functions can be approximated by continuous ones and limits of measurable functions are measurable, in order to prove Lemma 4.7.4 it is enough to show that $D_t f$ is measurable whenever f is continuous.

Given $R > 0$, $t > 0$, x and y in the same leaf L, define the truncated heat kernel $p(x,y,t,R)$ as $p(x,y,t)$ when the leaf distance between x and y is not greater than R and as 0 otherwise. Also, put

$$D_t^R f(x) = \int_{L_x} f(y)p(x,y,t,R)\Omega_{L_x}(y). \tag{4.7.11}$$

Clearly, $D_t f(x) = \lim_{R\to\infty} D_t^R f(x)$ for any $x \in M$. Therefore, it suffices to show that all the functions $D_t^R f$ are continuous for continuous f.

Continuity along any leaf is standard. So, let x and z be two nearby points such that $L_x \neq L_z$. Since H_{L_x}, the holonomy group of L_x is trivial, for any compact region K on L_x there exists a small transverse disc T ($x \in T$) such that the product foliation of $T\times K$, projects diffeomorphically to $\mathcal{F}$ on a neighbourhood of K. Let $K = \bar{B}_{L_x}(x, R) \subset L_x$ be the closed ball of radius R. Observe that, since the closed balls are compact and — as one can learn from many books on partial differential equations, for instance those mentioned recently — the heat kernel can be expressed locally in terms of the derivatives of order ≤ 2 of the coefficients of $\Delta_{\mathcal{F}}$, given $\varepsilon > 0$, there exists $\delta > 0$ such that

$$|p(x, y, t, R) - p(z, \phi_z(y), t, R)| < \varepsilon \tag{4.7.12}$$

whenever $d_M(x, z) < \delta$; here, $\phi_z : K \to L_z$ is the natural projection obtained from the product foliation of $T \times K$. It follows that

$$|D_t^R f(x) - D_t^R f(z)| \leq \varepsilon \cdot \max|f| \cdot A \cdot \max_{y\in M} \operatorname{vol} B_{L_y}(y, R), \tag{4.7.13}$$

where A is a constant independent of ε (namely, A is the maximum of the Jacobians of the maps ϕ_z for z close to x).

Inequality (4.7.13) proves the continuity of $D_t^R f$.

Now, since all the functions $D_t^R f$, $R > 0$, are continuous, in order to prove Lemma 4.7.5 it suffices to show that $D_t^R f \to D_t f$ uniformly when $R \to \infty$.

It is known (see, for example [**McK**], p. 93) that for any Riemannian manifold L with bounded geometry, any $t > 0$ and any $\varepsilon > 0$ there exists a constant $R(\varepsilon) > 0$ which depends on the geometry bounds only and satisfies the inequality

$$\int_{B(x,R(\varepsilon))} p_L(x, y, t)\Omega_L \geq 1 - \varepsilon, \quad x \in L, \tag{4.7.14}$$

where p_L is the heat kernel on L. Since the leaves of $\mathcal{F}$ have uniformly bounded geometry, we can find $R(\varepsilon)$ good for all the leaves. For $R > R(\varepsilon)$ we get

$$|D_t f(x) - D_t^R f(x)| \leq \varepsilon \cdot \max|f|, \quad x \in M. \tag{4.7.15}$$

Clearly, (4.7.15) implies the required convergence and ends the proof of Lemmas 4.7.4 and 4.7.5 for foliations without holonomy.

For arbitrary foliations, the situation is more complicated. Readers familiar with Connes algebras, Dirac operators etc. can consult [**Ro1**], where a short elegant proof of Lemma 4.7.5 (which implies Lemma 4.7.4 easily) can be found. Here, we shall describe briefly a proof, written down recently by Candel in an unpublished manuscript [**Ca2**] (see also [**CaC2**]) and based on a classical Yosida theorem. Some technical details will be left to the reader. Note that Candel's proof works also for more general differential operators on foliated manifolds. He calls all of them

Laplace operators and defines them as arbitrary elliptic differential operators (of order 2) which vanish on constant functions. Every Laplace operator A in this sense can be written in the form

$$A = \Delta + X,$$

where Δ is a foliated Laplace–Beltrami operator (with respect to some Riemannian metric) and X is a vector field.

Let us recall first the following Yosida theorem.

Theorem 4.7.6. [Yo] *Let A be a continuous linear operator defined on a subspace F of a Banach space E. If F is dense in E, A is closed and all the operators $\lambda \cdot I - A$ $(\lambda > 0)$ are surjective and have norms $\geq \lambda$, then there exists a one-parameter family D_t, $t \geq 0$, of bounded linear operators on E such that $D_0 = I$, $D_{t+s} = D_t \circ D_s$, $\|D_t\| \leq 1$ $(s, t \geq 0)$, $\lim_{t\to 0} \|D_t e - e\| = 0$ and*

$$Ae = \lim_{t\to 0} \frac{1}{t}(D_t e - e)$$

for all $e \in F$. □

We are going to apply the above to $E = C(M)$, the space of continuous functions on M equipped with the supremum norm, $\|f\| = \max\{|f(x)|; x \in M\}$, and a suitable extension A of $\Delta_{\mathcal{F}}$. (The operator $\Delta_{\mathcal{F}}$ itself is — in general — not closed and therefore does not satisfy the conditions of Theorem 4.7.6.) In order to fulfil this program, observe first that $\Delta_{\mathcal{F}} f(x_0)$ is non-negative (resp., non-positive) at points of minimum (resp., maximum) of a continuous function f and that this implies inequalities

$$\min\{\Delta_{\mathcal{F}} f(x); x \in M\} \leq 0 \leq \max\{\Delta_{\mathcal{F}} f(x); x \in M\}$$

and

$$\min\{\lambda f(x) - \Delta_{\mathcal{F}} f(x); x \in M\} \leq \lambda f(y) \leq \max\{\lambda f(x) - \Delta_{\mathcal{F}} f(x); x \in M\} \quad (4.7.16)$$

for all $\lambda > 0$, $y \in M$ and those $f \in C(M)$ which are C^2-differentiable along the leaves. This allows us to show that the norm of $\lambda \cdot I - \Delta_{\mathcal{F}}$ satisfies the required condition.

Next, we want to show that $\lambda \cdot I - \Delta_{\mathcal{F}}$ has dense image in $C(M)$. This follows from the following:

$(*)$ *For any Hölder continuous function g on M there exists a unique function $f \in C(M)$ which is C^2-differentiable along the leaves and satisfies the condition $\lambda f - \Delta_{\mathcal{F}} f = g$.*

The proof of $(*)$ consists of a few relatively simple steps which are listed below. Some details are left to the reader familiar with partial differential equations.

Step 1. If U is a distinguished chart which belongs to a nice foliated atlas on M and g is Hölder continuous on $\bar{U}$, then there exists a function f which is continuous on $\bar{U}$, is C^2-differentialble on each plaque $P \subset U$ and satisfies $\lambda f - \Delta_{\mathcal{F}} f = g$ on U. (The same holds whenever $\mathcal{F}$ is the product foliation on $D \times T$, T being a manifold, for instance a transversal for our original foliation, and D a manifold with the smooth boundary, for instance a closed regular domain contained in a leaf L of this foliation.)

Hint. Choose a sufficiently regular function ϕ on the boundary ∂U and solve on each plaque $P \subset U$ the Dirichlet problem for $\lambda f - \Delta_{\mathcal{F}} f - g = 0$ with the boundary condition $f|\partial P = \phi|\partial P$, and show the continuity of the solution.

Step 2. Choose a constant κ such that $g \geq \lambda\kappa$ (resp., $g \leq \lambda\kappa$) on M, define the function $(f_1)_L$ (resp., $(f_2)_L$), L being any leaf of $\mathcal{F}$, as the smallest upper bound (resp., the largest lower bound) of the set of functions f_D solving the Dirichlet problem for $\lambda f - \Delta_{\mathcal{F}} f - g = 0$ with the boundary condition $f|\partial D = \kappa$, where D ranges over all bounded regular domains $D \subset L$. Let $f_i : M \to \mathbb{R}$ $(i = 1, 2)$ be functions satisfying $f_i|L = (f_1)_L$ for any leaf L. Show that f_1 is upper semicontinuous while f_2 is lower semicontinuous.

Hint. By the Maximum Principle for elliptic operators, f_1 and f_2 are finite. To show that $\liminf_{y \to x} f_1(y) \geq f_1(y)$ consider the covering $\tilde{L}$ of L, the leaf of $\mathcal{F}$ through x, corresponding to the holonomy group H_L and use the fact that, given a bounded domain $D \subset \tilde{L}$, the foliation induced by $\mathcal{F}$ on the normal bundle $N(\tilde{L})$ of $\tilde{L}$ coincides with a product foliation in a neighbourhood of the zero section of $N(\tilde{L})$. (This is a bit technical but quite elementary and related to the arguments used above in the case of foliations without holonomy.)

Step 3. Consider $h = f_2 - f_1$, show that it is bounded, non-negative, upper semicontinuous and C^2-differentiable along the leaves. Therefore, h attains its global maximum at a point x_0. At this point, $\Delta_{\mathcal{F}} h(x_0) \leq 0$. Since h satisfies the equality $\Delta_{\mathcal{F}} h = \lambda h$, it follows that $h(x_0) \leq 0$. This implies that $h \equiv 0$ and $f_1 = f_2$ everywhere on M. Consequently, $f = f_1$ is continuous and satisfies the conditions of $(*)$.

Now, let us observe that if (f_n) is any sequence of continuous functions such that $\|f_n\| \to 0$ and $\|\Delta_{\mathcal{F}} f_n - g\| \to 0$ as $n \to \infty$, then $g \equiv 0$ on M. Indeed, if $g > 0$ on a closed disc D contained in a leaf and $h \geq 0$ is a function which is smooth and has a non-empty compact support contained in D, then — by Green's Theorem —

$$\int_D h \Delta_{\mathcal{F}} f_n = \int_D f_n \Delta_{\mathcal{F}} h \to 0$$

and — by our assumptions —

$$\int_D h \Delta_{\mathcal{F}} f_n \to \int_D hg > 0,$$

a contradiction. This observation shows that the operator $\Delta_{\mathcal{F}}$ has an extension A defined on a subspace F of $C(M)$ by the condition

$$\int_D \phi \cdot Af = \int_D f \cdot \Delta_{\mathcal{F}} \phi \tag{4.7.17}$$

for every regular bounded domain D contained in a leaf of $\mathcal{F}$ and every compactly supported function $\phi \in C^2(D)$. The space F can be chosen to contain the space of all the continuous functions C^2-differentiable along the leaves. Such an F is dense in $C(M)$. It is not too hard to show that the extension A is bounded and closed. The condition $(*)$ shows that the images of operators $\lambda I - \Delta_{\mathcal{F}}$ are dense in $C(M)$. By (4.7.16), these operators are injective, their inverses have finite norms ($\leq 1/\lambda$), so can be extended all over $C(M)$. It follows that the operators $\lambda I - A$ are surjective. Therefore, Yosida's Theorem 4.7.6 can be applied to A. Denote by $\tilde{D}_t$, $t \geq 0$, the operators on F determined by A. We leave to the reader the proof of the identity $\tilde{D}_t = D_t$ which holds on $C(M)$. This will complete the proof of Lemmas 4.7.4 and 4.7.5 in full generality.

A function $f : M \to \mathbb{R}$ is said to be *diffusion invariant* when

$$D_t f = f, \quad t \geq 0. \tag{4.7.18}$$

It follows directly from (4.7.8) that any diffusion invariant function f is leafwise harmonic (i.e., $\Delta_{\mathcal{F}} f = 0$). The converse is also true: any bounded leafwise harmonic function f satisfies (4.7.18). Roughly speaking, the argument goes as follows: (4.7.8) yields

$$\frac{d}{dt} D_t|_s = D_s \circ \Delta_{\mathcal{F}} = \Delta_{\mathcal{F}} \circ D_s. \tag{4.7.19}$$

Therefore, if f is leafwise harmonic, then

$$D_s f - f = \int_0^s \frac{d}{dt} D_t|_u f du = \int_0^s D_u \Delta_{\mathcal{F}} f du = 0$$

for any $s \geq 0$.

Definition 4.7.7. If μ is a Borel probability measure on M, then the formula

$$\int_M f d\mu_t = \int_M D_t f d\mu, \tag{4.7.20}$$

where f is any bounded measurable function on M, defines the *diffused measure* μ_t (denoted hereafter by $D_t\mu$), $t \geq 0$, on M. A measure μ is said to be *diffusion invariant* when

$$D_t \mu = \mu \tag{4.7.21}$$

for all $t \geq 0$.

It is not too hard to show that a measure μ is diffusion invariant if and only if $D_t\mu = \mu$ for some $t > 0$.

Lemma 4.7.8. *If $D_t\mu = \mu$, then D_t acts in a natural way on the space $L^1(M, \mu)$.*

Proof. Let f be a non-negative integrable function on (M, μ). Put

$$f_n(x) = \min\{f(x), n\}$$

for all $n \in \mathbb{N}$ and $x \in M$. The functions f_n are bounded, $f_n \nearrow f$ and $(D_t f_n)$ is an increasing sequence of bounded measurable functions. Put

$$D_t f = \lim_{n \to \infty} D_t f_n. \tag{4.7.22}$$

By our assumption,

$$\int_M D_t f_n d\mu = \int_M f_n d\mu \leq \int_M f d\mu.$$

This shows that $D_t f$ is integrable. Moreover, if $f = 0$ μ-a.e., then $f_n = 0$ μ-a.e. for all $n \in \mathbb{N}$. Also, $D_t f_n \geq 0$ and — by invariance of μ under D_t —

$$\int_M D_t f_n d\mu = \int_M f_n dD_t \mu = \int_M f_n d\mu = 0.$$

Therefore, $D_t f_n = 0$ and $D_t f = 0$ μ-a.e. This shows that formula (4.7.22) defines properly the action of D_t on $L^1(M, \mu)$. □

Harmonic measures can be characterized locally as follows.

Proposition 4.7.9. *A measure μ on M is harmonic if and only if it disintegrates locally, in any distinguished by $\mathcal{F}$ chart U, as*

$$\mu | U = \int_T h_x \Omega_x d\sigma(x), \tag{4.7.23}$$

where $T \subset U$ is a transversal, σ is a Borel measure on T, Ω_x is the Riemannian volume on the plaque $P_x \subset U$ through x and h_x, $x \in T$, is a harmonic function on P_x.

Proof. If μ is given by (4.7.23) with harmonic coefficients h_y, then — by (4.7.3) —

$$\int_{P_x} \Delta_{\mathcal{F}} f h_x \Omega_{L_x} = 0$$

for any plaque P_x and $\int_U \Delta_{\mathcal{F}} f d\mu = 0$ for any distinguished chart U and any bounded measurable C^2-differentiable along the leaves function f supported in U. Since M is compact, it can be covered by finitely many distinguished charts and any f on M can be decomposed into a finite sum of functions supported in these charts. Consequently, μ is harmonic. This proves the implication "$\Leftarrow$" in our statement.

Conversely, any measure ν on U disintegrates as

$$\nu = \int_{T_U} \nu_x d\sigma(x)$$

for some measures ν_x on the plaques P_x, $x \in T_U$, and some measure σ on T_U. If $\nu = \mu | U$, μ is harmonic on $(M, \mathcal{F})$ and f is a C^2-differentiable function compactly supported in U, then

$$\int_{P_x} \Delta_{\mathcal{F}} f d\nu_x = 0$$

for σ-almost all $x \in U$. Applying the above to a countable set (f_j) dense in the space of C^2-functions supported in U we can find a set $A \subset T_U$ of full σ-measure and such that the above equality holds for any C^2-function f supported in U and any $x \in A$. Therefore, ν_x is a weak solution for $\Delta_{\mathcal{F}} h = 0$ on P_x. By the regularity theorem for elliptic differential equations (see [**War**], for instance), weak solutions are smooth, i.e., $\nu_x = h_x \Omega_{L_x}$ for some harmonic function h_x. This establishes the implication "$\Rightarrow$" and completes the proof of the proposition. □

Now, assume that σ is a transverse invariant measure on $(M, \mathcal{F})$, take any nice covering $\mathcal{U}$ by distinguished charts and equip M with a Riemannian structure. For any $U \in \mathcal{U}$ put

$$\mu_\sigma | U = \int_{T_U} \Omega_x d\sigma(x), \tag{4.7.24}$$

where, as before, Ω_x is the Riemannian volume on the plaque $P_x \subset U$ through x. Since σ is invariant under the holonomy pseudogroup, formula (4.7.24) defines properly a finite Borel measure μ_σ on M. The measure μ_σ is harmonic since it is of the form (4.7.23) with $h_x \equiv 1$ for all x. In this way, we have established the following.

Corollary 4.7.10. *The space of transverse invariant measures on a compact foliated Riemannian manifold embeds naturally into the space of harmonic measures.* □

Another, this time global, characterization of harmonicity is provided by

Proposition 4.7.11. *A measure μ on $(M, \mathcal{F})$ is harmonic if and only if it is diffusion invariant.*

Proof. The implication "$\Leftarrow$" follows directly from (4.7.21) and (4.7.8). To get the converse, assume that μ is harmonic.

We have to prove that the equality

$$\int_M D_t f d\mu = \int_M f d\mu \tag{4.7.25}$$

holds for all t and f. Again, we shall provide the reader with two different proofs of this equality. First, we prove it in the case of foliations without holonomy following more elementary arguments of [**Gar1**]. Then, we prove it in general applying the extension of diffusion operators established in the course of proof of Lemmas 4.7.4 and 4.7.5 by use of Yosida's Theorem 4.7.6.

So, assume first that all the leaves of $\mathcal{F}$ have trivial holonomy groups and observe that it is enough to show that (4.7.25) holds for $t \leq 1$ and f non-negative and supported in a chart U distinguished by $\mathcal{F}$. To this end, it is sufficient to show that for any $\varepsilon > 0$ there exists $R_0 > 0$ such that

$$\int_M f d\mu - \int_M D_t^R f d\mu \leq \varepsilon \int_M f d\mu \tag{4.7.26}$$

for all such f and $R \geq R_0$, where D_t^R denotes, as before, the truncated heat diffusion along the leaves.

Given R, we can "slice" U into a finite number of "thin" distinguished charts $U_1, \ldots, U_m$ such that the foliations $\mathcal{F}|U_i(R)$, $U_i(R)$ being the set of all the points $y \in M$ which can be connected to some points of U_i by a leaf curve of length $\leq R$, are trivial. Clearly, it is enough to get (4.7.26) for f supported in one of U_i's.

By Proposition 4.7.9, $\mu|U_i(R)$ is of the form (4.7.23) with some measure σ and some harmonic functions h_y, $y \in T_i$, the transversal of U_i. Uniform geometry bounds for the leaves of $\mathcal{F}$ imply that the volumes of plaques in $U_i(R)$ grow at most exponentially with R. By the Harnack Inequality (see [**Mo**] or [**LSU**], for example), the values $h_y(z)$ grow also at most exponentially with R and $d_{\mathcal{F}}(z, U_i)$, $d_{\mathcal{F}}$ being the distance along the leaves. On the other hand, given any $t \leq 1$ and $K > 0$, the heat kernel $p(z, x, t)$ tends to 0 faster than $\exp(-K \cdot d_{\mathcal{F}}(x, z))$ when $d_{\mathcal{F}}(x, z) \to \infty$. Therefore, there exists R_0 such that

$$h_x(z) - \int_{L_x} h_x(y) p(z, y, t, R) \Omega_{L_x}(y) \leq \varepsilon h_x(z) \tag{4.7.27}$$

for all $R \geq R_0$, $z \in U_i \cap L_x$ and $x \in T_i$.

Denote by P_x and $P_x(R)$, respectively, the plaques of U_i and $U_i(R)$ passing through a point x of T_i. Since

$$\int_M D_t^R f d\mu = \int_{T_i} \int_{P_x(R))} \int_{P_x} f(y) p(z, y, t, R) \Omega_{L_x}(y) h_x(z) \Omega_{L_x}(z) d\sigma(x)$$

and the functions $(z, y) \mapsto p(z, y, t, R) h_x(y)$ are integrable, we can apply the Fubini Theorem and the symmetry of the heat kernel to get

$$\int_M D_t^R f d\mu = \int_{T_i} \int_{P_x} f(y) \int_{P_x(R)} h_x(z) p(y, z, t, R) \Omega_{L_x}(z) \Omega_{L_x}(y) d\sigma(x). \tag{4.7.28}$$

On the other hand,

$$\int_M f d\mu = \int_{T_i} \int_{P_x} f(y) h_x(z) \Omega_{L_x}(z) d\sigma(x). \tag{4.7.29}$$

Comparing (4.7.27), (4.7.28) and (4.7.29) yields (4.7.26) as required.

Now, in the general case, denote by $\tilde{D}_t : C(M) \to C(M)$ the diffusion operators for the extension A of $\Delta_{\mathcal{F}}$ established by use of Yosida's Theorem. By the Riesz Representation Theorem, it is sufficient to get (4.7.25) for all f in a dense subset of $C(M)$, for instance for all $f \in C(M)$ which are C^2-differentiable along the leaves. For such an f we have $\tilde{D}_t f = D_t f$ and $(d/dt) D_t f|_{t=s} = \Delta_{\mathcal{F}}(D_s f)$ (compare (4.7.19)). Consequently,

$$\int_M D_t f d\mu - \int_M f d\mu = \int_0^1 \left(\int_M \Delta_{\mathcal{F}}(D_s f) d\mu \right) ds = 0. \qquad \square$$

Remark 4.7.12. From the theory of diffusion and Brownian motion ([**Fe**], [**IW**], etc.) it follows that the above proposition can be interpreted in terms of Wiener measures. To explain this interpretation let us consider first an arbitrary complete Riemannian manifold N, fix $x_0 \in N$ and denote by $\mathcal{P}_{x_0}$ the space of all continuous paths $\gamma : [0, +\infty) \to N$ originated at x_0. Equip $\mathcal{P}_{x_0}$ with the topology of uniform convergence on compact sets. For arbitrary Borel subsets $E_1, \dots, E_m$ of N and arbitrary positive numbers $t_1 \leq t_2 \leq \dots \leq t_m$ denote by $C(\{t_i, E_i\})$ the Borel cylinder in $\mathcal{P}_{x_0}$ defined by

$$C(\{t_i, E_i\}) = \{\gamma \in \mathcal{P}_{x_0}; \gamma(t_i) \in E_i, i = 1, \dots, m\}.$$

Clearly, all the Borel cylinders generate the whole σ-algebra of Borel subsets of $\mathcal{P}_{x_0}$. Let ρ be the Riemannian distance function on N and ϕ the standard Gaussian distribution given by

$$\phi(t, u) = \frac{1}{(2\pi t)^{n/2}} \cdot e^{-u^2/(2t)}, \quad u \in \mathbb{R}, \quad t > 0,$$

where $n = \dim N$. Put

$$\omega(C(\{t_i, E_i\})) = \int_{E_1} \dots \int_{E_m} \phi(t_1, \rho(x_0, x_1)) \cdot \dots \cdot \phi(t_m - t_{m-1}, \rho(x_{m-1}, x_m)) \Omega_N(x_m) \dots \Omega_N(x_1),$$

Ω_N being the Riemannian volume element on N. The standard extension theorem shows that ω can be extended uniquely to a Borel measure $\omega = \omega_{x_0}$ on $\mathcal{P}_{x_0}$; ω_{x_0} is called the *Wiener measure* at x_0. Given any probability measure μ on N one can define $\bar{\mu}$, a measure on $\mathcal{P}$, the space of all the paths $\gamma : [0, \infty) \to N$, by

$$\bar{\mu} = \int_N \omega_x d\mu(x). \tag{4.7.30}$$

Such a measure $\bar{\mu}$ is called the *Wiener measure with initial distribution* μ.

Coming back to our foliated manifold $(M, \mathcal{F})$ one can perform a similar construction of a Wiener measure with initial distribution μ by formula (4.7.30), where ω_x is the Wiener measure at x for the leaf L_x of $\mathcal{F}$ through x. In this case, $\bar{\mu}$ becomes a Borel probability measure on $\mathcal{P}_{\mathcal{F}}$, the space of all the leaf paths $\gamma : [0, \infty) \to L$, $L \in \mathcal{F}$. For any $\tau > 0$ one has a natural transformation S_τ of $\mathcal{P}_{\mathcal{F}}$ given by

$$S_\tau(\gamma)(t) = \gamma(t + \tau). \tag{4.7.31}$$

With this notation Proposition 4.7.11 can be expressed as

Proposition 4.7.13. *A measure μ on $(M, \mathcal{F})$ is harmonic if and only if $\bar{\mu}$ is S_τ-invariant for all (equivalently, some) $\tau > 0$.* □

Coming back to the main line of our exposition we shall provide another proof of the existence of harmonic measures.

For any $t \geq 0$, D_t is a continuous linear operator on $C(M)$, the space of continuous real functions on M (Lemma 4.7.5). By the Riesz Representation Theorem, the functionals

$$C(M) \ni f \mapsto \frac{1}{t}\int_0^t \left(\int_M D_s f d\mu\right) ds, \quad t > 0,$$

where μ is any fixed Borel probability measure on M, represent elements μ_t of $\mathcal{M}(M)$, the space of all Borel probability measures on M. Since $\mathcal{M}(M)$ is compact, $\mu_{t_n} \to \mu_\infty \in \mathcal{M}(M)$ for some $t_n \to \infty$. μ_∞ is diffusion invariant, hence, by Proposition 4.7.11, harmonic. □

As $\mathcal{M}(X, \mathcal{G})$, the set of all Borel probability measures on X which are invariant under the action of a pseudogroup $\mathcal{G}$, is always convex in $\mathcal{M}(X)$ (Section 4.1), so is the set $\mathcal{H}(M, \mathcal{F})$ of all the harmonic probability measures on M in $\mathcal{M}(M)$. Also, one can define ergodicity of measures on $(M, \mathcal{F})$ as follows.

Definition 4.7.14. A Borel probability measure μ on a compact foliated manifold $(M, \mathcal{F})$ is *ergodic* when any $\mathcal{F}$-saturated Borel set $A \subset M$ has measure either 0 or 1.

It is easy to see that the harmonic measure μ_σ coming from a transverse invariant measure σ becomes ergodic (after normalization) if and only if $\sigma \in \mathcal{E}(T, \mathcal{H})$, where T is a complete transversal, $\mathcal{H}$ — the holonomy pseudogroup of $\mathcal{F}$ acting on T and $\mathcal{E}(T, \mathcal{H})$ — the set of all ergodic $\mathcal{H}$-invariant measures on T. Therefore, the above definition is coherent with that of ergodic invariant measures for pseudogroups (Definition 4.1.4). As mentioned at the beginning of this section, the advantage of this approach is that harmonic and ergodic measures in the above sense always exist. To see this more clearly, we shall prove a result analogous to Proposition 4.1.6. To this end, we need first the following ergodic theorem:

Theorem 4.7.15. *If μ is a harmonic measure on $(M, \mathcal{F})$ and $f : M \to \mathbb{R}$ is μ-integrable, then there exists a μ-integrable function $f^* : M \to \mathbb{R}$ constant along the leaves and such that*

$$f^*(x) = \lim_{n\to\infty} \frac{1}{n}\sum_{k=0}^{n-1} D_k f(x) \quad \text{for } \mu\text{-a.a. } x \tag{4.7.32}$$

and

$$\int_M f^* d\mu = \int_M f d\mu. \tag{4.7.33}$$

Proof. The existence of a D_1-invariant function f^* satisfying (4.7.32) and (4.7.33) follows directly from the individual ergodic theorem in [**Yo**] (Theorem 6, Chapter XIII.2). It remains to show that such an f^* is constant along almost all the leaves.

Since f^* is D_1-invariant, it is harmonic along the leaves. So, it is enough to show that any bounded Borel leaf-harmonic function h on M must be constant along μ-almost all the leaves; this means (here and after) just that the set of all the $x \in M$ for which h is constant on L_x has μ- measure 1.

To this end, observe first that any set A with $\mu(A) = 0$ has to satisfy the following: for μ-almost all the points x of M, the Riemannian volume of $A \cap L_x$ equals zero. Then, set $g_c = \min\{h, c\}$ for any $c > 0$. Clearly, $|g_c| \leq |h|$. Since $g_c \leq h$ and $g_c \leq c$, $D_1 g_c \leq D_1 h = h$ and $D_1 g_c \leq c$; consequently, $D_1 g_c \leq g_c$. It follows that both g_c and $D_1 g_c$ are μ-integrable. Moreover, since μ is harmonic, $\int_M D_1 g_c d\mu = \int_M g_c d\mu$ and $D_1 g_c = g_c$ μ-a.e. Therefore, there exists a saturated set Z_c such that $\mu(Z_c) = 1$ and $D_1 g_c = g_c$ almost everywhere (with respect to the Riemannian measure on the leaves) for every leaf contained in Z_c.

If $x \in Z_c$, then $D_1 h(x) > c$ if and only if $h > c$ on a set of positive Riemannian measure in L_x, the leaf through x. If so, $g_c = c$ on this set and, since $D_1 g_c = g_c$, $g_c = c$ and $h \geq c$ everywhere on L_x. (In fact, by one of the properties of diffusion operators mentioned at the beginning of this section, $D_1 g_c$ were everywhere strictly less than c if $h(y)$ would be strictly smaller than c at a point y of L_x.)

Let $Z = \bigcap_c Z_c$, where c ranges over the set of all positive rationals. The argument above applied to both h and $-h$ shows that h is constant on the leaves contained in Z. Obviously, $\mu(Z) = 1$. □

Next, let us prove the following Lebesgue decomposition for harmonic measures.

Theorem 4.7.16. *If μ and ν are two harmonic probability measures on $(M, \mathcal{F})$, then ν can be decomposed as $\nu = \nu_1 + \nu_2$, where ν_1 and ν_2 are harmonic, ν_1 is absolutely continuous with respect to μ while μ and ν_2 are mutually singular. Moreover, ν_1 and ν_2 are concentrated on disjoint saturated sets.*

Proof. By the standard Lebesgue decomposition theorem ([**Rud**], Thm. 6.9), there exists a set $A \subset M$ with $\mu(A) = 0$ and such that $\nu|(M \setminus A)$ is absolutely continuous with respect to μ. Let B be the union of all the leaves L of $\mathcal{F}$ for which the leaf volume of $L \cap A$ is positive. (The set B could be considered as the *essential saturation* of A.) Let $C = M \setminus B$, $\nu_1 = \nu|C$ and $\nu_2 = \nu|B$.

Obviously, $\nu = \nu_1 + \nu_2$, ν_1 and ν_2 are harmonic. Since $\mu(A) = 0$ and B is its essential saturation, then — by the structure of harmonic measures described above — $\mu(B) = 0$ and $\mu(C) = 1$. Therefore, ν_2 and μ are mutually singular. It remains to show that ν_1 is absolutely continuous with respect to μ.

Let $E \subset M$ and $\mu(E) = 0$. Since $\nu|(M \setminus A)$ is absolutely continuous with respect to μ, $\nu(E \setminus A) = 0$. Let E' be the essential saturation of $E \setminus A$, i.e., E' is the union of all the leaves L such that the leaf volume of $L \cap (E \setminus A)$ is positive. As before, by the structure of harmonic measures, $\nu(E') = 0$. Let E'' be the union of all the leaves L for which the volume of $L \cap C \cap E$ is positive. If $L \subset E''$, then the leaf volume of $L \cap E$ is positive and also the leaf volume of $L \cap (E \setminus A)$ is positive, consequently $E'' \subset E'$. It follows that $\nu(E'') = 0$ and — again since E'' is the essential saturation of $E \cap C$ — $\nu(E \cap C) = 0$. Finally, $\nu_1(E) = 0$. □

Now, we are able to prove the above-mentioned characterization of ergodic harmonic measures.

Theorem 4.7.17. *A harmonic probability measure μ on $(M, \mathcal{F})$ is ergodic if and only if it is an extreme point of $\mathcal{H}(M, \mathcal{F})$.*

Proof. If μ is not ergodic, then $M = A \cup B$, A and B being disjoint saturated Borel sets of positive measure. The measures μ_1 and μ_2 defined by $\mu_1(E) = \mu(A \cap E)/\mu(A)$ and $\mu_2(E) = \mu(E \cap B)/\mu(B)$ for all Borel sets $E \subset M$ are harmonic and provide the decomposition $\mu = \mu(A)\mu_1 + \mu(B)\mu_2$ which shows that μ is not extreme in $\mathcal{H}(M, \mathcal{F})$.

Conversely, if μ is ergodic, μ_1 and μ_2 are harmonic and $\mu = t\mu_1 + (1-t)\mu_2$ for some $t \in (0,1)$, then we can decompose μ_1 into the sum $\mu_1 = \nu_1 + \nu_2$, ν_1 and ν_2 being harmonic, ν_1 — absolutely continuous with respect to μ_2, ν_2 and μ_2 — mutually singular. Ergodicity of μ implies that $\nu_2 = 0$, i.e., μ_1 is absolutely continuous with respect to μ_2. Similarly, μ_2 is absolutely continuous with respect to μ_1. Consequently, μ, μ_1 and μ_2 have the same sets of measure zero. Now, if $E \subset M$ is an arbitrary Borel set, then — by Theorem 4.7.15 —

$$\mu_1(E) = \int_M \chi_E d\mu_1 = \int_M \chi_E^* d\mu_1 = \mu(E) \int_M d\mu_1 = \mu(E),$$

since — by ergodicity of μ — the function χ_E^* is constant and equal to $\mu(E)$ a.e. on M. So, $\mu_1 = \mu$ and, in the same way, $\mu_2 = \mu$. This shows that μ is an extreme point of $\mathcal{H}(M, \mathcal{F})$. □

Finally, we shall prove a result analogous to Proposition 3.1.6 in the classical ergodic theory. To this end we accept the following.

Definition 4.7.18. A non-compact proper leaf is said to be *eventually wandering.* A leaf is *eventually non-wandering* when it does not eventually wander.

More precisely, a leaf L of a foliation $\mathcal{F}$ is eventually wandering if and only if it is non-compact and each point x of L admits a transversal T such that $L \cap T = \{x\}$.

Theorem 4.7.19. *If μ is a harmonic measure on a compact foliated manifold $(M, \mathcal{F})$ and W is the union of all the eventually wandering leaves of $\mathcal{F}$, then $\mu(W) = 0$.*

In other words, this theorem says that any harmonic measure is supported in the union of eventually non-wandering leaves.

To prove the theorem, let us choose a countable set $\{f_i, i \in \mathbb{N}\}$ dense in the space of real continuous functions on M. For any $x \in M$ define the diffused Dirac measure δ_x^* by

$$\int_M f d\delta_x^* = f^*(x), \tag{4.7.34}$$

where f is any continuous function on M and f^* satisfies the conditions of Theorem 4.7.15. The measures δ_x^* are harmonic. We call a point $x \in M$ *regular* when δ_x^* is ergodic and $x \in \operatorname{supp} \delta_x^*$.

Lemma 4.7.20. *There exists a Borel saturated set $R \subset M$ such that all the points of R are regular, $\mu(R) = 1$ for any harmonic measure on M and $\delta_x^* = \delta_y^*$ whenever x and y belong to the same leaf $L \subset R$.*

Proof. By Theorem 4.7.15, for any $i \in \mathbb{N}$ there exists a saturated set $R_i \subset M$ such that f_i^* is constant along each leaf $L \subset R_i$ and $\mu(R_i) = 1$ for any $\mu \in \mathcal{H}(M, \mathcal{F})$. The set $R = \bigcap_i R_i$ has all the required properties. □

Lemma 4.7.21. *If $\mu \in \mathcal{H}(M, \mathcal{F})$ and f is a bounded Borel function on M, then*

$$\int_M f d\mu = \int_R \left(\int_M f d\delta_x^* \right) d\mu(x). \tag{4.7.35}$$

Proof. By Theorem 4.7.15, equation (4.7.34) and the properties of the set R we have

$$\int_M f d\mu = \int_M f^* d\mu = \int_R f^*(x) d\mu(x) = \int_R \left(\int_M f d\delta_x^* \right) d\mu(x)$$

for continuous f. Since any Borel measure on M is regular, equality (4.7.35) for bounded Borel functions follows from the above by the Monotone Convergence Theorem, first for characteristic functions of open and closed sets, then for characteristic functions of arbitrary Borel sets, etc. □

Passing to the proof of Theorem 4.7.19 let us observe that it is enough to show that points of eventually wandering leaves cannot belong to R. In fact, if $W \subset M \setminus R$, then $0 \leq \mu(W) \leq 1 - \mu(R) = 0$ for any harmonic μ.

So, assume that $x \in L \cap R$ and L is an eventually wandering leaf. Let U be a distinguished chart around x for which $U \cap L$ consists of a single plaque P. Since $x \in R$, $\delta_x^*(U) > 0$. Also, if $f : M \to \mathbb{R}$ is a continuous function which is positive on $U \setminus P$ and vanishes everywhere else, then $f \equiv 0$ on L and

$$\int_M f d\delta_x^* = 0.$$

It follows that $\delta_x^*(U \setminus P) = 0$ and $\delta_x^*(P) > 0$. Ergodicity of δ_x^* implies that $\delta_x^*(L) = 1$ and, therefore, L admits a harmonic probability measure. The structure of harmonic measures on Riemannian manifolds (see [**War**], for instance) implies the existence of a positive harmonic function $h : L \to \mathbb{R}$ integrable with respect to the Riemannian volume on L. Since the geometry of L is bounded, such a function has to be constant. This yields a contradiction since the Riemannian volume of L is infinite. In fact, L contains an infinite family of pairwise disjoint plaques of a nice covering $\mathcal{U}$ and the volumes of all the plaques are bounded from below by a positive constant. □

Harmonic measures on foliated manifolds have been successfully applied to the study of foliations. For example, Ghys obtained a kind of Gauss–Bonnet theorem for the average (with respect to a harmonic measures) of the scalar curvature of a 2-dimensional foliation ([**Gh2**], see also [**Ro2**]) and characterized generic (with respect to a harmonic measure again) leaves of 2-dimensional foliations (in fact, laminations) of compact spaces [**Gh3**]. For the first result we refer the reader to the original papers. The second one is described in Chapter 6. Here, we shall complete the section just by a remark concerning geometric entropy of foliations and a recent result concerning geometric entropy and non-wandering leaves.

First, let us say that it seems to be interesting (and important !) to search for a good definition of a measure theoretic entropy for foliations which could provide a kind of variational principle for the geometric entropy. Since holonomy invariant measures need not exist, the class of harmonic measures seems to be suitable here. If so, a variational principle and Theorem 4.7.19 would imply that the geometric entropy is always supported in the closure of the union of all the eventually non-wandering leaves. This was conjectured in [**GLW2**], where the authors suggested the reader to search for a "reasonable" definition of such an entropy referring to the notion considered in [**GLW1**] for arbitrary linear operators acting on the space of integrable functions. This idea could be probably applied to the heat diffusion operator D_1 on a compact foliated Riemannian manifold $(M, \mathcal{F})$. However, Frings [**Fri**] has shown that this entropy of D_1 vanishes when $\dim \mathcal{F} = 1$. Since also other attempts known to us ([**Fri**], [**Hu1**], [**Hu2**] etc.) of defining measure theoretic entropies for foliations (as well as for pseudogroups and groups) are not fully satisfactory from this point of view, the problem of defining good measure theoretic entropies for these systems stays open. Note that even in the relatively simple case of a commutative group, existing definitions of measure and topological entropies do not provide satisfactory variational principles. A variational principle for $\mathbb{Z}^n$ was proved by Ruelle [**Ru1**] (see also [**Mis1**]) but the entropies involved there vanish when generating maps are of finite entropies, for instance when $\mathbb{Z}^n$ acts smoothly on a compact manifold (see [**Con**]).

Finally, let us mention a recent Hurder's [**Hu4**] result relating entropies of codimension-one foliations and their restrictions to the set of all non-wandering points. The result does not solve the conjecture mentioned above in every respect, but shows that its affirmative solution can be plausibly expected.

Let $\mathcal{F}$ be a codimension-one C^1-foliation of a compact Riemannian manifold (M, g). Recall that a leaf of $\mathcal{F}$ is eventually wandering whenever it is non-compact and proper. A point $x \in M$ and the leaf L through x are said to be *wandering* when x has an open neighbourhood U (the domain of a distinguished chart) such that the intersection of any leaf L' and U consists of at most one plaque. Otherwise, x and L are called *non-wandering*. Clearly, the set $\Omega(\mathcal{F})$ of all the non-wandering points is closed and the relative entropy (see Remark 3.3.2(iii)) $h(\mathcal{F}, \Omega(\mathcal{F}), g)$ can be considered.

Theorem 4.7.22. *If* $\operatorname{codim} \mathcal{F} = 1$ *and* $h(\mathcal{F}, g) > 0$*, then* $h(\mathcal{F}, \Omega(\mathcal{F}), g) > 0$.

To begin with, we shall prove first the following.

Theorem 4.7.23. *If* $\operatorname{codim} \mathcal{F} = 1$ *and* μ *is a* ϕ*-invariant* $\pitchfork$*-hyperbolic measure on* $V = T^1\mathcal{F}$*, then the* $\pitchfork$*-support of* μ *is contained in* $\Omega(\mathcal{F})$.

Here, as in Section 4.6, $\phi = (\phi_t)$ is the geodesic flow of $\mathcal{F}$.

Proof. Assume first that μ is ergodic. Results of Section 4.6 show that a generic point $v \in \operatorname{supp} \mu$ provides us with a contracting holonomy map $h : I = (z - \varepsilon_1, z + \varepsilon_1) \to I$, where $\varepsilon_1 > 0$, z belongs to the ⫛-support of μ and can be chosen arbitrarily close to $x = \pi(v)$. Since the leaves L_y, $y \in I \setminus \{z\}$, intersect I infinitely many times and the points of intersection cluster on z, $z \in \Omega(\mathcal{F})$. Since x can be expressed as the limit of a sequence (z_n) of such points $z_n \in \Omega(\mathcal{F})$, $x \in \Omega(\mathcal{F})$. Finally, since the orbit $t \to \phi_t(v)$ is dense in $\operatorname{supp} \mu$ and its projection to M is entirely contained in $\Omega(\mathcal{F})$, $\pi(\operatorname{supp} \mu) \subset \Omega(\mathcal{F})$.

If μ is arbitrary (but ⫛-hyperbolic), then τ_μ-almost all measures m in the ergodic decomposition (4.6.9) are ⫛-hyperbolic. By the previous argument, almost all supports $\operatorname{supp} \mu$ are contained in $\pi^{-1}(\Omega(\mathcal{F}))$. If $f : V \to \mathbb{R}$ is supported outside $\pi^{-1}(\Omega(\mathcal{F}))$, then $\int_V f dm = 0$ for each such m,

$$\int_V f d\mu = \int_{\mathcal{E}(V,\phi)} \left(\int_V f dm \right) d\tau_\mu = 0$$

and therefore μ is supported in $\pi^{-1}(\Omega(\mathcal{F}))$. □

Finally, let us complete this section with the proof of Theorem 4.7.22.

To this end, observe that the result on the existence of resilient leaves (Theorem 4.6.1) shows that the assumption $h(\mathcal{F}, g) > 0$ implies the existence of a transverse segment I and two contracting holonomy maps $h_1, h_2 : I \to I$ such that $h_1(I) \cap h_2(I) = \emptyset$. Each of them has a fixed point $x_i = h_i(x_i) \in I$. The leaves passing through points y ($y \neq x_i$) arbitrarily close to x_i are not proper, therefore the leaves L_i through x_i, $i = 1, 2$, are contained in $\Omega(\mathcal{F})$. Let $K = \overline{L_1 \cup L_2}$. Then $K \subset \Omega(\mathcal{F})$ is compact and saturated, and the holonomy pseudogroup $\mathcal{H}_K$ of $\mathcal{F}|K$ contains a subpseudogroup $\mathcal{G}$ generated $h_1|K$ and $h_2|K$. By arguments used in the proofs of Theorems 3.4.1 and 3.6.1,

$$h(\mathcal{F}, \Omega(\mathcal{F}), g) \geq h(\mathcal{F}, K, g) \geq c \cdot h(\mathcal{G}, \mathcal{G}_1) > 0,$$

where c is a positive real number and $\mathcal{G}_1 = \{\mathrm{id}_K, h_1|K, h_2|K, (h_1|K)^{-1}, (h_2|K)^{-1}\}$ is the symmetric set generating $\mathcal{G}$.

This ends the proof of Theorem 4.7.22. □

4.8 Patterson–Sullivan measures

Let us recall once again that the set $\mathcal{M}(X, G)$ of G-invariant Borel probability measures for a finitely generated group G of transformations of a space X may be empty. For such a group G one may search for measures satisfying less restrictive conditions, for instance for measures μ such that the measures μ and $g^*\mu$, $g \in G$, are mutually absolutely continuous, i.e., for measures μ which are G-quasi-invariant in the sense of the following

Definition 4.8.1. A measure μ on X is *G-quasi-invariant* when

$$g^*\mu = \phi_g \cdot \mu, \quad g \in G, \tag{4.8.1}$$

for some positive measurable functions ϕ_g on X.

The best known and most geometric construction of quasi-invariant measures is due to Patterson [**Pat1**] and Sullivan ([**Su2**] and [**Su4**]) who dealt with discrete groups of Möbius transformations of the unit ball in $\mathbb{R}^n$. In this section, we recall this construction and collect some natural and important properties of Patterson–Sullivan measures. Our exposition is rather brief and condensed. The reader interested in more on this topic is referred to the original Patterson and Sullivan papers or to good and more elaborated expositions by Patterson [**Pat2**] and Nicholls [**Ni**]. Also, note that Patterson–Sullivan type measures on the boundary of arbitrary hyperbolic metric space have been defined and studied by Coornaert [**Coo**].

The group of transformations of $\mathbb{R}^n$ generated by the inversion $x \mapsto x/|x|^2$ and all affine transformations of the form $x \mapsto a \cdot Ax + b$, where $a \in \mathbb{R} \setminus \{0\}$, $b \in \mathbb{R}^n$ and $A \in \mathrm{O}(n)$, the orthogonal group, is called the *Möbius group*. If $B \subset \mathbb{R}^n$ is the unit ball and $M(B)$ is the subgroup of the Möbius group consisting of all the transformations preserving B, then any discrete subgroup Γ of $M(B)$ is called *Fuchsian* when $n = 2$ or *Kleinian* when $n \geq 3$. The terminology may vary from one reference to another. Some authors restrict the term "Kleinian" either to the case $n = 3$ or to Γ's with the limit set Λ (see below) satisfying some conditions (like $\Lambda \neq S$, S being the boundary of B).

Equip B with the hyperbolic Riemannian metric g,

$$g(x) = \frac{4}{(1-|x|^2)^2} \cdot dx^2, \quad x \in B.$$

The Riemannian manifold (B, g) has constant sectional curvature -1, so it becomes a hyperbolic metric space when considered with the Riemannian distance function d. Its boundary can be identified with the unit sphere $S = \{x; |x| = 1\}$. $M(B)$ coincides with the group of isometries of (B, g). Recall that

$$d(0, x) = \log \frac{1+|x|}{1-|x|} \tag{4.8.2}$$

and the formula for the distance of arbitrary points x and y of B can be derived from (4.8.2) by passing through any map of $M(B)$ which keeps x fixed and sends y to 0.

For any discrete group $\Gamma \subset M(B)$, points x, y of B and $r > 0$ put

$$N(x, y, r) = \#\{\gamma \in \Gamma; d(x, \gamma(y)) \leq r\}. \tag{4.8.3}$$

N is called the *orbital counting function* of Γ. One can check easily that its type of growth does not depend on the choice of x and y.

Lemma 4.8.2. *Given x, y and Γ, there exists a constant A such that*

$$N(x,y,r) \leq A \cdot e^{(n-1)r} \tag{4.8.4}$$

for all $r > 0$.

Proof. Let $B(\varepsilon)$ be the ball of centre y and radius $\varepsilon > 0$ so small that $d(y, \gamma(y)) > 2\varepsilon$ for all $\gamma \in \Gamma$, $\gamma \neq e$. Such ε exists since Γ is discrete. Let $V(r)$ denote the volume of the hyperbolic ball of radius r. Then

$$N(x,y,r) \cdot V(\varepsilon) \leq V(r+\varepsilon) \leq \frac{1}{2^{n-1}} \int_0^{r+\varepsilon} e^{(n-1)t}\, dt$$

(compare (2.2.22)). Therefore, equality (4.8.4) holds with

$$A = \frac{e^{(n-1)\varepsilon}}{2^{n-1}(n-1)V(\varepsilon)}.$$

□

Again, since Γ is discrete, orbits of Γ have no points of accumulation in B.

Definition 4.8.3. The limit set Λ of Γ is defined as the set of all $x_\infty \in S$ such that the orbit $\Gamma(x) = \{\gamma(x); \gamma \in \Gamma\}$ accumulates on x_∞ for some (equivalently, all) $x \in B$. The group Γ is of the *first kind* when $\Lambda = S$, of the *second kind* otherwise.

All the transformations of $M(B)$ act on S. Our goal is to construct a probability measure μ on S which is Γ-quasi-invariant,

$$\gamma^* \mu = |\gamma'|^\delta \mu, \tag{4.8.5}$$

where $|\gamma'|$ is the linear distortion of $\gamma \in \Gamma$ and δ is an exponent defined by some geometric data. We begin by finding a suitable exponent δ.

Definition 4.8.4. The *Poincaré series* $P_s(x,y)$, $x, y \in B$, of Γ is given by

$$P_s(x,y) = \sum_{\gamma \in \Gamma} e^{-s \cdot d(x, \gamma(y))}. \tag{4.8.6}$$

The triangle inequality implies that the convergence of P_s does not depend on x and y. Also, there exists a positive number δ (called the *critical exponent*) such that P_s converges for all $s > \delta$ and diverges for all $s < \delta$. Lemma 4.8.2 yields the inequality $\delta \leq n-1$. The group Γ is of *convergent* (resp., *divergent*) type when P_δ converges (resp., diverges).

The critical exponent can be defined for many abstract series $\sum_m a_m$ of positive numbers a_m. It equals δ whenever the series $\sum_m a_m^{-s}$ converges for all $s > \delta$ and diverges for all $s < \delta$. The construction of Patterson–Sullivan measures μ relies on the following elementary fact.

Lemma 4.8.5. *Let δ be the critical exponent of a series $\sum_{m=1}^{\infty} a_m^{-s}$. Then, there exists an increasing function $k : [0,\infty) \to (0,\infty)$ such that the series*

$$\sum_{m=1}^{\infty} k(a_m) a_m^{-s} \tag{4.8.7}$$

has the same exponent of convergence δ and diverges at δ. Moreover, for any $\varepsilon > 0$, there exists $t_0 > 0$ such that

$$k(rt) \leq r^{\varepsilon} k(t) \tag{4.8.8}$$

for any $t \geq t_0$ and $r \geq 1$.

Proof. The existence of the critical exponent implies that we can reorder the sequence (a_m) in such a way that the entries a_m increase. Then $a_m \to \infty$ since $\delta > 0$. Let (ε_m) be a convergent to zero decreasing sequence of positive reals. First, take $t_1 = 1$ and put $k(t) = 1$ on $[0, t_1]$. Then, suppose that we already defined $t_1 < t_2 < \cdots < t_m$ and k on $[0, t_m]$. Choose t_{m+1} large enough to provide the inequality

$$k(t_m) \cdot \sum_{j \in (t_m, t_{m+1}]} a_j^{-(\delta - \varepsilon_m)} \geq t_m^{\varepsilon_m}.$$

This can be done since the series $\sum_j a_j^{-(\delta-\varepsilon_m)}$ diverges. Define k on $[t_m, t_{m+1}]$ by

$$k(t) = k(t_m) \cdot \left(\frac{t}{t_m}\right)^{\varepsilon_m}.$$

Obviously, the function k is positive and increases.

Since

$$\sum_{j=1}^{\infty} k(a_j) a_j^{-\delta} = \sum_{m=1}^{\infty} \sum_{j \in (t_m, t_{m+1}]} k(t_m)(a_j/t_m)^{\varepsilon_m} a_j^{-\delta} \geq \sum_{m=1}^{\infty} 1 = \infty,$$

the series in (4.8.7) diverges.

Choose $\varepsilon > 0$ and $m_0 \in \mathbb{N}$ such that $\varepsilon_m < \varepsilon$ for all $n \geq m_0$. It is not too hard to see that (4.8.8) holds for $t \geq t_{m_0}$. Finally, if $s > \delta$ and $\varepsilon = (s - \delta)/2$, then, by (4.8.8), the sequence $(k(a_m)/a_m^{\varepsilon})$ is bounded and

$$k(a_m) a_m^{-s} \leq c \cdot a_m^{-\frac{s+\delta}{2}} \leq c \cdot 2^{-\frac{s+\delta}{2}}$$

for all m large enough and some $c > 0$. This shows that the series $\sum_m k(a_m) a_m^{-s}$ converges when $s > \delta$ and ends the proof. □

Now, adapt a function k of Lemma 4.8.5 to the Poincaré series in (4.8.6). For any $s > \delta$ define the measure $\mu_{s,x,y}$ on B by the formula

$$\mu_{s,x,y} = \frac{1}{P_s^*(x,y)} \cdot \sum_{\gamma \in \Gamma} k(e^{d(x,\gamma(y))}) e^{-sd(x,\gamma(y))} \delta_{\gamma(y)}, \tag{4.8.9}$$

where δ_z denotes, as before, the Dirac measure at a point $z \in B$ and

$$P_s^*(x,y) = \sum_{\gamma\in\Gamma} k(e^{d(x,\gamma(y))})e^{-sd(x,\gamma(y))}. \tag{4.8.10}$$

For any x, y and s, $\mu_{s,x,y}$ is a probability measure on the compact space $\bar{B} = B \cup S$, so one can find a sequence (s_j) such that $s_j \searrow \delta$ and $\mu_{s_j,x,y} \to \mu$, μ being a probability measure on $\bar{B}$. Since $P_s^*(x,y) \to \infty$ when $s \searrow \delta$ and any compact subset K of B contains only a finite number of points $\gamma(y)$, $\gamma \in \Gamma$, μ is supported in S, in fact in Λ, the limit set of Γ.

Definition 4.8.6. Any measure μ constructed as above is called a *Patterson–Sullivan measure* (for the group Γ).

To begin with properties of Patterson–Sullivan measures, let us see first what happens when we change the "base point" x.

Proposition 4.8.7. *For all $s > \delta$, $x, y \in B$ and $\tau \in \Gamma$ the equality*

$$\mu_{s,\tau(x),y} = \tau^*\mu_{s,x,y} \tag{4.8.11}$$

holds. Therefore, we can always choose Patterson–Sullivan measures $\mu_{x,y}$ in such a way that

$$\mu_{\tau(x),y} = \tau^*\mu_{x,y} \tag{4.8.12}$$

for all $\tau \in \Gamma$.

Proof. Since every $\tau \in \Gamma$ is an isometry,

$$\begin{aligned} P_s(\tau(x),y) &= \sum_{\gamma\in\Gamma} e^{-sd(\tau(x),\gamma(y))} = \sum_{\gamma\in\Gamma} e^{-sd(x,(\tau^{-1}\circ\gamma)(y))} \\ &= \sum_{\eta\in\Gamma} e^{-sd(x,\eta(y))} = P_s(x,y) \end{aligned}$$

and the same holds for $P_s^*(x,y)$. It follows that

$$\begin{aligned} \mu_{s,\tau(x),y} &= \frac{1}{P_s^*(\tau(x),y)} \cdot \sum_{\gamma\in\Gamma} k(e^{d(\tau(x),\gamma(y))})e^{-sd(\tau(x),\gamma(y))}\delta_{\gamma(y)} \\ &= \frac{1}{P_s^*(x,y)} \cdot \sum_{\eta\in\Gamma} k(e^{d(x,\eta(y))})e^{-sd(x,\eta(y))}\delta_{\tau\eta(y)} = \tau^*\mu_{s,x,y}. \end{aligned}$$

Here again, we used the substitution $\tau^{-1}\circ\gamma \mapsto \eta$. Therefore, if $(\mu_{s_j,x,y})$ converges to $\mu_{x,y}$, then $(\mu_{s_j,\tau(x),y})$ converges to $\tau^*\mu_{x,y}$ for all $\tau \in \Gamma$. $\square$

To prove that our measures satisfy (4.8.5) we have to define precisely the measure $|\gamma'|$ of *distortion* of $\gamma \in \Gamma$ on S. First, denote by d_0 the standard distance function on the unit sphere S. Then, choose $x \in B$ and a Möbius transformation τ such that $\tau(x) = 0$. Let

$$d_x(y_\infty, z_\infty) = d_0(\tau(y_\infty), \tau(z_\infty)). \tag{4.8.13}$$

Clearly, d_x does not depend on the choice of τ and one can interpret $\alpha = d_x(y_\infty, z_\infty)$ as follows: an eye situated at x and equipped with a hyperbolic telescope sees the segment $[y_\infty, z_\infty]$ at the angle α. Finally, set

$$|\gamma'_x(y_\infty)| = \lim_{z_\infty \to y_\infty} \frac{d_x(\gamma(y_\infty), \gamma(z_\infty))}{d_x(y_\infty, z_\infty)} \tag{4.8.14}$$

for $x \in B$, $y_\infty \in S$ and $\gamma \in \Gamma$.

Lemma 4.8.8. *If*

$$p(x, y_\infty) = (1 - |x|^2)/|x - y_\infty|^2, \quad x \in B,\ y_\infty \in S,$$

is the Poisson kernel, then

$$|\gamma'_x(y_\infty)| = \frac{p(\gamma^{-1}(x), y_\infty)}{p(x, y_\infty)}, \tag{4.8.15}$$

where $|a - b|$ denotes the Euclidean distance of points a and b of $\mathbb{R}^n$.

Proof. If a Möbius transformation $\tau \in M(B)$ maps x to the origin and τ' is its differential, then, for any $z \in \bar{B}$, $\tau'(z)$ is a conformal matrix, so $\tau'(z) = |\tau'(z)| \cdot A(z)$ for some orthogonal matrix $A(z)$ and some scalar coefficient $|\tau'(z)|$.

It is not too hard to show that

$$|\eta(z) - \eta(y)| = |\eta'(z)|^{1/2} \cdot |\eta'(y)|^{1/2} \cdot |z - y| \tag{4.8.16}$$

for all y and z of $\bar{B}$ and any $\eta \in M(B)$. (In fact, this is a matter of a straightforward calculation when η is the inversion with respect to the unit sphere and can be derived for any η by the Chain Rule.) In particular,

$$|\tau(z_\infty) - \tau(y_\infty)| = |\tau'(y_\infty)|^{1/2} \cdot |\tau'(z_\infty)|^{1/2} \cdot |z_\infty - y_\infty|$$

and

$$|(\tau \circ \gamma)(y_\infty) - (\tau \circ \gamma)(z_\infty)) = |(\tau \circ \gamma)'(y_\infty)|^{1/2} \cdot |(\tau \circ \gamma)'(z_\infty)|^{1/2} \cdot |z_\infty - y_\infty|.$$

Since, by an elementary geometric argument, the quotient

$$d_x(y_\infty, z_\infty)/|\tau(z_\infty) - \tau(y_\infty)|$$

converges to 1 when $z_\infty \to y_\infty$ and the same holds for the points $\gamma(y_\infty)$ and $\gamma(z_\infty)$, it follows that

$$|\gamma'_x(y_\infty)| = \frac{|(\tau \circ \gamma)'(y_\infty)|}{|\tau'(y_\infty)|}.$$

Since τ maps x to the origin, it can be expressed as the composition of a Euclidean isometry with the inversion ι with respect to the sphere of radius $r = (1-|x|^2)^{1/2}/|x|$ centred at $a = x/|x|^2$. Therefore, $|\tau'(y_\infty)| = |\iota'(y_\infty)|$ and again this is a matter of a straightforward calculation to derive the formula

$$|\iota'(y_\infty)| = p(x, y_\infty)$$

from the equation

$$\iota(z) = a^* + (|a^* - 1)(z - a^*)^*,$$

where $u^* = u/|u|^2$ is the image of u under the inversion with respect to the unit sphere.

The same argument applied to $\tau \circ \gamma$ yields (4.8.15). □

Now, we are in a position to prove

Theorem 4.8.9. *For any Borel set $E \subset S$, any $x \in B$ and $\gamma \in \Gamma$ the equality*

$$\mu_{\gamma(x),y}(E) = \int_E |\gamma_x'(z_\infty)|^\delta d\mu_{x,y}(z_\infty) \tag{4.8.17}$$

holds.

Proof. Formula (4.8.2) defining the hyperbolic distance d implies the equality

$$\lim_{z \to z_\infty} \frac{e^{-d(x',z)}}{e^{-d(x,z)}} = \frac{p(x', z_\infty)}{p(x, z_\infty)}.$$

Also, property (4.8.8) of the function k in Lemma 4.8.5 implies that for t in any bounded interval $I \subset \mathbb{R}$,

$$k(e^{r+t})/k(e^r) \to 1$$

uniformly when $r \to \infty$. These observations and equation (4.8.9) defining Patterson–Sullivan measures imply that

$$\lim_{t \to 0} \frac{\mu_{s,x,y}(E(t))}{\mu_{s,x',y}(E(t))} = \left(\frac{p(x, y_\infty)}{p(x', y_\infty)}\right)^s,$$

where $E(t)$ is the set of all the points $z_\infty \in E$ within the Euclidean distance t from y_∞ and $x' = \gamma^{-1}(x)$. More precisely, for any $\varepsilon > 0$ there exists $t(\varepsilon) > 0$ such that

$$(1-\varepsilon)\mu_{s,x,y}(E(t)) \left[\left(\frac{p(x, y_\infty)}{p(x', y_\infty)}\right)^s - \varepsilon\right] \le \mu_{s,x',y}(E(t))$$
$$\le (1+\varepsilon)\mu_{s,x,y}(E(t)) \left[\left(\frac{p(x, y_\infty)}{p(x', y_\infty)}\right)^s + \varepsilon\right]$$

for all $t \in (0, t(\varepsilon))$ and $s \in (\delta, \delta+1)$. If (s_j) is such a sequence that $\mu_{s_j,x,y} \to \mu_{x,y}$ and $\mu_{s_j,x',y} \to \mu_{x',y}$, then the above inequalities written for s_j in place of s yield, after passing to the limit as $j \to \infty$,

$$(1-\varepsilon)\mu_{x,y}(E(t)) \left[\left(\frac{p(x,y_\infty)}{p(x',y_\infty)}\right)^\delta - \varepsilon\right] \le \mu_{x',y}(E(t))$$

$$\le (1+\varepsilon)\mu_{x,y}(E(t)) \left[\left(\frac{p(x,y_\infty)}{p(x',y_\infty)}\right)^\delta + \varepsilon\right].$$

This shows that the measures $\mu_{x,y}$ and $\mu_{x',y}$ are absolutely continuous each with respect to the other and that the Radon–Nikodym derivative $d\mu_{x',y}/d\mu_{x,y}$ is given by

$$\frac{d\mu_{x',y}}{d\mu_{x,y}}(z_\infty) = \left(\frac{p(x',z_\infty)}{p(x,z_\infty)}\right)^\delta. \tag{4.8.18}$$

Clearly, (4.8.17) follows from (4.8.15) and (4.8.18). □

Proposition 4.8.7 and the theorem above imply immediately the following.

Corollary 4.8.10. *For any Kleinian group* Γ *there exist Patterson–Sullivan measures* μ *satisfying condition* (4.8.5) *with* $\delta > 0$ *being the critical exponent of the Poincaré series of* Γ. □

We shall complete this section by proving some ergodicity property of Patterson–Sullivan measures. To this end, we need the following

Definition 4.8.11. A limit point $x_\infty \in \Lambda$ is called *conical* when the quotients

$$\frac{|x_\infty - \gamma_j(x)|}{1-|\gamma_j(x)|}, \quad x \in B, \tag{4.8.19}$$

remain bounded as $j \to \infty$ for some $\gamma_j \in \Gamma$ such that $\gamma_j(x) \to x_\infty$. All the conical limit points form the *conical limit set* Λ_c of the group Γ.

Clearly, Λ_c is Γ-invariant. The condition (4.8.19) can be expressed more geometrically in the following way: there exists a Euclidean cone C with vertex at x_∞ such that $C \subset B$ and almost all points $\gamma_j(x)$ belong to C.

Recall also (Section 1.5) that finite groups and finite extensions of a cyclic group $\mathbb{Z}$ are called elementary.

Theorem 4.8.12. *If* Γ *is non-elementary, then any Patterson–Sullivan measure* $\mu_{x,y}$ *is ergodic on* Λ_c, *i.e., if a Borel set* $E \subset \Lambda_c$ *is* Γ*-invariant, then either* $\mu_{x,y}(E) = 0$ *or* $\mu_{x,y}(E) = \mu_{x,y}(\Lambda_c)$.

Proof. Without loss of generality, we may assume that $x = y = 0$. Let us write μ_0 instead of $\mu_{0,0}$.

Suppose that a set $E \subset \Lambda_c$ is Γ-invariant and $\mu_0(E) > 0$. We have to prove that

$$\mu_0(E) = \mu_0(\Lambda_c).$$

We split the argument into two parts. In the first one, we show that

$$\mu_0(E) \geq \mu_0(\Lambda_c) - m_0, \tag{4.8.20}$$

where $m_0 = \max\{\mu_0(\{z_\infty\}); z_\infty \in \Lambda\}$ is the largest mass of an atom of μ_0. In the second, we show that

$$m_0 = 0,$$

that is μ_0 has no atoms at all.

Hereafter, $D(x, y, r)$ ($x, y \in B$, $r \in \mathbb{R}_+$) denotes the hyperbolic shadow of a hyperbolic ball $B(y, r)$ observed on S by an eye situated at x. That is, $D(x, y, r)$ is the ball in the metric space (S, d_x) consisting of all the points $z_\infty \in S$ for which the hyperbolic geodesic joining x to z_∞ intersects the ball $B(y, r)$.

Part 1. Let x_∞ be the density point of E and (γ_j) a sequence of elements of Γ for which $\gamma_j(0) \to x_\infty$ and the quotients in (4.8.19) stay bounded. Let $C \subset \bar{B}$ be a Euclidean cone at x_∞ containing almost all the points $\gamma_j(0)$, $j \in \mathbb{N}$.

By definition of density points, given $r > 0$,

$$\frac{\mu_0(E \cap D(0, \gamma_j(0), r))}{\mu_0(D(0, \gamma_j(0), r))} \to 1 \tag{4.8.21}$$

as $j \to \infty$.

Fix $\varepsilon > 0$. There exists $j_0 \in \mathbb{N}$ and $r_0 \in \mathbb{R}_+$ such that for all $j > j_0$ and $r > r_0$ the inequality

$$\mu_0(D(\gamma_j^{-1}(0), r)) > \mu_0(\Lambda) - m_0 - \varepsilon,$$

holds. In fact, otherwise we were able to find sequences (j_k) and (r_k) converging to infinity and such that

$$\mu_0(\Lambda \setminus D(\gamma_{j_k}^{-1}(0), 0, r_k)) \geq m_0 + \varepsilon$$

for all k. Since $r_k \to \infty$, the diameters of the sets $\Lambda \setminus D(\gamma_{j_k}^{-1}(0), 0, r_k)$ converge to 0 and these sets are contained in some balls $B_k = B(*, \eta_k) \subset S$, where $\eta_k \to 0$ as $k \to \infty$. Passing to a subsequence if necessary, we may assume that the balls B_k are descending to a point z_∞ of S. For this point we obtain

$$\mu_0(\{z_\infty\}) = \lim_{k \to \infty} \mu_0(B_k) \geq m_0 + \varepsilon,$$

a contradiction.

Since $D(\gamma_j^{-1}(0), 0, r) = \gamma_j^{-1}(D(0, \gamma_j(0), r)$ and E is Γ-invariant, we obtain from (4.8.18) the equalities

$$\begin{aligned}\frac{\mu_0(D(\gamma_j^{-1}(0),0,r)\cap E)}{\mu_0(D(\gamma_j^{-1}(0),0,r))} &= \frac{\mu_{\gamma_j(0)}(D(0,\gamma_j(0),r)\cap E)}{\mu_{\gamma_j(0)}(D(0,\gamma_j(0),r))} \\ &= \frac{\int_{D(0,\gamma_j(0),r)\cap E} p(\gamma_j(0),y_\infty)^\delta d\mu_0(y_\infty)}{\int_{D(0,\gamma_j(0),r)} p(\gamma_j(0),y_\infty)^\delta d\mu_0(y_\infty)} \\ &= 1 - \frac{\int_{D(0,\gamma_j(0),r)\setminus E} p(\gamma_j(0),y_\infty)^\delta d\mu_0(y_\infty)}{\int_{D(0,\gamma_j(0),r)} p(\gamma_j(0),y_\infty)^\delta d\mu_0(y_\infty)}.\end{aligned}$$

An elementary calculation shows that for any $r>0$ there exists a constant $a>0$ such that

$$a < p(x,y_\infty)\cdot(1-|x|) < 2 \tag{4.8.22}$$

whenever $y_\infty \in D(0,x,r)$ and $x \in C$. Moreover, the upper bound in (4.8.22) holds trivially for all x and y_∞. Applying these estimates to the integrals above we obtain the inequality

$$\frac{\mu_0(D(\gamma_j^{-1}(0),0,r)\cap E)}{\mu_0(D(\gamma_j^{-1}(0),0,r))} \geq 1 - \left(\frac{2}{a}\right)^\delta \cdot \frac{\mu_0(D(0,\gamma_j(0),r)\setminus E)}{\mu_0(D(0,\gamma_j(0),r))} \geq 1-\varepsilon$$

for j and r large enough. The last inequality follows from (4.8.21). Consequently,

$$\begin{aligned}\mu_0(E) &\geq \mu_0(D(\gamma_j^{-1}(0),0,r)\cap E) \geq (1-\varepsilon)\mu_0(D(\gamma_j^{-1}(0),0,r)) \\ &\geq (1-\varepsilon)(\mu_0(\Lambda) - m_0 - \varepsilon).\end{aligned}$$

Letting $\varepsilon \to 0$ we obtain (4.8.20).

Part 2. To prove that μ_0 has no atoms, observe first that it cannot have exactly one atom. If so, the atom, say $z_\infty \in S$, would be fixed by all $\gamma \in \Gamma$, so Γ would be elementary. If μ_0 has at least two atoms, then, by (4.8.20), some of them lie in E. So, Λ_c would contain an atom of μ_0. To complete the proof we shall show that this is not the case.

For this purpose, it is more convenient to work with the upper half space model of the hyperbolic space. Let $\mathbb{R}^n_+ = \{(x_1,\ldots,x_n); x_n > 0\}$ and η be a Möbius transformation mapping B onto $\mathbb{R}^n_+$. Then, η maps $S = \partial B$ onto $\mathbb{R}^{n-1} \cup \{\infty\}$, where $\mathbb{R}^{n-1}$ is identified with the set of all $(x_1,\ldots,x_n) \in \mathbb{R}^n$ for which $x_n = 0$. Conjugating elements of Γ with η we get a group, denoted by Γ again, of transformations of $\mathbb{R}^n_+$. The Poisson kernel p assumes the form

$$p(x,\infty) = x_n \quad \text{and} \quad p(x,y_\infty) = x_n/|x-y_\infty|^2,$$

where $x = (x_1,\ldots,x_n) \in \mathbb{R}^n_+$ and $y_\infty \in \mathbb{R}^{n-1}$.

Suppose that μ_0 has a conical atom. Without loss of generality we may assume that η maps it onto ∞.

First, consider the stabilizer Γ_∞ of ∞: $\gamma \in \Gamma_\infty$ iff $\gamma \in \Gamma$ and $\gamma(\infty) = \infty$. If $\gamma \in \Gamma_\infty$, then for any $x \in \mathbb{R}^n_+$ the measure $\mu_x = \mu_{x,\infty}$ satisfies $\mu_x(\{\gamma^{-1}(\infty)\}) = \mu_x(\{\infty\})$ and $\mu_x(\{\infty\}) = \mu_{\gamma(x)}(\{\infty\})$. The transformation rule (4.8.18) for Patterson–Sullivan measures yields

$$\frac{y_n}{x_n} = \frac{p(\gamma(x), \infty)}{p(x, \infty)} = \left(\frac{d\mu_{\gamma(x)}}{d\mu_x}(\infty)\right)^{1/\delta} = 1,$$

where $y = (y_1, \ldots, y_n) = \gamma(x)$ and $x \in \mathbb{R}^n_+$. This shows that the elements of Γ_∞ preserve the hyperplanes $x_n =$ const. (that is, the horospheres at ∞ of our hyperbolic space).

Next, consider the quotient $\Delta = \Gamma/\Gamma_\infty$ and the series

$$\sum_{[\gamma]\in\Delta} ([\gamma(y)]_n)^\delta, \tag{4.8.23}$$

where $[z]_n$ denotes the n-th coordinate of $z \in \mathbb{R}^n$. The quantity in (4.8.23) is well defined since — as was observed above — elements of Γ_∞ preserve horizontal hyperplanes. Moreover, series (4.8.23) converges. In fact, since Patterson–Sullivan measures are finite and points $\gamma^{-1}(\infty)$ for different classes $[\gamma]$ in Δ are different, we have

$$\sum_{[\gamma]\in\Delta} \mu_x(\{\gamma^{-1}(\infty)\}) < \infty \quad \text{and} \quad \sum_{[\gamma]\in\Delta} \mu_{\gamma(x)}(\{\infty\}) < \infty.$$

The last inequality and the transformation rule (4.8.17) yield the inequality

$$\sum_{[\gamma]\in\Delta} \left(\frac{p(\gamma(x), \infty)}{p(x, \infty)}\right)^\delta \cdot \mu_x(\{\infty\}) < \infty$$

which is equivalent to the convergence of (4.8.23).

Finally, since ∞ is a conical limit point, we can find a sequence (γ_j) of elements of Γ and a point $y \in \mathbb{R}^n_+$ such that $[\gamma_j(y)]_n \nearrow \infty$. Since $[\gamma(y)]_n = [y]_n$ when $\gamma \in \Gamma_\infty$, all the cosets $[\gamma_j]$, $j \in \mathbb{N}$, are different and

$$\sum_{[\gamma]\in\Delta} ([\gamma(y)]_n)^\delta \geq \sum_{j=1}^{m} ([\gamma_j(y)]_n)^\delta \to \infty$$

as $m \to \infty$, a contradiction.

This completes the proof of Theorem 4.8.12. □

Some application of Patterson–Sullivan measures will be given in the next chapter (Section 5.5), where we shall deal with the Hausdorff dimension of the limit set Λ.

Let us complete this section with the following remark. As far as we know, the existence of Patterson–Sullivan measures on boundaries of leaves in compact foliated manifolds is an open question. Certainly, a suitable notion of a leaf boundary is needed for a reasonable formulation of this question. "Good" boundaries ∂X of complete metric spaces X should allow homeomorphic extension (to ∂X) of any quasi-isometry of X, therefore ∂X should depend (up to a homeomorphism) on the quasi-isometry class of X only. In the literature, there exists the notion of Higson [**Hig**] boundary which has been shown to be "good" for complete open Riemannian manifolds and adapted (by Higson and Roe [**HR**]) to complete metric spaces. This idea can be applied to the leaves of foliations of compact manifolds. It can be explained in terms of K-theory and certain C*-algebras but this goes far beyond the limits put by the author in this book. A reader interested in these ideas should consult [**Hu3**] and references there.

Chapter 5

Hausdorff dimension

Since the late 1970s of last century, a large interest in geometry of fractals can be observed. Fractals can be defined in different ways with the use of various notions of dimensions, usually as sets or spaces with non-integer dimension. The most classical notion of dimension is due to Hausdorff [**Hau**], see [**Ed**] and [**Fa**] for a modern exposition. Here, we calculate or estimate the Hausdorff dimension of several fractal-like sets which appear, for instance, as limit sets of group actions (Section 5.5) or exceptional minimal sets on foliated manifolds.

The main results concerning dimensions in the context of foliated manifolds are presented in Section 5.3, where we show that:

(i) the Hausdorff dimension of the Markov invariant set of any C^2-Markov system on $\mathbb{R}$ is always strictly less than 1 (Theorem 5.3.5) and

(ii) the transverse Hausdorff dimension (defined in Section 5.1 of this Chapter) of a codimension-one C^2-foliation $\mathcal{F}$ is equal to the standard Hausdorff dimension of the intersection of a complete transversal T and the union $C(\mathcal{F})$ of all the compact leaves provided that the leaves of $\mathcal{F}$ are not dense (Theorem 5.3.10).

Since the present interest in fractal geometry came from the theory of complex dynamical systems, we decided to devote some pages to Hausdorff dimension of Julia sets, subsets of the Riemann sphere $\mathbb{C}_\infty$ which carry the major part of dynamics of complex transformations of $\mathbb{C}_\infty$ (Section 5.2) even if these pages are the only ones in this book which are related to complex dynamics.

5.1 Definitions and basic facts

Let (X,d) be a metric space. For any $\varepsilon > 0$ denote by $\mathcal{C}_\varepsilon(X)$ the family of all countable covers of X by sets of diameter $\leq \varepsilon$. For any $s > 0$ and a countable cover $\mathcal{A}$ of X put

$$H_s(\mathcal{A}) = \sum_{A\in\mathcal{A}} (\operatorname{diam} A)^s, \quad H^s_\varepsilon(X) = \inf\{H_s(\mathcal{A});\ \mathcal{A} \in \mathcal{C}_\varepsilon(X)\} \tag{5.1.1}$$

and

$$H^s(X) = \lim_{\varepsilon \to 0^+} H^s_\varepsilon(X). \tag{5.1.2}$$

By convention, $\inf \emptyset = +\infty$. Clearly, the limit in (5.1.2) exists and $H^s(X) \in [0, +\infty]$.

The function $X \supset Z \mapsto H^s(Z)$ is an outer measure on X. In $X = \mathbb{R}^1$, H^1 coincides with the 1-dimensional Lebesgue measure. In $X = \mathbb{R}^n$, H^n is proportional (by a constant factor) to the n-dimensional Lebesgue measure. Clearly, $H^s(X) = 0$ for all $s > 0$ when X is either finite or countable.

The following fact is crucial for defining the Hausdorff dimension.

Lemma 5.1.1. *If $0 < H^t(X) < \infty$ and $s < t < u$, then $H^s(X) = +\infty$ and $H^u(X) = 0$.*

Proof. If $s < t < u$, then for any cover $\mathcal{A} \in \mathcal{C}_\varepsilon(X)$ we have

$$\varepsilon^{t-s} \sum_{A \in \mathcal{A}} (\operatorname{diam} A)^s \geq \sum_{A \in \mathcal{A}} (\operatorname{diam} A)^t \geq \varepsilon^{t-u} \sum_{A \in \mathcal{A}} (\operatorname{diam} A)^u.$$

This implies the inequalities

$$\varepsilon^{t-s} H^s_\varepsilon(X) \geq H^t_\varepsilon(X) \geq \varepsilon^{t-u} H^u_\varepsilon(X).$$

Passing to limits as $\varepsilon \to 0^+$ yields the statement. □

Definition 5.1.2. The *Hausdorff dimension* $\dim_H X$ of a metric space (X, d) is given by

$$\dim_H X = \inf\{s; H^s(X) = 0\} = \sup\{s; H^s(X) = +\infty\}. \tag{5.1.3}$$

From the previous remarks it follows that $\dim_H U = n$ when U is a non-empty open subset of $\mathbb{R}^n$ (or, of any n-dimensional manifold M). Obviously, $\dim_H X = 0$ when X is either finite or countable and $\dim_H Y \leq \dim_H X$ when Y is a subspace of X. Moreover, $\dim_H X = \max\{\dim_H Y_1, \dim_H Y_2\}$ when $X = Y_1 \cup Y_2$.

Proposition 5.1.3. *If $f : X \to Y$ is a Lipschitz map, then*

$$\dim_H f(Z) \leq \dim_H Z \tag{5.1.4}$$

for any $Z \subset X$.

Proof. If $c > 0$ is a Lipschitz constant for f and $\mathcal{A} \in \mathcal{C}_\varepsilon(Z)$, then $f(\mathcal{A}) = \{f(A); A \in \mathcal{A}\} \in \mathcal{C}_{c\varepsilon}(f(Z))$ and $\operatorname{diam} f(A) \leq c \cdot \operatorname{diam} A$ for any $A \in \mathcal{A}$. It follows that $H^s_{c\varepsilon}(f(Z)) \leq H^s_\varepsilon(Z)$ for all s and $\varepsilon > 0$. Consequently, for any $s > \dim_H Z$ we have $H^s(Z) = 0$, $H^s(f(Z)) = 0$ and $s \geq \dim_H f(Z)$. □

Corollary 5.1.4. *If two metric spaces X and Y are Lipschitz homeomorphic, then $\dim_X H = \dim_H Y$.* □

Note that Hausdorff dimensions of quasi-isometric spaces may differ. For instance, any compact metric space is quasi-isometric to a single point but it can have Hausdorff dimension as large as we want (including $+\infty$).

As we could observe while proving Proposition 5.1.3 it is relatively easy to estimate the Hausdorff dimension from above. To this end, it is sufficient to find covers by sets of small diameters and to estimate the suitable sum in (5.1.1). Estimating Hausdorff dimension from below is, in general, more difficult but it becomes easier when we can find a probability measure with good properties.

Definition 5.1.5. A Borel probability measure μ on X is said to be *s-continuous* ($s > 0$) when there exist numbers $r_0 > 0$ and $c > 0$ such that

$$\mu(B(x,r)) \leq cr^s \tag{5.1.5}$$

for any $x \in X$ and $r \in (0, r_0)$.

Clearly, the Lebesgue measure on $\mathbb{R}^n$ (and on any Riemannian manifold) is n-continuous, s-continuous measures have no atoms and s-continuous measures are t-continuous for all $t \leq s$.

Proposition 5.1.6. *If (X,d) admits an s-continuous Borel probability measure μ, then* $\dim_H X \geq s$.

Proof. If $\varepsilon > 0$ is small enough and $\mathcal{A} \in \mathcal{C}_\varepsilon(X)$, then

$$1 \leq \sum_{A \in \mathcal{A}} \mu(A) \leq c \cdot \sum_{A \in \mathcal{A}} (\operatorname{diam} A)^s.$$

Therefore, $H^s_\varepsilon(X) \geq c^{-1}$, $H^s(X) \geq c^{-1} > 0$ and $\dim_H X \geq s$. □

The following example is very classical and well known but it is so instructive that we decided to describe it scrupulously.

Example 5.1.7. Let $Z \subset [0,1]$ be the classical ("triadic") Cantor set considered here already in Section 1.4 (Example 1.4.4 (b)). For any $n \in \mathbb{N}$ we can cover Z by 2^n sets of the form $I_{k,n} = Z \cap [k3^{-n}, (k+1)3^{-n}]$ and of diameter $\varepsilon_n = 3^{-n}$. Therefore, $H^s_{\varepsilon_n}(Z) \leq 2^n \cdot 3^{-ns}$ and $H^s(Z) = 0$ when $s > \log 2/\log 3$. This shows that $\dim_H Z \leq \log 2/\log 3$. On the other hand, let μ_n be the Borel probability measure on $[0,1]$ equal to the sum of all Lebesgue measures on the intervals $I_{k,n}$ multiplied by the factor $(\frac{3}{2})^n$. That is, $\mu_n(A) = \sum_k (\frac{3}{2})^n \cdot \lambda(A \cap I_{k,n})$ for any Borel set $A \subset [0,1]$, where λ is the Lebesgue measure on the real line $\mathbb{R}$. The sequence (μ_n) converges to a probability measure μ on $[0,1]$ which is supported in Z. If $r = 3^{-n}$ and $x \in Z$, then the Z-ball $B = B(x,r)$ is contained in one of the components $I_{k,n}$. Similarly, B is contained in the union of 2^m components $I_{k,n+m}$.

Therefore, $\mu_{n+m}(B) \leq (\frac{3}{2})^{n+m} \cdot 2^m \cdot 3^{-(n+m)} = r^{\log 2/\log 3}$ for all $m \geq 0$ and $\mu(B) \leq r^{\log 2/\log 3}$. Finally, if $3^{-(n+1)} \leq r \leq 3^{-n}$, then $\mu(B(x,r) \leq \mu(B(x,3^{-n}) \leq 3^{\log 2/\log 3} \cdot r^{\log 2/\log 3}$. This shows that the measure μ is $(\log 2/\log 3)$-continuous and implies the inequality $\dim_H Z \geq \log 2/\log 3$. Consequently,

$$\dim_H Z = \frac{\log 2}{\log 3}.$$

The argument used in the above example can be generalized to arbitrary subsets of $\mathbb{R}^n$ which are self-similar and satisfy the so-called Moran's condition. If $X \subset \mathbb{R}^n$ can be decomposed as $X = \bigcup_{i=1}^k f_i(X)$, where $f_i : \mathbb{R}^n \to \mathbb{R}^n$ is a similarity with ratio $r_i \in (0,1)$, then $\dim_H X \leq s$, where s is the unique solution of the equation

$$\sum_{i=1}^{k} r_i^s = 1. \tag{5.1.6}$$

If, moreover, the similarities $f_1, \ldots, f_k$ satisfy Moran's condition

(*) there exists a non-empty open set U such that $f_i(U) \subset U$ and $f_i(U) \cap f_j(U) = \emptyset$ for all i and $j \leq k$, $j \neq i$,

then $\dim_H X = s$.

This allows us to compute dimensions of several classical fractal sets such as, among many others, Koch curve K, Sierpiński gasket S and Menger sponge M (see, for example, [**Ed**] and Figure 5.1 below): $K = f_1(K) \cup f_2(K)$ with $r_1 = r_2 = 1/\sqrt{3}$, so $\dim_H K = 2\log 2/\log 3$; $S = f_1(S) \cup f_2(S) \cup f_3(S)$ with $r_1 = r_2 = r_3 = 1/2$, so $\dim_H S = \log 3/\log 2$; $M = \bigcup_{i=1}^{20} f_i(M)$ with $r_1 = \cdots = r_{20} = 1/3$, so $\dim_H M = \log 20/\log 3$.

Figure 5.1: Fractals: a) Koch curve, b) Sierpiński gasket, c) Menger sponge.

Now, let us consider a finitely generated pseudogroup $\mathcal{G}$ of locally Lipschitz homeomorphisms between open subsets of a compact metric space (X,d). For any $\varepsilon > 0$ denote by $\mathcal{C}_\varepsilon(\mathcal{G})$ the set of all finite generating (not necessarily symmetric) sets $\mathcal{G}_1$ consisting of maps g with domains D_g of diameter $\leq \varepsilon$. Analogously to (5.1.1) and (5.1.2) put

$$H_\varepsilon^s(\mathcal{G}) = \inf\{\sum_{g \in \mathcal{G}_1} (\operatorname{diam} D_g)^s ; \mathcal{G}_1 \in \mathcal{C}_\varepsilon(\mathcal{G})\} \tag{5.1.7}$$

and

$$H^s(\mathcal{G}) = \lim_{\varepsilon \to 0+} H^s_\varepsilon(\mathcal{G}). \tag{5.1.8}$$

The function $s \to H^s(\mathcal{G})$ has a property similar to that described in Lemma 5.1.1: if $0 < H^t(\mathcal{G}) < \infty$ and $s < t < u$, then $H^t(\mathcal{G}) = +\infty$ and $H^u(\mathcal{G}) = 0$. Therefore, we may accept the following.

Definition 5.1.8. [**Wal2**] The *Hausdorff dimension* $\dim_H \mathcal{G}$ of a pseudogroup $\mathcal{G}$ is given by

$$\dim_H \mathcal{G} = \inf\{s; H^s(\mathcal{G}) = 0\} = \sup\{s; H^s(\mathcal{G}) = +\infty\}. \tag{5.1.9}$$

Comparing the above definitions of Hausdorff dimensions for spaces and pseudogroups one can easily see that $\dim_H X = \dim_H \mathcal{G}(\mathrm{id}_X)$ for any X. Before providing more examples we shall prove some basic properties of the dimension of pseudogroups.

Proposition 5.1.9. (i) *If* $\mathcal{G} \subset \mathrm{Homeo}\, X$, *then* $\dim_H \mathcal{G} \leq \dim_H X$.

(ii) *If* $\mathcal{G}'$ *is a subpseudogroup of* $\mathcal{G}$, *then* $\dim_H \mathcal{G}' \geq \dim_H \mathcal{G}$.

(iii) *If* $Y \subset X$ *is* $\mathcal{G}$*-invariant, then* $\dim_H \mathcal{G}|Y \leq \dim_H \mathcal{G}$.

Proof. (i) Let $\mathcal{G}_1 = \{g_1, \dots, g_k\}$ be any finite set generating $\mathcal{G}$. For any cover $\mathcal{A} \in \mathcal{C}_\varepsilon(X)$ put $\mathcal{G}_1|\mathcal{A} = \{g_i|A; A \in \mathcal{A}, i \leq k\}$. Without losing generality we may assume that all the members of $\mathcal{A}$ are open. Then, $\mathcal{G}_1|\mathcal{A} \in \mathcal{C}_\varepsilon(\mathcal{G})$ and

$$\sum_{g \in \mathcal{G}_1|\mathcal{A}} (\mathrm{diam}\, D_g)^s \leq k \cdot \sum_{A \in \mathcal{A}} (\mathrm{diam}\, A)^s.$$

This implies the inequality $H^s_\varepsilon(\mathcal{G}) \leq k \cdot H^s_\varepsilon(X)$ and yields the required relation between Hausdorff dimensions of $\mathcal{G}$ and X.

(ii) The proof is analogous to that of (i). However, one has to replace $\mathcal{A}$ by $\mathcal{G}'_1 \in \mathcal{C}_\varepsilon(\mathcal{G}')$, assume that $\mathrm{id}_X \in \mathcal{G}_1$, put $\mathcal{G}''_1 = \{g \circ g'; g \in \mathcal{G}_1, g' \in \mathcal{G}'_1\}$ and observe that $\mathcal{G}''_1$ generates $\mathcal{G}$. Since $D_{g \circ g'} \subset D_{g'}$, $\mathrm{diam}\, D_{g \circ g'} \leq \mathrm{diam}\, D_{g'}$, $\mathcal{G}''_1 \in \mathcal{C}_\varepsilon(\mathcal{G})$ and

$$\sum_{g \in \mathcal{G}''_1} (\mathrm{diam}\, D_g)^s \leq k \cdot \sum_{g \in \mathcal{G}'_1} (\mathrm{diam}\, D_g)^s.$$

As before, this yields the inequality $H^s_\varepsilon(\mathcal{G}) \leq k \cdot H^s_\varepsilon(\mathcal{G}')$ and the required relation between the dimensions.

(iii) If $\mathcal{G}_1 \in \mathcal{C}_\varepsilon(\mathcal{G})$, then $\mathcal{G}_1|Y = \{g|Y; g \in \mathcal{G}_1\} \in \mathcal{C}_\varepsilon(\mathcal{G}|Y)$ and $\mathrm{diam}\, D_{g|Y} \leq \mathrm{diam}\, D_g$ for any $g \in \mathcal{G}_1$. □

The following result is analogous to Proposition 5.1.6.

Proposition 5.1.10. *If* X *carries a* $\mathcal{G}$*-invariant* s*-continuous measure* μ, *then*

$$\dim_H \mathcal{G} \geq s.$$

Proof. If $\varepsilon > 0$ is small enough, μ satisfies (5.1.5) and $\mathcal{G}_1 \in \mathcal{C}_\varepsilon(\mathcal{G})$, then the condition

$$X = \bigcup_{g \in \mathcal{G}_1} (D_g \cup R_g)$$

implies the inequalities

$$1 \leq \sum_{g \in \mathcal{G}_1} (\mu(D_g) \cup \mu(R_g)) \leq 2 \sum_g \mu(D_g) \leq 2c \cdot \sum_g (\operatorname{diam} D_g)^s.$$

The rest is immediate. □

Now, let us collect more simple examples.

Example 5.1.11. (i) If $X = S^1$, $f(z) = z^2$ and $\mathcal{G} = \mathcal{G}(f)$, then for any $\varepsilon > 0$ there exist open arcs $U_1, \dots, U_4$ of length $\leq \varepsilon$ and a number $k \in \mathbb{N}$ such that the set $\{f^k|U_i, f^{k+1}|U_i; i = 1, \dots, 4\}$ generates $\mathcal{G}$. Therefore, $H^s_\varepsilon(\mathcal{G}) \leq 8\varepsilon^s$ and $H^s(\mathcal{G}) = 0$ for any $s > 0$. This shows that $\dim_H \mathcal{G} = 0$.

(ii) If $\mathcal{G} = \mathcal{G}(\Gamma)$, $\Gamma = \{f_1, \dots, f_m\}$ being a finite group of Lipschitz homeomorphisms, then $\dim_H \mathcal{G} = \dim_H X$. In fact, if $\mathcal{A}_k$, $k = 1, \dots, m$, are open coverings of X and $\mathcal{G}_1 = \{f_k|U; U \in \mathcal{A}_k\} \in \mathcal{C}_\varepsilon(\mathcal{G})$, then all the sets of the form $f_j(U)$, where $U \in \mathcal{A}_k$, $j \leq m$ and $k \leq m$, cover X and

$$\begin{aligned}\sum_{g \in \mathcal{G}_1} (\operatorname{diam} D_g)^s &= \sum_{k=1}^m \sum_{U \in \mathcal{A}_k} (\operatorname{diam} U)^s \geq \frac{1}{mc^s} \sum_{j,k=1}^m \sum_{U \in \mathcal{A}_k} (\operatorname{diam} f_j(U))^s \\ &\geq \frac{1}{mc^s} H^s_{c\varepsilon}(X),\end{aligned}$$

where c is the maximum of Lipschitz constants for all f_k's. Therefore,

$$H^s_\varepsilon(\mathcal{G}) \geq \frac{1}{mc^s} H^s_{c\varepsilon}(X) \quad \text{and} \quad H^s(\mathcal{G}) \geq \frac{1}{mc^s} H^s(X).$$

This yields the inequality $\dim_H \mathcal{G} \geq \dim_H X$. The opposite one follows from Proposition 5.1.9 (i).

(iii) If $\mathcal{G}$ is a pseudogroup of local isometries of a compact Riemannian manifold M, then $\dim_H \mathcal{G} = \dim M$. In fact, $\mathcal{G}$ preserves the Riemannian volume element on M and $\operatorname{vol} B(x, r) \leq c r^{\dim M}$ for some $c > 0$, all $x \in M$ and all $r > 0$ sufficiently small.

Finally, let us prove an analogue of Corollary 5.1.4.

Proposition 5.1.12. *If finitely generated pseudogroups $\mathcal{G}$ on X and $\mathcal{H}$ on Y are Lipschitz isomorphic, then $\dim_H \mathcal{G} = \dim_H \mathcal{H}$.*

Proof. Let $\Phi : \mathcal{G} \to \mathcal{H}$ be an isomorphism consisting of Lipschitz maps $\phi : D_\phi \subset X \to R_\phi \subset Y$. Let $\Phi_0 \subset \Phi$ be a finite set generating Φ. If $\mathcal{G}_1 \in \mathcal{C}_\varepsilon(\mathcal{G})$, then $\mathcal{H}_1 = \{\phi \circ g \circ \psi^{-1}; g \in \mathcal{G}_1, \phi, \psi \in \Phi_0\} \in \mathcal{C}_{c\varepsilon}(\mathcal{H})$, where c is the maximum of

Lipschitz constants for all the members of Φ_0. Therefore, $H^s_\varepsilon(\mathcal{H}) \leq m^2c^s \cdot H^s_{c\varepsilon}(\mathcal{G})$ with $m = \#\Phi_0$ and this yields the inequality $\dim_H \mathcal{H} \leq \dim_H \mathcal{G}$. The converse is similar. □

Applying the above and Proposition 1.3.6 to a C^1-foliation of a compact manifold we obtain the following.

Corollary 5.1.13. *If T_1 and T_2 are complete transversals of a C^1-foliation $\mathcal{F}$ of a compact manifold M, then $\dim_H \mathcal{H}_1 = \dim_H \mathcal{H}_2$, where $\mathcal{H}_i$ is the holonomy pseudogroup of $\mathcal{F}$ acting on T_i.* □

This allows us to define the transverse dimension of $\mathcal{F}$ as the dimension of its holonomy pseudogroup.

Definition 5.1.14. The *transverse Hausdorff dimension* $\dim_H^{\pitchfork} \mathcal{F}$ of a C^1-foliation $\mathcal{F}$ of a compact manifold M equals $\dim_H \mathcal{H}$, $\mathcal{H}$ being the holonomy pseudogroup of $\mathcal{F}$ acting on any complete transversal T.

From Examples 5.1.11 (i) and (iii) it follows that the transverse Hausdorff dimension of the Hirsch foliation equals 0 while the transverse Hausdorff dimension of any Riemannian foliation $\mathcal{F}$ equals $\operatorname{codim} \mathcal{F}$. In general, $\dim_H^{\pitchfork} \mathcal{F} \leq \operatorname{codim} \mathcal{F}$. Later on, we shall establish some relations between $\dim_H^{\pitchfork} \mathcal{F}$ and the structure of $\mathcal{F}$ in the case of codimension 1.

5.2 Julia sets

As we could see in Section 5.1 it is quite easy to compute the Hausdorff dimension of self-similar sets in $\mathbb{R}^n$. Such sets appear often in the theory of dynamical systems but there are many sets important for dynamics and much more complicated ones. Calculating (or, estimating) dimensions of such sets is an important topic in the theory of dynamical systems.

One good example of such sets, a Julia set, is provided by complex dynamical systems defined by rational maps $f = P/Q$ of the Riemann sphere $\mathbb{C}_\infty = \mathbb{C} \cup \{\infty\}$. Here, P and Q are complex polynomials of one variable of degrees $n = \deg P$ and $m = \deg Q$. The number $\deg f = \max\{m, n\}$ is called the degree of f. The dynamics of rational maps of degree 1 (that is, of homographies $z \mapsto (az+b)/(cz+d)$ with $a, b, c, d \in \mathbb{C}$ satisfying $ad - bc \neq 0$) is rather simple, so we shall assume throughout this section that $\deg f \geq 2$.

Given f, $\mathbb{C}_\infty$ can be decomposed into the union of two disjoint sets, $\mathbb{C}_\infty = J_f \cup F_f$, called the *Julia set* and the *Fatou set* of f, respectively. To define them, let us recall that a point $z \in \mathbb{C}_\infty$ is said to be *periodic* when $f^k(z) = z$ for some $k \geq 1$. Such a periodic point z is called *repelling* when $|(f^k)'(z)| > 1$. Then, J_f can be defined as the closure of the set of all repelling periodic points of f. So, J_f is compact while F_f, the complement of J_f in $\mathbb{C}_\infty$, is open. Also, it is almost clear from the definition that the both sets J_f and F_f are completely f-invariant: $f(J_f) = f^{-1}(J_f) = J_f$ and $f(F_f)) = f^{-1}(F_f) = F_f$. Since the Fatou set can be defined also as the largest open subset U of $\mathbb{C}_\infty$ for which all the iterations f^k, $k \in \mathbb{N}$, are equicontinuous (i.e., for any $\varepsilon > 0$ and $z_0 \in U$ there exists $\delta > 0$ such that $|f^k(z) - f^k(z_0)| < \varepsilon$ for all $k \in \mathbb{N}$ and $z \in B(z_0, \delta)$), the most chaotic part of the dynamics of f is carried by J_f and this is why Julia sets of rational maps are of great interest. This does not mean that the behaviour of the Fatou set is of little interest. One of the most remarkable results of the 1980s in complex dynamics was due to Sullivan [**Su5**] who proved that every connected component F_0 of F_f is eventually periodic, i.e., that there exist k and l in $\mathbb{N}$ such that $f^k(f^l(F_0)) = f^l(F_0)$.

Note that the set F_f may be empty while the Julia set J_f is always non-empty (in fact, infinite and perfect, therefore uncountable) when $\deg f \geq 2$. Indeed, if $J_f = \emptyset$, then the family (f^n) is equicontinuous on the whole Riemann sphere $\mathbb{C}_\infty$ and — by the Arzelá–Ascoli Theorem — contains a subsequence (f^{n_k}) converging uniformly to a rational map f_∞. Since — by the classical Rouché Theorem — the map $f \mapsto \deg(f)$ is continuous with respect to the topology of uniform convergence,

$$(\deg f)^{n_k} = \deg(f^{n_k}) \to \deg f_\infty$$

as $k \to \infty$ and $\deg f = 1$ contradicting our assumption. Now, if J_f were finite and $z \in J_f$, then — by complete invariance of J_f — the union $\bigcup_{n\in\mathbb{Z}} f^n(\{z\})$ would be finite and — by elementary arguments — z would lie in F_f, a contradiction. Therefore, J_f is infinite. Similar arguments show that J_f contains no isolated points.

In this section, we restrict ourselves to some results about the Hausdorff dimension of the Julia set. For more results concerning the sets J_f and F_f, as well as for the detailed discussion of the structure of components of F_f leading to their classification used and discussed briefly below, we refer to [**Be2**] and several references there.

First, let us list some examples showing how different Julia sets can be.

Example 5.2.1. (i) If $f(z) = z^n$ for some $n > 1$, then $J_f = S^1$, the unit circle.

(ii) If f is the n-th Tchebychev polynomial defined by $f(\cos z) = \cos nz$, then $J_f = [-1, 1]$. In fact, $f^k(\cos z) = \cos(n^k z)$ and this implies that $f^k(w) \to \infty$ when $w \notin [-1, 1]$. Moreover, if $x = \cos y \in [-1, 1]$ is periodic, then $y = \frac{2l}{n^k \pm 1} \cdot \pi$ for some $l \in \mathbb{Z}$ and $k \in \mathbb{N}$, and $|(f^k)'(x)| > 1$ whenever $l \neq 0$ and k is sufficiently large.

(iii) If $f(z) = z^2 + az^{-3}$, where $a > 0$ is small, then J_f is the uncountable union of Jordan curves ("a Cantor set of circles").

(iv) If $f(z) = (z-2)^2/z^2$, then $F_f = \emptyset$ and $J_f = \mathbb{C}_\infty$. (For a justification of Examples (iii) and (iv) we refer to [**Be2**], Sections 11.8 and 11.9.)

The last example shows that we cannot expect any non-trivial estimate of the Hausdorff dimension of Julia sets from above. To estimate $\dim_H J_f$ from below we have to find first a point $z_0 \in F_f$ with some particular properties.

Lemma 5.2.2. *There exists z_0 in F_f such that the sets $f^{-n}(\{z_0\})$, $n \in \mathbb{N}$, accumulate on J_f and $f^n(z_0)$ is a regular value of f for any $n = 1, 2, \ldots$.*

Proof. The existence of such z_0 follows from the classification of periodic connected components of the Fatou set. If F_0 is such a component, then either

(i) F_0 is parabolic, i.e., it contains an attracting periodic point of f, or $f^{kn}(z) \to w_0$ as $n \to \infty$ for some $w_0 \in J_f$, $k \in \mathbb{N}$ and all $z \in F_0$, or

(ii) F_0 is a Siegel disc, i.e., $f|F_0$ is analytically conjugate to a rotation of a disc, or

(iii) F_0 is a Hermann ring, i.e., $f|F_0$ is analytically conjugate to a rotation of an annulus.

In all these cases, $f^{-1}(F_0)$ contains a closed disc D such that no sequence $(f^n(z))$, $z \in F_f$, accumulates on D. It follows that for any open neighbourhood U of J_f there exists $n_0 \in \mathbb{N}$ such that $f^{-n}(D) \subset U$ for all $n \geq n_0$. For, otherwise we could find a sequence $n_1 < n_2 < \cdots$ and points $z_j \in f^{-n_j}(D) \setminus U$, $j = 1, 2, \ldots$. Passing to a subsequence, if necessary, we could assume that $z_j \to z_\infty$ for some $z_\infty \in \mathbb{C}_\infty \setminus U$. Then $z_\infty \in F_f$, all the iterates f^n, $n = 1, 2, \ldots$, are equicontinuous in a neighbourhood of z_∞ and, given $\varepsilon > 0$, $|f^{n_j}(z_j) - f^{n_j}(z_\infty)| < \varepsilon$ for j large enough. Since $f^{n_j}(z_j) \in D$ for all j, this would show that the sequence $(f^n(z_\infty))$ accumulates on D, a contradiction.

Now, the union of all the orbits of critical points of f is at most countable, so D contains a point z_0 satisfying all the requirements. □

To continue, assume that $m = \deg f \geq 2$ and $\infty \in F_f$. Then, the Julia set J_f is contained in $\mathbb{C}$ and there exists $\eta > 0$ such that the η-neighbourhood $N_\eta(J_f) = \{z; d(z, J_f) < \eta\}$ of the Julia set contains no poles of f. Let

$$M = M(\eta) = \max\{|f'(z)|; z \in N_\eta(J_f)\}. \tag{5.2.1}$$

Since J_f contains a repelling periodic point w_0, $|(f^k)'(w_0)| > 1$ for some $k \geq 1$ and $M(\eta) \geq M(0) = \max_{J_f} |f'| > 1$. Choose z_0 satisfying conditions of Lemma 5.2.2. Since we may replace z_0 by any point of $f^{-j}(\{z_0\})$, we may assume that

$$\bigcup_{n=0}^{\infty} f^{-n}(\{z_0\}) \subset N_\eta(J_f). \tag{5.2.2}$$

With these assumptions and notation we have

Theorem 5.2.3. [Ga] $\dim_H J_f \geq \log m / \log M$.

Let $z_0 \in F_f$ be as above. Set

$$d_n = \operatorname{dist}(J_f, f^{-n}(\{z_0\})) \quad \text{and} \quad \mu_n = m^{-n} \cdot \sum_{z \in f^{-n}(\{z_0\})} \delta_z, \tag{5.2.3}$$

δ_z being the Dirac measure supported by z. Then (μ_n) is a sequence of probability measures on $\mathbb{C}_\infty$. It contains a converging subsequence (μ_{n_k}). Let $\mu = \lim_{k\to\infty} \mu_{n_k}$.

The theorem will follow from Proposition 5.1.6 if we prove that the measure μ is s-continuous with $s = \log m / \log M$. Since s-continuity of μ follows from s-continuity of the measures μ_n, it is sufficient to show that the measures (5.2.3) satisfy

$$\mu_n(D(z,r)) \le c \cdot r^s \tag{5.2.4}$$

for all $n \in \mathbb{N}$, $z \in \mathbb{C}_\infty$, $r > 0$ small enough and some $c > 0$. For this purpose, we shall prove two lemmas.

Lemma 5.2.4. $d_n \le M \cdot d_{n+1}$.

Proof. Take any $w \in J_f$ and $z \in f^{-(n+1)}(z_0)$. If $|w - z| \ge \eta$, then (5.2.2) yields inequalities $d_n \le \eta < M\eta \le M|w - z|$. If $|w - z| < \eta$, then the segment $[w, z]$ is entirely contained in $N_\eta(J_f)$, $|f'| \le M$ along it and $d_n \le |f(w) - f(z)| \le M|w - z|$ since $f(w) \in J_f$ and $f(z) \in f^{-n}(\{z_0\})$. □

Next, let $C = \{c_1, \ldots, c_l\}$ be the set of all the critical points of f in J_f. None of the points of C is periodic. Indeed, if $f'(c_j) = 0$, c_j would be attracting and, if periodic, would belong to the Fatou set F_f. Therefore, we can choose an integer $p > 0$ such that $f^p(C) \cap C = \emptyset$. Next, we can find $\rho > 0$ such that the open discs $D(c_j, 2\rho)$ are pairwise disjoint, $|(f^p)'| < 1$ on each of them and $d(C, f^p(C)) > 4\rho$. Let

$$J_f^\rho = J_f \setminus \bigcup_{j=1}^{l} D(c_j, \rho).$$

Since the set J_f^ρ is compact, we can choose $\tau > 0$ such that f is univalent (that is, injective) on any disc $D(z, \tau)$ with $z \in J_f^\rho$ and an integer $q \ge 1$ such that $d_0/M^q < \min\{\eta, \rho, \tau\}$.

Lemma 5.2.5. *The inequality*

$$\#[D(z, d_0/M^k) \cap f^{-n}(\{z_0\})] \le m^{n+q-k} \tag{5.2.5}$$

holds for all $z \in J_f$, $n \ge 1$ and $k \ge q$.

Proof. Since $\#f^{-n}(\{z_0\}) \le m^n$, it is clear that (5.2.5) holds for $k = q$ and all $n \ge 1$. Also, if $k > q$ and $1 \le n \le k+1$, then by Lemma 5.2.4,

$$d_0/M^{k+1} \le d_0/M^n \le d_n, \quad D(z, d_0/M^{k+1}) \cap f^{-n}(\{z_0\}) = \emptyset$$

and inequality (5.2.5) is satisfied trivially.

We shall continue by induction on k. We assume that our inequality holds for some k, all $z \in J_f$ and all $n \ge 1$ and we prove that it holds for $k+1$, all z and all $n > k+1$. We split the inductive step into two cases: (i) $z \in J_f^\rho$ and (ii) $z \in \bigcup_j D(c_j, \rho)$.

In case (i), since $d_0/M^{k+1} \le d_0/M^q < \min\{\eta, \tau\}$, f is univalent on the disc $D(z, d_0/M^{k+1})$ and $|f'| \le M$ there. It follows that f maps this disc into $D(f(z), d_0/M^k)$ and therefore, by inductive assumption,

$$\begin{aligned}\#[D(z, \frac{d_0}{M^{k+1}}) \cap f^{-n}(\{z_0\})] &\le \#[D(f(z), \frac{d_0}{M^k}) \cap f^{-(n-1)}(\{z_0\})] \\ &\le m^{(n-1)+q-k} = m^{n+q-(k+1)}.\end{aligned}$$

In case (ii), since $d_0/M^{k+1} \le d_0/M^q < \rho$, we see that $|(f^p)'| \le 1$ on the disc $D(z, d_0/M^{k+1})$ and that f^p maps this disc into $D(f^p(z), d_0/M^{k+1})$. Clearly, $\#f^{-p}(\{w\}) \le m^p$ for any $w \in D(f^p(z), d_0/M^{k+1})$. Also, $f^p(z)$ lies in J_f^ρ, so (5.2.5) holds there by case (i). It follows that

$$\begin{aligned}\#[D(z, \frac{d_0}{M^{k+1}}) \cap f^{-n}(\{z_0\})] &\le m^p \cdot \#[D(f^p(z), \frac{d_0}{M^{k+1}}) \cap f^{-(n-p)}(\{z_0\})] \\ &\le m^p m^{(n-p)+q-(k+1)} = m^{n+q-(k+1)}.\end{aligned}$$

This completes the proof. □

We can interpret (5.2.5) in terms of measures μ_n as

$$\mu_n(D(z, d_0/M^k)) \le m^q/m^k. \tag{5.2.6}$$

Passing to the limit when $n \to \infty$, we obtain (5.2.6) with μ in place of μ_n. Finally, if $d_0/M^{k+1} \le r \le d_0/M^k$, then (5.2.6) yields

$$\mu(D(z,r)) \le m^{q+1} d_0^{-s} r^s$$

with $s = \log m / \log M$. This ends the proof of Theorem 5.2.3. □

The result presented above is rather simple and general. The past 20 years have brought at least 50 papers devoted to Hausdorff dimension and related measures for Julia sets, where several more specific and deep results have been obtained. Among them, Sullivan [**Su3**] has shown that Julia sets behave quite analogously to the limit sets of Fuchsian groups and that, in some cases, the Hausdorff dimension of J_f depends analytically on the coefficients of rational maps f, a similar result was obtained by Ruelle [**Ru2**] (see [**DS**] for a short proof), Garnett [**Gar2**] provided a computer algorithm for computing $\dim_H J_f$, Kotus ([**Ko1**], [**Ko2**]) and Barański [**Bar**] received some estimates of $\dim_H J_f$ for several families of non-rational meromorphic maps f, Bodart and Zinsmeister [**BZ**] proved a continuity result for the function $c \mapsto \dim_H J_{f_c}$, f_c being the quadratic polynomial $z \mapsto z^2 + c$. An interested reader is referred also to [**DU1**], [**DU2**], [**JM**], [**Ur1**], [**Zd**] and comprehensive surveys [**EL**] and [**Ur2**].

5.3 Dimension in foliated manifolds

In this section, we are going to estimate Hausdorff dimension of Markov invariant sets and Markov exceptional minimal sets of a codimension-one foliation $\mathcal{F}$ and to show that Hausdorff dimension of the holonomy pseudogroup of such an $\mathcal{F}$ coincides with the dimension of the trace of the union of all the compact leaves of $\mathcal{F}$ (if any) on a complete transversal. These results are related to the following, old but still open question in the theory of foliations.

Is the Lebesgue measure of any exceptional minimal set of a codimension-one C^2-foliation of a compact manifold equal to zero (see **[La]**)?

Theorem 5.3.5 below shows that the answer is "yes" in the Markovian situation. In general, the answer is unknown and there exist examples of non-Markovian exceptional minimal sets (see **[CC5]**).

To begin with, take a Markov system $\mathcal{S} = \{h_1, \ldots, h_m\}$, $h_i : D_i \to R_i$, on the real line $\mathbb{R}$. Throughout this section we keep the notation used in Section 1.4. So, D_i and R_i are open intervals, each interval R_i contains a compact non-degenerated subinterval K_i, $R_i \cap R_j = \emptyset$ when $i \neq j$ and either $K_i \cap D_j = \emptyset$ or $K_i \subset Q_j = h_j^{-1}(K_j)$. Let $P = [p_{ij}]$ be the transition matrix of $\mathcal{S}$. We shall see that the dimension $\dim_H Z_0$ of the Markov invariant set Z_0 depends on the dynamics of the shift σ_P defined by P and of the strength of contraction of the maps h_i. Our first result is analogous to that for Julia sets (Theorem 5.2.3).

Theorem 5.3.1. *If P satisfies condition* (1.4.12) *of Section* 1.4 *and $\mathcal{S}$ consists of C^1-maps, then the Markov invariant set Z_0 has positive Hausdorff dimension. More precisely, if λ is the largest positive eigenvalue of P and*

$$A = \max\{|h_i'(x)|, ||(h_i^{-1})'(y)|; \ x \in Q_i, \ y \in K_i, \ i = 1, \ldots, m\}, \tag{5.3.1}$$

then

$$\dim_H Z_0 \geq \frac{\log \lambda}{\log A}. \tag{5.3.2}$$

Proof. Let us recall again that we keep the notation of Section 1.4 and define probability measures on $\mathbb{R}$ by

$$\mu_n = \frac{1}{2p_n} \cdot \sum_{x \in Z_n} \delta_x, \tag{5.3.3}$$

where $Z_n = \bigcup_{g \in \mathcal{S}_n} \partial K_g$, $p_n = \frac{1}{2}\#Z_n = \sum_{i,j=1}^m p_{ij}^{(n)}$ and δ_x is the Dirac measure at x. Our goal is to estimate μ_n-measures of neighbourhoods of points of Z_0.

First, put

$$d_{n,k} = \max_{x \in Z_0} \#[Z_n \cap (x - \frac{1}{A^k}, x + \frac{1}{A^k})]. \tag{5.3.4}$$

Then

$$d_{n+l,k} \leq p_l \cdot (d_{n,k} + 6) \tag{5.3.5}$$

and

$$d_{n+1,k+1} \le d_{n,k} \tag{5.3.6}$$

when k is large enough to satisfy the inequality

$$A^{-k} < \min_{i,j \le m} \operatorname{dist}(K_i, K_j). \tag{5.3.7}$$

To prove (5.3.5) take any $z \in Z_{n+l} \cap (x - \frac{1}{A^k}, x + \frac{1}{A^k})$ such that $z \in \partial K_g$, where $g = g_1 \circ g_2$, $g_1 \in \mathcal{S}_n$ and $g_2 \in \mathcal{S}_l$. Since $K_g \subset K_{g_1}$, we have three possibilities:

(1) $K_{g_1} \subset (x - \frac{1}{A^k}, x + \frac{1}{A^k})$,

(2) $\#[\partial K_{g_1} \cap (x - \frac{1}{A^k}, x + \frac{1}{A^k})] = 1$,

(3) $(x - \frac{1}{A^k}, x + \frac{1}{A^k}) \subset K_{g_1}$.

By definition, the sets K_{g_1}, $g_1 \in \mathcal{S}_n$, are pairwise disjoint, therefore the number of $g_1 \in \mathcal{S}_n$ satisfying (1) is at most $d_{n,k}$, (2) — at most 2 and (3) — at most 1. Since the number of all $g_2 \in \mathcal{S}_l$ is p_l, (5.3.5) follows.

To prove (5.3.6) just note that if k satisfies (5.3.7), then the Markov shift τ (which is defined locally as the inverse of one of the h_i's) maps $(x - \frac{1}{A^{k+1}}, x + \frac{1}{A^{k+1}}) \cap Z_{n+1}$ into $(x - \frac{1}{A^k}, x + \frac{1}{A^k}) \cap Z_n$ and it is one-to-one there.

Inequalities (5.3.5) and (5.3.6) imply

$$d_{n,k} \le d_{n+k_0-k,k_0} \le (d_{k_0,k_0} + 6) \cdot p_{n-k} \tag{5.3.8}$$

for all $k > k_0$ and $n > k$, where k_0 is any number satisfying (5.3.7). Condition (5.3.8) can be read as

$$\mu_n((x - \frac{1}{A^k}, x + \frac{1}{A^k})) \le c_1 \cdot \frac{p_{n-k}}{p_n}, \tag{5.3.9}$$

where $c_1 = \frac{1}{2}(d_{k_0,k_0} + 6)$. Also, since p_n's are defined in terms of $p_{ij}^{(n)}$'s, the entries of the n-th power of the matrix P, there exists a constant $c_2 > 0$ such that $p_{n-k}/p_n \le c_2 \cdot \lambda^{-k}$ when n is sufficiently large. Therefore, (5.3.9) implies

$$\mu_n((x - \frac{1}{A^k}, x + \frac{1}{A^k})) \le c_1 c_2 \lambda^{-k} \tag{5.3.10}$$

for large n and k.

Finally, if μ is any point of accumulation of the sequence (μ_n), k is large enough and $A^{-(k+1)} \le r \le A^{-k}$, then μ is supported in Z_0 and (5.3.10) yields

$$\mu((x - r, x + r) \cap Z_0) \le c_1 c_2 r^s,$$

where

$$s = \frac{k}{k+1} \cdot \frac{\log \lambda}{\log A}.$$

The theorem follows from Proposition 5.1.6. □

The estimate in (5.3.2) is sharp. For instance, the equality holds in the case of the standard 1/3-Cantor set discussed earlier in Examples 1.4.4 (b) and 5.1.7. Also, a simple modification (replace, for instance, homotheties of scale 1/3 by these of scale $1/(2+\varepsilon)$, where ε is positive and small) of this example provides Markov invariant sets of dimension arbitrarily close to 1 (and equal to $\log 2/\log(2+\varepsilon)$ in the situation just described). We shall show that the dimension 1 cannot be reached by Markov systems $\mathcal{S} = \{h_1, \ldots, h_m\}$ consisting of C^2-maps. The core of the proof of this fact consists in estimating carefully sizes of gaps in the sets $Z_n = \bigcap_{m=1}^n \bigcup\{K_g; g \in S_m\}$ "converging" to the Markov set Z_0 (in the sense that — compare Definition 1.4.2 — $Z_0 = Z \setminus \operatorname{int} Z$, where $Z = \bigcap_n Z_n$). To this end, we begin by a series of lemmas coming from elementary differential calculus, the first of them being due to Sacksteder [**Sa**].

Lemma 5.3.2. *If an interval $[x,y]$, $x, y \in \mathbb{R}$, is contained in the domain of a map $g_n = h_{i_n} \circ \cdots \circ h_{i_1}$, $g_k = h_{i_k} \circ \cdots \circ h_{i_1}$ for $k = 1, \ldots, n-1$, $g_0 = \mathrm{id}$ and*

$$\theta = \max\{|(h_i^{\pm 1})''(z)|\} \cdot \max\{|(h_i^{\pm 1})'(z)|\}, \tag{5.3.11}$$

where z ranges over Q_i or K_i and i over the set $\{1, \ldots, m\}$, then

$$|g_n'(x)| \le |g_n'(y)| \cdot \exp\left(\theta \sum_{k=1}^{n-1} |g_k(x) - g_k(y)|\right). \tag{5.3.12}$$

Proof. Since $g_n' = \prod_k h_{i_k}' \circ g_{k-1}$, $\log g_n' = \sum_k \log(h_{i_k}' \circ g_{k-1})$ and, by the Mean Value Theorem,

$$\log g_n'(x) - \log g_n'(y) = \sum_k \frac{h_{i_k}''(t_k)}{h_{i_k}'(t_k)} (g_{k-1}(x) - g_{k-1}(y))$$

for some $t_k \in (x,y)$. Therefore,

$$|\log g_n'(x) - \log g_n'(y)| \le \theta \cdot \sum_k |g_{k-1}(x) - g_{k-1}(y)|$$

and this implies (5.3.12). □

The next two lemmas provide some consequences of the previous one. Essentially, they are due to Matsumoto [**Mat1**].

Lemma 5.3.3. *Under the notation above, if J is a gap of Z_0 (i.e., J is a connected component of $\mathbb{R} \setminus Z_0$) and $x_0 \in \partial J$, then there exist positive constants σ and δ such that*

$$\sum_{k=0}^{n} |g_k'(y_k)| \le \sigma \tag{5.3.13}$$

whenever $|y_k - x_0| < \delta$ and $g = g_n$ satisfies the condition

(*) *$(x_0 - \delta, x_0 + \delta) \cup J \subset D_g$ and the intervals $J, g_1(J), \ldots, g_n(J)$ are mutually disjoint.*

Proof. Let A be the length of a compact segment containing all the intervals Q_i and K_i, $i \leq m$. Then, by the Mean Value Theorem and Lemma 5.3.2,

$$A \geq \sum_k |g_k(J)| = |J| \cdot \sum_k |g_k'(t_k)| \geq |J| \cdot \sum_k |g_k'(x_0)| e^{-\theta \sum_{j<k} |g_j(x_0) - g_j(t_k)|}$$
$$\geq |J| e^{-\theta A} \sum_k |g_k'(x_0)|$$

for some elements t_k, $k = 1, \dots, n$, of J. Hereafter, $|K|$ denotes the length of an interval K.

Let $\sigma = 2Ae^{A\theta}/|J|$ and choose δ so small that $e^{\theta\sigma\delta} < 2$. The last inequality reads

$$\sum_k |g_k'(x_0)| \leq \frac{1}{2}\sigma. \tag{5.3.14}$$

Moreover, (5.3.12) implies that

$$|g_k'(y)| \leq 2|g_k'(x_0)| \tag{5.3.15}$$

for all $y \in (x_0 - \delta, x_0 + \delta)$ and $k \leq n$. In fact, since $\sigma \geq 1$, (5.3.15) holds for $k = 1$ and if it holds for all $k < l$, then

$$|g_l'(y)| \leq |g_l'(x_0)| \cdot \exp\left(\theta \sum_{k<l} |g_k(y) - g_k(x_0)|\right)$$
$$\leq |g_l'(x_0)| \cdot \exp\left(\theta \sum_{k<l} |g_k'(t_k)||y - x_0|\right) \leq |g_l'(x_0)| \cdot e^{\theta\sigma\delta} \leq 2|g_l'(x_0)|$$

with some t_k's in the interval with end points x_0 and y. Clearly, (5.3.14) and (5.3.15) yield (5.3.13). □

Lemma 5.3.4. *If $s \leq 1$, x_0 and δ satisfy conditions of the previous lemma, $K \subset (x_0 - \delta, x_0 + \delta)$ is a closed non-degenerate interval and $\mathcal{I} = \{I_1, \dots, I_p\}$, $p \geq 2$, is a finite family of pairwise disjoint intervals contained in K, then there exists $\lambda > 0$ such that for any g satisfying $(*)$ the inequality*

$$\sum_{k=1}^{p} |g(I_k)|^s \leq \lambda \cdot |g(K)|^s \tag{5.3.16}$$

holds. If $s > s_0$, where

$$s_0 = \frac{\log p}{\log(p\alpha + p - 1) - \log \alpha} \tag{5.3.17}$$

for some $\alpha > 0$, then λ can be chosen in such a way that $\lambda < 1$.

Certainly, given g, (5.3.16) holds for some $\lambda < 1$. The essence of the lemma consists in that λ can be chosen uniformly, the same for all maps g satisfying $(*)$.

Proof. For any $k = 1, \ldots, p$ choose a gap J_k of $K \setminus \bigcup \mathcal{I}$ adjacent to I_k. Let $\alpha_k = |I_k| \cdot |J_k|^{-1} \cdot \exp(\theta\sigma|K|)$ and $\alpha = \max \alpha_k$. Inequalities (5.3.12) and (5.3.13) together with the Mean Value Theorem imply the relations

$$\begin{aligned}\frac{|g(I_k)|}{|g(J_k)|} &= \frac{|I_k|}{|J_k|} \cdot \frac{|g'(x_k)|}{|g'(y_k)|}\\ &\le \frac{|I_k|}{|J_k|} \cdot \exp\left(\theta \sum_j |g_j(x_k) - g_j(y_k)|\right)\\ &\le \frac{|I_k|}{|J_k|} \cdot \exp\left(\theta \sum_j |g_j(K)|\right) \le \frac{|I_k|}{|J_k|} \cdot \exp\left(\theta |K| \sum_j |g_j'(z_j)|\right)\\ &\le \frac{|I_k|}{|J_k|} \cdot \exp(\theta\sigma|K|) = \alpha_k \le \alpha\end{aligned}$$

for some points x_k, y_k and z_j of K. Since the number of gaps of $K \setminus \mathcal{I}$ lies between $p-1$ and $p+1$, it follows that

$$\sum_k |g(I_k)| \le |g(K)| \cdot \frac{p\alpha}{p(\alpha+1)-1}. \tag{5.3.18}$$

Any function ϕ of the form

$$\phi(s) = \frac{1}{p} \sum_{k=1}^{p} a_k^s$$

with $a_1, \ldots, a_p > 0$ is convex on $\mathbb{R}$. Therefore, (5.3.18) implies (5.3.16) with

$$\lambda = \lambda(s) = p \cdot \left(\frac{\alpha}{p(\alpha+1)-1}\right)^s. \tag{5.3.19}$$

The last statement follows from the equality $\lambda(s_0) = 1$ and the property $\lambda'(s) < 0$, $s \in \mathbb{R}$, of the function λ in (5.3.19). □

The constant λ in inequality (5.3.16) is "universal" in the following sense: if K' is a subinterval of K and I_k', $k = 1, \ldots, p$, are subintervals of $I_k \cap K'$, then this inequality remains true with the same λ, K and I_k's being replaced respectively by K' and I_k''s. Indeed, since the corresponding gaps J_k' of $K' \setminus \bigcup_k I_k'$ are larger than these of $K \setminus \bigcup \mathcal{I}$, $\alpha' = \max_k(|I_k'| \exp(\theta\sigma|K'|)/|J_k'|) \le \alpha$, $s_0' = \log p/(\log(p\alpha' + p - 1) - \log \alpha') \le s_0$ and $\lambda' = p \cdot (\alpha'/(p(\alpha'+1)-1) \le \lambda$.

It is easy to check that $s_0 < 1$ for any $\alpha > 0$ in (5.3.17), so we always have (5.3.16) with some $\lambda < 1$ and $s < 1$. This observation is crucial for the proof of

Theorem 5.3.5. [**Wal1**] *For any C^2-Markov system $\mathcal{S} = \{h_1, \ldots, h_m\}$, $m > 1$, the Markov invariant set Z_0 satisfies the inequality*

$$\dim_H Z_0 < 1. \tag{5.3.20}$$

Corollary 5.3.6. *Under the same assumptions, Z_0 has 1-dimensional Lebesgue measure zero.* □

The proof of the theorem is essentially due to Cantwell and Conlon [**CC5**] who proved the above corollary. To get the theorem one needs some modifications consisting first and foremost in the use Lemmas 5.3.3 and 5.3.4.

Throughout the proof we shall denote by A_∞, $A \subset Z_0$, the set of all the points x of Z_0 such that $\tau^n(x) \in A$ for infinitely many n in $\mathbb{N}$. A point $x \in Z_0$ will be called *good* when it admits a neighbourhood V open in Z_0 and such that $\dim_H V_\infty < 1$. G will denote the set of all good points while $B = Z_0 \setminus G$ the set of all *bad* points. By definition, G is open in Z_0 while B is closed there.

Lemma 5.3.7. *If $A \subset G$ is compact, then* $\dim_H A_\infty < 1$.

Proof. There exist open sets $V_1, \ldots, V_n$ covering A and such that $\dim_H (V_i)_\infty < 1$ for $i = 1, \ldots, n$. Since $A_\infty \subset (V_1 \cup \cdots \cup V_n)_\infty = (V_1)_\infty \cup \cdots \cup (V_n)_\infty$, $\dim_H A_\infty \leq \dim_H (V_1 \cup \cdots \cup V_n)_\infty \leq \max_i \dim_H (V_i)_\infty < 1$. □

This Lemma shows that for the proof of the theorem it is sufficient to show that $B = \emptyset$. If so, $Z_0 = G$ is compact, $(Z_0)_\infty = Z_0$ and $\dim_H Z_0 < 1$ as stated.

So, we have to show that any point x_0 of Z_0 is good. Evidently, all the isolated points of Z_0 are good. Therefore, we have to consider two cases:

(i) x_0 is a one-sided point of accumulation of Z_0,

(ii) x_0 is a two-sided point of accumulation of Z_0.

Case (i). A one sided-point x_0 belongs to ∂J for some gap J of Z_0. Assume, for instance, that $x_0 = \inf J$. (The case $x_0 = \sup J$ is quite similar.) Let δ and σ satisfy conditions of Lemma 5.3.3. Since Z_0 has empty interior, we can find $\delta' < \delta$ such that $x_0 - \delta' \notin Z_0$ and put $V = (x_0 - \delta', x_0] \cap Z_0$.

Assume first that x_0 is non-cyclic, i.e., that $\tau^k(x_0) \neq x_0$ for all $k > 0$. This allows us to apply Lemma 5.3.4 to any segment K of the form $K_g \cap (x_0 - \delta', x_0]$ with $g \in \mathcal{S}_{n_0}$, where $n_0 \in \mathbb{N}$ is fixed and sufficiently large. Let $\mathcal{K}$ be the family of all such segments K. For each $K \in \mathcal{K}$ denote by $\mathcal{I}_K$ the family of all the segments of the form $K \cap K_h$ with $h \in \mathcal{S}_{n_0+1}$. Then, choose positive constants $s_K < 1$ and $\lambda_K < 1$ satisfying (5.3.16). Put $s = \max s_K$ and $\lambda = \max \lambda_K$. For any $n \in \mathbb{N}$ the families $\mathcal{A}_n = \{g(K) \cap V_\infty; g \in \mathcal{S}_l, l \geq n \text{ and } K \in \mathcal{K}\}$ and $\mathcal{A}'_n = \{g(I) \cap V_\infty; g \in \mathcal{S}_l, l \geq n, I \in \mathcal{I}_K \text{ and } K \in \mathcal{K}\}$ cover V_∞ and, by inductive application of Lemma 5.3.4,

$$\sum_{A' \in \mathcal{A}'_n} (\operatorname{diam} A')^s \leq \lambda \sum_{A \in \mathcal{A}_n} (\operatorname{diam} A)^s < \infty.$$

If n_0 is large enough, all the sets K_g ($g \in \mathcal{S}_{n_0}$) are either contained in or disjoint from $(x_0 - \delta', x_0]$ and then, by the Markovian property of our system $\mathcal{S}$, $\mathcal{A}'_n = \mathcal{A}_{n+1}$. Consequently, $H^s(\mathcal{A}_n) \to 0$ and $\operatorname{diam} \mathcal{A}_n \to 0$ as $n \to \infty$. This implies that $H^s(V_\infty) = 0$, $\dim_H V_\infty \leq s < 1$ and $x_0 \in G$.

Next, assume that x_0 is cyclic, $\tau^{p+1}(x_0) = x_0$, $p \geq 0$ being the smallest possible. Let $X = \{x_0, x_1, \dots, x_p\}$, where $x_k = \tau^k(x_0)$. Also, let $\{y_1, \dots, y_r\} = \tau^{-1}(X) \setminus X$ and for each $j \leq r$ find $g_j \in S_{n_j}$ such that $g_j(x_0) = y_j$ and $n_j \in \mathbb{N}$ is the smallest possible. Choose an open interval U and $l \in \{1, \dots, m\}$ such that $x_0 \in V = U \cap Z_0 \cap K_l$.

Lemma 5.3.8. $V_\infty = \bigcup_{j=1}^r (g_j V)_\infty \cup A$, *where* A *is a countable set.*

Proof. Find numbers $j_0, \dots, j_p \in \{1, \dots, m\}$ such that $h_{j_i}(x_i) = x_{i+1}$ and put $g_0 = h_{j_p} \circ \cdots \circ h_{j_0}$. Then, $g_0(x_0) = x_0$ and — since points of Z_0 fixed by g_0 are τ-periodic and τ is semiconjugate to the shift $\sigma_P : \mathcal{X}_P \to \mathcal{X}_P$ *via* a two-to-one map — there exists at most one other point $x'_0 \in Z_0 \cap K_l$ fixed by g_0. Let A be the union of the $\mathcal{G}(\mathcal{S})$-orbits of x_0 and x'_0. Clearly, A is countable and

$$A \cup (g_1 V)_\infty \cup \cdots \cup (g_r V)_\infty \subset V_\infty.$$

To complete the proof take any $y \in V_\infty$ and let $n_1 < n_2 < \cdots$ be such a sequence of natural numbers that $\tau^{n_j}(y) \in V$ for all j. Set $V_0 = V$ and $V_i = h_{j_{i-1}} \circ \cdots \circ h_{j_0}(V)$ when $i \geq 1$. Since $V = U \cap Z_0 \cap K_l$ and U is an interval, $V_{p+1} \subset V_0$. For any j, $\tau^{n_j - 1}(y) \in \tau^{-1}(V) = V \cup g_{t_1}(V) \cup \cdots \cup g_{t_q}(V)$ for some $t_1, \dots, t_q \in \{1, \dots, r\}$. If $\tau^{n_j-1}(y) \in V_1$, then $\tau^{n_j-2}(y) \in \tau^{-1}(V_1) = V_2 \cup g_{s_1}(V) \cup \cdots \cup g_{s'_q}(V)$ for some $s_1, \dots, s_{q'} \in \{1, \dots, r\}$. Continue as long as possible.

If $\tau^n(y) \in \bigcup_j V_j$ for all large n, then there exists $n_0 \in \mathbb{N}$ such that $z = \tau^{n_0}(y) \in V$, $g_0^{-n}(z)$ is defined and lies in V for all $n \geq 0$. In other words, $z \in \bigcap_n g_0^n(K_l \cap Z_0) = \{x_0, x'_0\}$. This implies that $y \in A$. If $\tau^{m_k}(y) \in g_1(V) \cup \cdots \cup g_r(V)$ for some $m_1 < m_2 < \cdots$, then $y \in \bigcup_j (g_j V)_\infty$. □

The points y_j are not periodic. By previous arguments, one can shrink V, if necessary, in such a way that $\dim_H (g_j V)_\infty < 1$ for all j. Since $\dim_H A = 0$, the lemma above implies that $\dim_H V_\infty \leq \max_j \dim_H (g_j V)_\infty < 1$ and $x \in G$.

In this way, we have proved that all the one-sided points of accumulation of Z_0 are good.

Case (ii). Assume that Z_0 accumulates on a point y from both sides and that y is bad. Then, y lies in the interior of K_l for some $l \in \{1, \dots, m\}$ and there exist a gap $J = (a, b)$ of Z_0 and an admissible sequence $j = (j_1, \dots, j_{n_0})$ such that $K_g = [c, d] \subset K_l$ and $a < b < c < y_0 < d$, where $g = h_{j_1} \circ \cdots \circ h_{j_{n_0}}$ and $y_0 = \min(B \cap [b, +\infty))$.

Write $\mathcal{S}_{n_0} = \{f_1, \dots, f_{p+r}\}$, where the enumeration is chosen in such a way that $K_{f_i} \cap (a, c) = \emptyset$ if and only if $i \leq p$. Express each f_i in the form $f_i = h_{\alpha_i} \circ g_i$, where $\alpha_i \in \{1, \dots, m\}$ and $g_i \in \mathcal{S}_{n_0 - 1}$. Let $f_i^* = h_{\alpha_i} | K_{g_i}$ and denote by $\mathcal{G}^*$ the pseudogroup generated by the Markov system $\mathcal{S}^* = \{f_1^*, \dots, f_p^*\}$. Since $\mathcal{S}^*$ consists of restrictions of all the elements of $\mathcal{S}_{n_0}$ which contribute to Z_0, the corresponding Markov invariant set Z_0^* is sparser than Z_0 and therefore J is contained in a gap $J^* = (a^*, b^*)$ of Z_0^*. By definition of Z_0^*, $(a,c) \subset J^*$ and $b^* \in K_g$ (see Figure 5.2). The corresponding shift τ^* defines the operation $Z_0^* \supset A \mapsto A_{\infty *} = \bigcap_{n\geq 0}\bigcup_{k\geq n}(\tau^*)^{-k}(A)$ analogous to the correspondence $Z_0 \ni A \mapsto A_\infty$ considered before.

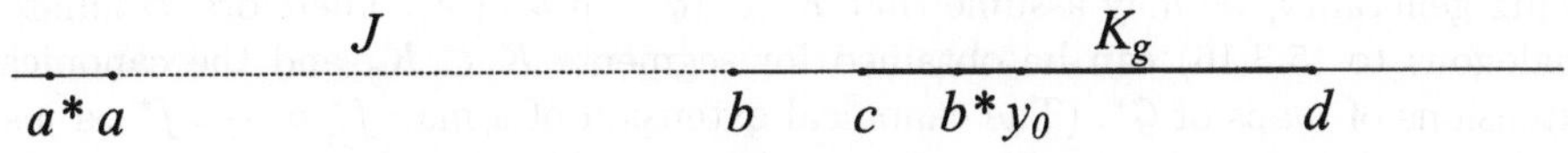

Figure 5.2: Gaps of Z_0 and Z_0^*.

Lemma 5.3.9. *For any $V \subset Z_0$ one has*

$$V_\infty \subset \bigcup_{i=1}^{r}(A_i)_\infty \cup \bigcup_{k=0}^{\infty}\tau^{-k}((V\cap Z_0^*)_{\infty *}), \tag{5.3.21}$$

where $A_i = K_{f_{p+i}}$.

Note that from the definition of A_∞ it follows directly that $\tau^{-k}(A_\infty) \subset A_\infty$ for all $k \geq 0$ and A, but here we deal with the operation $(\cdot)_{\infty *}$, so the statement of the lemma is not trivial.

Proof. Let $x \in V_\infty$. If $\tau^k(x) \in Z_0^*$ for some $k \geq 0$, then one can find infinitely many $n \geq k$ for which $\tau^n(x) = (\tau^*)^{n-k}(\tau^k(x)) \in V$. Then, $\tau^k(x) \in (V \cap Z_0^*)_{\infty *}$ and $x \in \tau^{-k}((V\cap Z_0^*)_{\infty *})$.

If $\tau^k(x) \notin Z_0^*$ for all $k \geq 0$, then for any $k \geq 0$ there exist $n \geq k$ such that $\tau^{n+k}(x) \notin \bigcup_{i=1}^{p} K_{f_i}$. Also, $\tau^n(x)$ lies in the union of all K_{f_i}'s, $i \leq p+r$, just because x lies in Z_0. Therefore, there exist infinitely many natural numbers n for which $\tau^n(x) \in \bigcup_i^r K_{f_{p+i}} = \bigcup_i A_i$. This means that $x \in (A_1)_\infty \cup \cdots \cup (A_r)_\infty$. □

To complete the proof of the theorem we intend to consider all possible positions of b^* and to arrive at a contradiction in all the cases. This will show that the assumption $y \in B$ is false and B has to be empty.

(1) $b^* = d$. Then $(a,d)\cap Z_0^* = \emptyset$ and $V = (a,d)\cap Z_0$ satisfies, by (5.3.21),

$$V_\infty \subset \bigcup_{i=1}^{r}(A_i \cap (a,d))_\infty \subset \bigcup_{i=1}^{r}(A_i)_\infty.$$

By the construction, each A_i is compact and contained in (b,c), therefore in G. Consequently, $\dim_H((A_i)_\infty < 1$ and $\dim_H V_\infty < 1$. Since $y_0 \in V$, $y_0 \in G$, a contradiction.

(2) $b^* < d$ and $(\tau^*)^{-n}(\{b^*\}) = \emptyset$ for some $n \geq 1$. Then, any neighbourhood V of y_0 contained in $(c,d)\cap Z_0$ satisfies $(V\cap Z_0^*)_{\infty *} = \emptyset$ and, as before, $\dim_H V_\infty < 1$, $y_0 \in G$, a contradiction. In fact, since the set $(\tau^*)^{-n}(\{b^*\})$ is empty, b^* belongs to

the domains of only finitely many maps of G^*. Since $b^* \in K_l$, K_l intersects only finitely many such domains and $(K_l \cap Z_0^*)_{\infty *} = \emptyset$.

(3) $b^* < d$ and $(\tau^*)^{-n}(\{b^*\}) \neq \emptyset$ for all $n \in \mathbb{N}$. That is, b^* lies in the domain of infinitely many maps $g^* \in \mathcal{G}^*$. In this case, we have to apply Lemmas 5.3.3 and 5.3.4 to $\mathcal{S}^*$ and J^*. From the proof of Lemma 5.3.3 it follows that the corresponding constants δ^* and δ for J^* and J can be chosen in such a way that $\delta^* \geq \delta$. Without losing generality, we may assume that $K_g \subset (b^* - \delta, b^* + \delta)$. Therefore, estimates analogous to (5.3.16) can be obtained for segments $K \subset K_g$ and the canonical extensions of maps of $\mathcal{G}^*$. (The canonical extension of a map $f_{j_1}^* \circ \cdots \circ f_{j_k}^*$ is just $h_{\alpha_{j_1}} \circ \cdots \circ h_{\alpha_{j_k}}$.) Consequently,

$$\dim_H (V \cap Z_0^*)_{\infty *} < 1,$$

where $V = (a, d) \cap Z_0$. This inequality together with (5.3.21) shows that $\dim_H V_\infty < 1$ and that $y_0 \in G$, a contradiction.

This completes the proof of Theorem 5.3.5. □

The last result presented in this section concerns the transverse Hausdorff dimension of codimension-one foliations.

Theorem 5.3.10. [**IWa**] *If $\mathcal{F}$ is a transversely oriented codimension-one C^2-foliation of a compact manifold M and the leaves of $\mathcal{F}$ are not dense, then*

$$\dim_H^{\pitchfork} \mathcal{F} = \dim_H (C(\mathcal{F}) \cap T), \tag{5.3.22}$$

where $C(\mathcal{F})$ is the union of all the compact leaves of $\mathcal{F}$ and T is a compact complete transversal.

Proof. The inequality

$$\dim_H^{\pitchfork} \mathcal{F} \geq \dim_H (T \cap C(\mathcal{F}))$$

follows directly from Proposition 5.1.9 and the fact that the holonomy of $\mathcal{F}|C(\mathcal{F})$ is trivial.

To prove the converse, fix $s > \dim_H(C(\mathcal{F}) \cap T)$, $\eta > 0$ and $\varepsilon > 0$. Recall once again (compare Section 3.6) that the centre $Z(\mathcal{F}) = C(\mathcal{F}) \cup E_1 \cup \cdots \cup E_m$ is compact and contains finitely many exceptional minimal sets E_i. Moreover, each leaf L of $\mathcal{F}$ accumulates on $Z(\mathcal{F})$: $\bar{L} \cap Z(\mathcal{F}) \neq \emptyset$.

The Sacksteder Theorem [**Sa**] allows us to choose points $z_i \in T \cap E_i$, $i = 1, \ldots, m$, and contracting holonomy maps h_i defined in some transverse segments J_i containing z_i.

Since the set $C(\mathcal{F})$ is compact, we can cover it by a finite number of mutually disjoint foliated I-bundles (some of them degenerating to single leaves isolated in $C(\mathcal{F})$) $C_1, \ldots, C_n$ bounded by closed leaves L_j^+ and L_j^- semi-isolated in $C(\mathcal{F})$. Choose fibres $I_j = [x_j, y_j] \subset T$ of C_j and extend them to slightly larger segments $I_j' = [x_j', y_j']$ such that the segments $[x_j', x_j]$ and $[y_j, y_j']$ are attracted to x_j and y_j by the global holonomy groups Γ_j of $\mathcal{F}|C_j$. More precisely, for any $j = 1, \ldots n$ and $\delta > 0$ there should exist holonomy maps f_j and f_j' in Γ_j which extend to I_j'

and bring x'_j (resp., y'_j) to within distance δ of x_j (resp., y_j). Shrink segments I'_j, if necessary, to $I''_j = [x''_j, y''_j]$ in such a way that $x'_j < x''_j < x_j$, $y_j < y''_j < y'_j$ and $I''_j \cap I''_k = \emptyset$ whenever $j \neq k$. Set

$$T' = \bigcup_{i=1}^{m} J_i \cup \bigcup_{j=1}^{n} I''_j.$$

T' is a complete transversal for $\mathcal{F}$.

For any $j \leq n$, choose a finite symmetric set Γ_j^0 generating Γ_j. Without loss of generality we may assume that the previously chosen segments I'_j are so small that all the members of Γ_j^0 extend to holonomy maps defined on I'_j. Let $\mathcal{H}'$ be the subpseudogroup of the holonomy pseudogroup $\mathcal{H}_{T'}$ generated by the contractions h_i, $i \leq m$, and the holonomy maps of Γ_j^0 (extended to I''_j), $j \leq n$.

Since $s > \dim_H(C(\mathcal{F}) \cap T)$, we can cover $T' \cap C(\mathcal{F})$ by segments $K_1, \ldots, K_N$ with endpoints in $C(\mathcal{F})$ and lengths $|K_i| \leq \varepsilon$, and such that

$$\sum_{i=1}^{N} |K_i|^s < \eta.$$

Put $\delta = \frac{1}{2}\varepsilon \max\{1, |K_1|, \ldots, |K_N|\}$. For any j, choose holonomy maps g_j^- (resp., g_j^+) in Γ_j which extend to I'_j and bring x'_j (resp., y'_j) to within distance δ of x_j (resp., y_j). Also, for any gap (i.e., a connected component) $U = (a, b)$ of $I_j \setminus \bigcup_i K_i$ choose — at the same time — points $c, d \in U$, $c < d$, and holonomy maps f_U^- (resp., f_U^+) which bring d (resp., c) to within distance δ of a (resp., b).

Take any $i \leq N$. If $K_i = [\alpha, \beta] \subset (x_j, y_j)$, then take the gaps U^- and U^+ adjacent to K_i, such that, say, $\alpha \in \bar{U}^-$ and $\beta \in \bar{U}^+$, denote by V^- (resp., V^+) the δ-neighbourhood of α (resp., β) and put

$$A_i = \{h|\operatorname{int} K_i,\ h \circ (f_{U^-}^+)^{-1}|V^-,\ h \circ (f_{U^+}^-)^{-1}|V^+;\ h \in \Gamma_j^0\}.$$

If $K_i = [x_j, \beta] \subset I_j$, then take U^+, V^+ and $f_{U^+}^-$ as before and put

$$A_i = \{h|\operatorname{int} K_i,\ h \circ g_j^{-1}|W_j,\ h \circ (f_{U^+}^-)^{-1}|V^+;\ h \in \Gamma_j^0\},$$

where $W_j = (x_j - \delta, x_j + \delta)$. Define A_i similarly when $K_i = [\alpha, y_j]$. Finally, for any $i \leq m$ choose an exponent $n_i \in \mathbb{N}$ such that the image $h_i^{n_i}(J_i)$ has diameter less than ε and put

$$A = \{h_i^{-n_i},\ h_i^{-(n_i+1)};\ i \leq m\} \cup A_1 \cup \cdots \cup A_N.$$

The set A generates $\mathcal{H}'$, the diameters of the domains of the maps in A do not exceed ε and

$$\sum_{h \in A} (\operatorname{diam} D_h)^s \leq 2m\varepsilon^s + \max_j \#\Gamma_j^0 \cdot (1 + 2\varepsilon^s)\eta.$$

It follows that $H^s(\mathcal{H}') < \infty$ and $s \geq \dim_H \mathcal{H}'$. Since $\dim_H \mathcal{H}' \geq \dim_H^{\pitchfork} \mathcal{F}$ by Proposition 5.1.9, we obtain the required inequality

$$\dim_H^{\pitchfork} \mathcal{F} \leq \dim_H(C(\mathcal{F}) \cap T). \qquad \square$$

Remark 5.3.11. (i) The argument above shows that $\dim_H^{\pitchfork} \mathcal{F} = 0$ when $\mathcal{F}$ is minimal with non-trivial holonomy. One has just to take as T' any arbitrarily short segment transverse to $\mathcal{F}$, choose any finite set A_0 generating $\mathcal{H}_T$, a point $x_0 \in \operatorname{int} T'$ and two holonomy maps h_1 and h_2 attracting the end points of T' to within distance ε of x_0. Then the maps $g \circ h_1^{-1}$ and $g \circ h_2^{-1}$, $g \in A_0$, will generate $\mathcal{H}_{T'}$ while their domains will have diameters at most 2ε. This will yield $H^s(\mathcal{H}_{T'}) = 0$ for any $s > 0$.

(ii) The Denjoy diffeomorphism f of S^1 (Example 1.2.2) admits a minimal invariant set X such that $f|X$ preserves the 1-dimensional Lebesgue measure λ. Since $\lambda(X) > 0$, $\dim_H X = 1$. Since f has no periodic points, its suspension $\mathcal{F}$ has no compact leaves, $\dim_H(C(\mathcal{F}) \cap T) = 0$ while $\dim_H^{\pitchfork} \mathcal{F} = 1$. Therefore, the assumption of C^2-differentiability in Theorem 5.3.10 is essential.

(iii) If $\mathcal{F}$ is minimal with trivial holonomy, then we are not able to determine its transverse Hausdorff dimension. For instance, such an $\mathcal{F}$ can be Riemannian and then $\dim_H^{\pitchfork} \mathcal{F} = \operatorname{codim} \mathcal{F}$. On the other hand, Arnold [**Ar**] produced an example of an analytic diffeomorphism f of S^1 with an irrational rotation number $\rho(f)$ obtained as the limit of a sequence of rationals with "small denominators". The suspension $\mathcal{F}$ of f is minimal and its leaves have trivial holonomy since they are simply connected. It is possible to show that $\dim_H^{\pitchfork} \mathcal{F} = 0$. See [**IWa**] for details. Till now, we do not know examples of holonomy free minimal foliations of codimension 1 for which the transverse Hausdorff dimension is strictly positive and strictly less than 1.

5.4 Dimension of a hyperbolic boundary

It is easy to imagine that the Hausdorff dimension of the boundary of a hyperbolic space may be infinite. However, the most natural hyperbolic spaces, Riemannian manifolds of negative sectional curvature and finitely generated hyperbolic groups, as well as metric spaces quasi-isometric to such objects, have boundaries of finite Hausdorff dimension. The main reason for that is the following: the "areas" of spheres in such spaces grow at most exponentially.

Definition 5.4.1. A hyperbolic metric space (X, d) *grows at most exponentially* when there exist positive constants α and β such that — for a fixed point $x_0 \in X$ — each sphere $S(r) = \{x \in X; d(x_0, x) = r\}$ of radius $r > 0$ and centre x_0 contains a finite set Z such that the balls $B(z, 1)$, $z \in Z$, cover $S(r)$ and

$$\# Z \leq \alpha \cdot e^{\beta r}. \qquad (5.4.1)$$

Example 5.4.2. (i) If G is a hyperbolic group, X is its Cayley graph determined by a finite symmetric generating set G_1, $n \in \mathbb{N}$ and $S(n) = \{g \in G; d(e,g) = n\}$, then $\#S(n) \leq C^n$, where $C = \#G_1$. This implies that X grows at most exponentially.

(ii) If M is a simply connected complete n-dimensional Riemannian manifold of negative sectional curvature K, $-b^2 \leq K \leq -a^2 < 0$, then the volumes $V(r)$ of balls of radius r in M satisfy inequalities of the form

$$c_1 e^{a_1 r} \leq V(r) \leq c_2 e^{a_2 r}$$

(compare (2.2.22) in Chapter 2) with some fixed positive constants a_1, a_2, c_1 and c_2. A maximal 1-separated set $Z \subset S(r)$ ($d(z,z') \geq 1$ when $z, z' \in Z$ and $z \neq z'$) is 1-dense: $S(r) \subset \bigcup_{z \in Z} B(z,1)$. For such Z, $\#Z \cdot V(1/2) \leq V(r+1) - V(r-1)$. More precisely,

$$\#Z \cdot c_1 e^{a_1/2} \leq c_2 e^{a_2(r+1)} - c_1 e^{a_1(r-1)} \leq c_2 e^{a_2(r+1)}.$$

Clearly, this yields an inequality of the form (5.4.1) and shows that M grows at most exponentially in the sense of the above definition.

Proposition 5.4.3. *If a hyperbolic space X grows at most exponentially, then*

$$\dim_H \partial X < \infty.$$

Of course, we think about ∂X equipped with the metric d_ε defined for some $\varepsilon > 0$ as in Section 1.5.

Proof. Let $\mathcal{B}(n) = \{B_z^{(n)}, z \in Z_n\}$, be a minimal cover by unit balls of the sphere $S(n)$ centred at a fixed point $x_0 \in X$. Define the cover $\mathcal{U}(n) = \{U_z^{(n)}; z \in Z_n\}$ of ∂X as follows: $x_\infty \in U_z^{(n)}$ if and only if there exists a geodesic ray γ originated at x_0, representing x_∞ and intersecting the ball $B_z^{(n)}$.

If x_∞ and y_∞ lie in $U_z^{(n)}$, then the Gromov product $(\cdot|\cdot)$ based at x_0 satisfies the inequality

$$(x_\infty | y_\infty) \geq n - 3.$$

In fact, let γ_0 and γ_1 be intersecting $B_z^{(n)}$ and originated at x_0 rays representing x_∞ and y_∞. Choose points $\gamma_0(t)$ and $\gamma_1(s)$ of $B_z^{(n)}$. Then, $n-2 \leq s, t \leq n+2$,

$$d(\gamma_0(n+i), \gamma_1(n+j)) \leq d(\gamma_0(n+i), \gamma_0(t)) + d(\gamma_1(n+j), \gamma_1(s)) + 2 \leq i + j + 6$$

and

$$(\gamma_0(n+i)|\gamma_1(n+j) \geq n - 3$$

for any i and j of $\mathbb{N}$.

It follows that $\operatorname{diam} U_z^{(n)} \leq \exp(3\varepsilon) \cdot \exp(-n\varepsilon)$. If $\beta - \varepsilon s < 0$, where α and β are constants involved in (5.4.1), then

$$\sum_{z \in Z_n} \left(\operatorname{diam} U_z^{(n)}\right)^s \leq \alpha e^{3s\varepsilon} e^{(\beta - s\varepsilon)n} \to 0$$

when $n \to \infty$. This shows that

$$\dim_H \partial X \le \beta/\varepsilon < \infty. \qquad \square$$

The next result concerns the Hausdorff dimension of the pseudogroup $\mathcal{G}(\Gamma)$ generated by a hyperbolic group Γ acting on the boundary ∂X of its Cayley graph $X = \mathcal{C}(\Gamma, \Gamma_1)$, Γ_1 being any finite generating set.

If Γ is non-elementary, then it contains a hyperbolic element h which keeps some points x_∞ and y_∞ of ∂X fixed. The force of one of them, say x_∞, is strictly less than 1. Then the force $\Phi(h, y_\infty)$ has to be strictly bigger than 1. Take any finite set $\Gamma_1 = \{g_1, \dots, g_N\}$ generating Γ, $k \in \mathbb{N}$ arbitrarily large and set $\Gamma_1' = \{h^k, g_1 h^k, \dots, g_N h^k\}$. Clearly, Γ_1' generates Γ. Since $\Phi(h^k, x_\infty) = \Phi(h, x_\infty)^k$ can be arbitrarily close to 0, the maps $g_i h^k$ and $(g_i h^k)^{-1}$, $i = 1, \dots, N$, contract some balls $B(x_\infty, R)$ and $B(y_\infty, R)$, respectively, into themselves. Therefore, each of the maps $g_i h^k$ has exactly two fixed points x_∞^i and y_∞^i in ∂X. By the classification of elements of Γ, Γ_1' consists of hyperbolic elements only.

In this way, we have obtained

Proposition 5.4.4. *Any non-elementary hyperbolic group can be generated by hyperbolic elements.* $\square$

Let us keep the above notation for a while. Take any $\varepsilon > 0$, let $U = B(x_\infty, \frac{\varepsilon}{2})$ and $V = B(y_\infty, \frac{\varepsilon}{2})$. Observe that, if k is large enough, h^k maps $\partial X \setminus U$ into V while h^{-k} maps $\partial X \setminus V$ into U. The maps

$$h^k|U,\ h^{k+1}|U,\ h^{-k}|V,\ h^{-(k+1)}|V,\ g_i h^k|U \quad \text{and} \quad g_i h^{-k}|V, \quad i = 1, \dots, N,$$

generate $\mathcal{G}(\Gamma)$ and have domains of diameter $\le \varepsilon$. It follows that

$$H^s_\varepsilon(\mathcal{G}(\Gamma)) \le (2N+4)\varepsilon^s \quad \text{and} \quad H^s(\mathcal{G}(\Gamma)) = 0$$

for all $s > 0$. This implies

Proposition 5.4.5. *If Γ is a non-elementary hyperbolic group, then*

$$\dim_H \mathcal{G}(\Gamma) = 0. \tag{5.4.2}$$

$\square$

The reader can easily see that what we really needed in the proof of this proposition was exactly the existence of a hyperbolic element. Therefore, this proposition can be formulated in the form of

Proposition 5.4.6. *If a finitely generated group Γ of isometries of a hyperbolic metric space X contains a hyperbolic element, then equality* (5.4.2) *holds.* $\square$

5.5 Dimension of a limit set

Let us consider again a discrete group Γ of Möbius transformations of the open unit ball $B \subset \mathbb{R}^n$. Recall that the limit set Λ of Γ consists of all the points $x_\infty \in S = \partial B$ for which $\gamma_j(x) \to x_\infty$ for some $\gamma_1, \gamma_2, \dots \in \Gamma$ and $x \in B$. We shall suppose that Γ is *convex co-compact*, i.e., that the action of Γ on the convex hull $C(\Lambda)$ of the limit set $\Lambda = \Lambda(\Gamma)$ has a compact fundamental domain D. That is, there exists a compact subset D of B such that the orbit of any point of $C(\Lambda)$ meets D. Our goal here is to prove

Theorem 5.5.1. [Su2] *If Γ is convex co-compact, then*

$$\dim_H \Lambda(\Gamma) = \delta(\Gamma), \tag{5.5.1}$$

where $\delta(\Gamma)$ is the exponent of convergence of Γ.

Recall (Section 4.8) that $\delta = \delta(\Gamma)$ is defined as the largest lower bound of the set of all $s > 0$ for which the Poincaré series

$$P_s(x, y) = \sum_{\gamma \in \Gamma} \exp(-sd(x, \gamma(y)))$$

of Γ converges for some (equivalently, any) x and y of B.

Theorem 5.5.1 follows directly from

Theorem 5.5.2. *If Γ is convex co-compact, then there exists a positive constant $a > 0$ such that*

$$\mu_0 = a \cdot H^\delta, \tag{5.5.2}$$

where μ_0 is the Patterson–Sullivan measure considered in Theorem 4.8.12 and H^δ is the δ-dimensional Hausdorff measure on Λ.

To prove Theorem 5.5.2 define the measure

$$m = \mu_0 + H^\delta$$

and observe that μ_0 is absolutely continuous with respect to m. Let

$$f = \frac{d\mu_0}{dm} \tag{5.5.3}$$

be the Radon–Nikodym derivative. Since μ_0 satisfies the transformation rule (4.8.17) and — by definition (4.8.14) of the distortion $|\gamma'|$ of $\gamma \in \Gamma$ — the similar holds for H^δ, f is Γ-invariant. By Theorem 4.8.12, f has to be constant a.e. on Λ_c, the conical limit set of Γ. To complete the proof we have to show that $\Lambda_c = \Lambda$ and that f cannot be identically 0 nor 1 on Λ_c. We shall do all of that in a series of lemmas and propositions.

Proposition 5.5.3. *If Γ is convex co-compact, then $\Lambda_c = \Lambda$.*

Proof. Let $x_\infty \in \Lambda$ and $x \in C(\Lambda)$. The geodesic ray σ joining x to x_∞ is entirely contained in $C(\Lambda)$. If D is a compact fundamental domain for Γ acting on $C(\Lambda)$, then the ray σ meets infinitely many images $\gamma_j(D)$ with $\gamma_j \in \Gamma$, $j = 1, 2\ldots.$ If $y \in D$, then the hyperbolic distances between $\gamma_j(y)$ and some points of σ, say $\sigma(t_j)$ with $t_j \nearrow \infty$, are bounded by the diameter of D. Since any tubular neighbourhood (of a fixed radius) of a geodesic ray is contained in a Euclidean cone with the vertex at the end point of the ray, this shows that $\gamma_j(y)$ converges to x_∞ within a Euclidean cone, i.e., x_∞ is conical. □

Next, let us recall that $D(x, y, r)$, $x, y \in B$, $r > 0$, is the set of all the points $z_\infty \in S$ for which the geodesic ray joining x to z_∞ meets the hyperbolic ball of radius r and centre y. $D(x, y, r)$ turns out to be a ball in (S, d_x), d_x being the distance function given by (4.8.13) in Chapter 4. Clearly, the centre of this ball coincides with the end point of the geodesic ray which originates at x and passes through y. An elementary calculation in hyperbolic trigonometry (see, for example, [**Be1**], p. 147) shows that the d_x-radius $\rho = \rho(x, y, r)$ of this ball satisfies the relation

$$\tan\rho \cdot \sinh d(x, y) = \tanh r,$$

where d denotes the hyperbolic distance in B. For $x \neq 0$ this yields

$$\tan\rho = \frac{\tanh r \cdot (1 - |x|^2)(1 - |y|^2)}{2|x||x - y||\iota(x) - y|}, \tag{5.5.4}$$

where ι is the inversion with respect to the unit sphere S. For $x = 0$, (5.5.4) reduces to

$$\tan\rho = \frac{\tanh r \cdot (1 - |y|^2)}{2|y|}. \tag{5.5.5}$$

Lemma 5.5.4. *For any $r_1 > r_2 > 0$ there exist positive constants c_1, c_2 and c_3 such that*

$$\rho(0, y, r_1 + r_2) \leq c_1 \rho(0, x, r_1) \quad \textit{and} \quad \rho(0, y, r_1 - r_2) \geq c_2 \rho(0, x, r_1) \tag{5.5.6}$$

whenever $x, y \in B$, $d(x, y) \leq r_2$, $d(0, x) \geq c_3$ and $d(0, y) \geq c_3$.

Proof. From (5.5.5) it follows that

$$\frac{\tan\rho(0, y, r_1 + r_2)}{\tan\rho(0, x, r_1)} = \frac{\tanh(r_1 + r_2)}{\tanh r_1} \cdot \frac{|x|}{|y|} \cdot \frac{1 - |y|^2}{1 - |x|^2}.$$

If $c_3 > 0$ is large enough and $d(0, x) \geq c_3$, then $\tan\rho(0, x, r_1) \leq 2\rho(0, x, r_1)$. Since $\tan t \geq t$ for all $t \in [0, \frac{\pi}{2})$, it follows that

$$\frac{\rho(0, y, r_1 + r_2)}{\rho(0, x, r_1)} \leq a \cdot \frac{1 - |y|}{1 - |x|},$$

where $a > 0$ is a positive constant which depends in fact on r_1 and r_2 only. Finally, by (4.8.2),

$$\frac{1-|y|}{1-|x|} \leq 2e^{d(0,x)-d(0,y)} \leq 2e^{d(x,y)} \leq 2e^{r_2}$$

and this yields the first inequality in (5.5.6). The proof of the second one is quite similar. □

Now, denote by $B(x_\infty, r)$ the common part of S and the Euclidean ball of radius $r > 0$ and centre $x_\infty \in S$. With this notation we have the following.

Lemma 5.5.5. *If Γ is convex co-compact, then there exist positive constants c, C and r_0 such that*

$$cr^\delta \leq \mu_0(B(x_\infty, r)) \leq Cr^\delta \tag{5.5.7}$$

whenever $x_\infty \in \Lambda_c$ and $r \leq r_0$.

Proof. Let D be a fundamental domain for Γ acting on $C(\Lambda)$ and D_η be the η-neighbourhood of D: $D_\eta = \{x \in B; d(x,D) \leq \eta\}$. Take η large enough to satisfy $0 \in D_\eta$. Let r_2 be the hyperbolic diameter of D_η. Since x_∞ lies in Λ_c, the geodesic ray joining 0 to x_∞ is covered by some images $\gamma(D_\eta)$, $\gamma \in \Gamma$. Also, given r small enough and $r_1 > r_2$, there exists a unique point y on this geodesic ray for which $B(x_\infty, r) = D(0, y, r_1)$. Then, $y \in \gamma(D_\eta)$ for some $\gamma \in \Gamma$, $d(y, \gamma(0)) \leq r_2$ and

$$D(0, \gamma(0), r_1 - r_2) \subset D(0, y, r_1) \subset D(0, \gamma(0), r_1 + r_2).$$

It follows that

$$\mu_0(D(0, \gamma(0), r_1 - r_2)) \leq \mu_0(B(x_\infty, r)) \leq \mu_0(D(0, \gamma(0), r_1 + r_2)).$$

If r is small enough, then $r \approx \rho(0, y, r_1)$. More precisely,

$$1 - \varepsilon \leq \frac{\rho(0, y, r_1)}{r} \leq 1 + \varepsilon$$

for some $\varepsilon > 0$. Dividing all the quantities in the previous inequalities by r^δ and applying the last ones together with Lemma 5.5.4 yields

$$\begin{aligned} c_2^\delta(1-\varepsilon)^\delta \frac{\mu_0(D(0,\gamma(0),r_1-r_2))}{\rho(0,\gamma(0),r_1-r_2)^\delta} &\leq \frac{\mu_0(B(x_\infty,r))}{r^\delta} \\ &\leq c_1^\delta(1+\varepsilon)^\delta \frac{\mu_0(D(0,\gamma(0),r_1+r_2))}{\rho(0,\gamma(0),r_1+r_2)^\delta}. \end{aligned} \tag{5.5.8}$$

Since Γ is convex co-compact, μ_0 cannot reduce to a single atom and, as in the proof of Theorem 4.8.12,

$$\mu_0(D(\tau(0), 0, R)) \geq \frac{1}{2}(\mu_0(\Lambda) - m_0) > 0$$

for almost all $\tau \in \Gamma$ and R large enough. Here again, m_0 denotes the largest mass of an atom of μ_0.

Also, moving our "eye" from 0 to $\tau(0)$ yields

$$\mu_0(D(\tau(0),0,R))=\mu_{\tau^{-1}(0)}(D(0,\tau^{-1}(0),R)=\int_{D(0,\tau^{-1}(0),R)} p(\tau^{-1}(0),y_\infty)^\delta d\mu_0(y_\infty),$$

and, by (4.8.22) in Chapter 4,

$$\begin{aligned}\frac{a_1}{(1-|\tau^{-1}(0)|)^\delta}\mu_0(D(0,\tau^{-1}(0),R)) &\le \mu_0(D(\tau(0),0,R))\\ &\le \frac{a_2}{(1-|\tau^{-1}(0)|)^\delta}\mu_0(D(0,\tau^{-1}(0),R))\end{aligned} \tag{5.5.9}$$

for some positive constants a_1 and a_2. When $|\tau^{-1}(0)|$ is close to 1, then (by (5.5.5)) $1-|\tau^{-1}(0)|$ is close to $\rho(0,\tau^{-1}(0),R)$ and (5.5.9) yields

$$b_1\rho(0,\tau^{-1}(0),R)^\delta \le \mu_0(D(0,\tau^{-1}(0),R) \le b_2\rho(0,\tau^{-1}(0),R)^\delta \tag{5.5.10}$$

for other positive constants b_1 and b_2. Recall that (5.5.10) holds for all but finitely many $\tau \in \Gamma$.

Comparing (5.5.8) with (5.5.10) written for $R = r_1 \pm r_2$ and $\tau = \gamma^{-1}$ yields (5.5.7). □

Inequalities (5.5.7) imply (by arguments similar to those in the proof of Proposition 5.1.6) that the Radon–Nikodym derivative f in (5.5.3) cannot be identically either 0 or 1. This completes the proof of Theorem 5.5.2. □

The same result, $\dim_H \Lambda = \delta(\Gamma)$, was obtained by Sullivan [**Su4**] and Tukia [**Tu**] for geometrically finite Kleinian groups. Quite recently, Bishop and Jones [**BJ**] proved the equality $\dim_H \Lambda_c = \delta(\Gamma)$, Λ_c being the conical limit set of arbitrary discrete group Γ of Möbius transformations of the 3-dimensional ball, by means of definitions and elementary properties of Möbius transformations only. Also, one can observe a great interest in relating different geometrical data like entropy of the geodesic flow, Hausdorff dimension of the limit set of the fundamental group, Hausdorff measures etc. on negatively (or, non-positively) curved manifolds. Presentation of these beautiful and important ideas goes beyond purposes of this book. Interested readers are referred to, among many other papers, the series of papers [**Ham1**]–[**Ham5**] (see also references there), the survey [**EHS**] and several books ([**BGS**], [**Ni**] etc.) devoted to these and related topics.

Chapter 6

Varia

In this chapter, we collect different results related to some topics (like entropy or harmonic measures) discussed before. In the first section, we follow beautiful ideas due to Attie and Hurder [**AH**] who have shown that geometry of Riemannian manifolds quasi-isometric to leaves on compact foliated manifolds cannot be very chaotic in the sense that, on such a manifold, the maximal number of non-quasi-isometric pieces of a given radius grows at most exponentially as a function of the radius. In Section 2, we introduce, after Inaba and Tsuchiya [**IT**], the notion of expansivity for pseudogroups and foliations and prove that expansive foliations of codimension 1 have strictly positive entropy and, therefore, contain some resilient leaves. In Section 3, we show how to calculate entropy of a group in terms of separated pseudo-orbits. This should provide a computational method of calculation (or, estimation) of entropies of groups, pseudogroups and foliations. Finally, Section 4 is devoted to the study of topology of generic, with respect to any harmonic measure, leaves. Following Ghys [**Gh3**] we show that there are just six topological types of such leaves in dimension 2.

6.1 Complexity growth

Let (X, d) be a metric space. Fix $R > 0$ and $\varepsilon \geq 1$ and define the number $N_X(R, \varepsilon)$ as the smallest number k of metric spaces $(Z_1, d_1), \ldots, (Z_k, d_k)$ such that $\operatorname{diam} Z_i \leq R$ and any subset Y of X with $\operatorname{diam} Y \leq R$ is ε-quasi-isometric to a subset of one of the spaces (Z_i, d_i). Also, put $N_X(R, \varepsilon) = \infty$ when such k's do not exist.

Definition 6.1.1. The type of growth of the function $R \mapsto N_X(R, \varepsilon)$ is called the type of *ε-complexity growth* of (X, d).

Let us begin with the simplest examples and properties of this notion.

Example 6.1.2. If X is compact, its complexity growth type is constant. The same holds when X is homogeneous, i.e., when any two points x and y of X admit an isometry f of X such that $f(x) = y$, or when X is a (metric) covering of a compact space. To produce an example with $N_X(R, \varepsilon) = \infty$, take a real line $\mathbb{R}$ and at each point $t = n^2$ ($n \in \mathbb{N}$) attach a large number k, say $k = 2^n$, of segments of large length l, say $l = n$. Given $\varepsilon \geq 1$ and $R > 0$, there exists $n_0 \in \mathbb{N}$ such that the balls $B(n^2, R)$ and $B(m^2, R)$ are not ε-quasi-isometric for all n and $m \geq n_0$.

Proposition 6.1.3. *If two metric spaces (X, d) and (X', d') are quasi-isometric, then for any $\varepsilon \geq 1$ there exists $\varepsilon' \geq 1$ such that the types of ε'-complexity growth of X' and ε-complexity growth of X are the same.*

Proof. If X and X' are λ-quasi-isometric *via* maps $f : X \to X'$ and $g : X' \to X$, $Y' \subset X'$ has diameter $\leq R$ and $Y = g(Y')$ is ε-quasi-isometric to a space Z of diameter $\leq (R + 1)\lambda$ *via* maps $\phi : Y \to Z$ and $\psi : Z \to Y$, then Y' is ε''-quasi-isometric to Z *via* the maps $\phi' = \phi \circ g$ and $\psi' = f \circ \psi$, where $\varepsilon'' = \lambda\varepsilon + \max\{\lambda, \varepsilon\}$. Multiplying the metric on Z by a constant factor $R/[(R+1)\lambda]$ we achieve diam $Z \leq R$. With this new metric, ϕ' and ψ' become ε'-quasi-isometries for some ε' which depends on ε and λ only. This shows that $N_{X'}(R, \varepsilon') \leq N_X(\lambda(R + 1), \varepsilon)$. In the same way, $N_X(R, \varepsilon') \leq N_{X'}(\lambda(R + 1), \varepsilon)$. □

The examples above show that we can easily produce metric spaces with arbitrarily large complexity. This is not true in the case of Riemannian manifolds of bounded geometry.

Proposition 6.1.4. *If (M, g) is a complete Riemannian manifold of bounded geometry, then, for any ε large enough, its ε-complexity growth is at most that of the function* exp ∘ exp.

Proof. Recall that "bounded geometry" means here that the sectional curvature K of M is bounded by some constants a and b, $-\infty < a \leq K \leq b < \infty$, and the radius of injectivity $r(x)$ of M is bounded from below by a positive constant r_0: $r(x) \geq r_0 > 0$ for all $x \in M$.

The curvature bound from below implies that the volumes of balls of radius R are bounded from above by a number $V(R) \leq ce^{a'R}$ where c and a' are positive constants which depend on the curvature bounds and the dimension on M. The other curvature bound together with the bound for injectivity radius imply that the volumes of balls of radius $\eta \leq r_0$ are bounded from below by a positive number $v(\eta)$. Therefore, there exists another constant $c' > 0$ such that if $Z = \{z_1, \ldots, z_k\} \subset B(x, R) \subset M$ is a maximal η-separated set, then $k = \#Z \leq V(R)/v(\eta/2) \leq c'e^{a'R}$. Also, if $\eta < \frac{1}{2}$, then $B = B(x, R)$ is 1-quasi-isometric to Z *via* the maps id : $Z \to B$ and $f : B \to Z$ defined by $f(y) = z_i$, where $i = \min\{j; d(y, z_j) < \eta\}$.

For any such Z define the pattern of intersection as

$$\{(z, z') \in Z \times Z; B(z, \eta) \cap B(z', \eta)\} \neq \emptyset.$$

If Z and Z' have the same cardinality and the same pattern of intersection, then there exists a bijection $\phi : Z \to Z'$ such that $B(\phi(z), \eta) \cap B(\phi(z'), \eta) \neq \emptyset$ if and only if $B(z, \eta) \cap B(z', \eta) \neq \emptyset$. The map ϕ is a 2-quasi-isometry. Since the number of possible patterns for sets consisting of k elements does not exceed the number of all possible subsets of the set $\{1, \ldots, k\} \times \{1, \ldots, k\}$ we can conclude by

$$N_M(R, \varepsilon) \leq 2^{c'^2 \cdot \exp(2a'R)}$$

for all R and all ε large enough. □

Later on, we shall show that the estimate in Proposition 6.1.4 is sharp, i.e., that there exist complete Riemannian manifolds of bounded geometry which have super-exponential type of complexity growth. First, however, we shall prove that this cannot happen in the case of leaves of differentiable foliations of compact manifolds.

Theorem 6.1.5. [**AH**] *If L is a simply connected leaf of a C^1-foliation $\mathcal{F}$ of a compact Riemannian manifold (M, g), then the type of complexity growth of (L, g_L), g_L being the Riemannian structure on L induced from M, is at most exponential.*

Proof. Fix a nice covering $\mathcal{U}$ of $(M, \mathcal{F})$ and a complete transversal T. Let $A \subset L \cap T$ be a finite $(4R, \varepsilon)$-spanning set, ε being small enough. This means that for any $y \in T \cap L$ there exists $x \in A$ such that any curve $\gamma : [0, 1] \to L$ originated at y and of length $l(\gamma) \leq 4R$ can be projected orthogonally to L in such a way that the projection $p_x(\gamma)$ is originated near x and $d_M(\gamma(t), p_x(\gamma)(t)) < \varepsilon$ for all $t \in [0, 1]$.

For any $z \in B(y, R)$, the leaf ball of radius R centred at y, we can define a projection $p(z) \in L$ by $p(z) = p_x(\gamma)(1)$, where $\gamma : [0, 1] \to L$ is any curve of length $\leq R$ joining $y = \gamma(0)$ to $z = \gamma(1)$. Since L is simply connected, it has no holonomy and this definition is correct. Also, since M is compact, there exists a constant $\lambda > 1$ such that the norm of any orthogonal projection between two plaques of $\mathcal{U}$ within distance $\leq \varepsilon$ is bounded by λ. Therefore, p maps $B(y, R)$ into $B(x, \lambda R + \delta)$, where δ is an upper bound for distances between the projection of any point of T to a plaque P (within distance $\leq \varepsilon$) and the closest point of intersection of P and T. Also, since any two points y_1 and y_2 admit a loop γ which is contained in L, passes through y, y_1 and y_2, has length $\leq 4R$ and can be projected onto another one passing nearby x and through $p(y_1)$ and $p(y_2)$, the projection p is a λ-quasi-isometry. After shrinking the image $p(B(y, R))$ into $B(x, R)$ if necessary we can see that the balls $B(x, R)$ and $B(y, R)$ are η-quasi-isometric with $\eta > 1$ which depends on the choice of ε, $\mathcal{U}$ and the geometry of M only. Therefore,

$$N_L(R, \eta) \leq r(4R, \varepsilon, L) \leq r(4R, \varepsilon, \mathcal{F}),$$

where $r(4R, \varepsilon, L)$ is the smallest cardinality of $(4R, \varepsilon)$-spanning subsets of $T \cap L$.
Now, the result follows from Proposition 2.4.7 and Remark 3.3.2. □

Applying Proposition 6.1.3 we get immediately

Corollary 6.1.6. *If a complete simply connected Riemannian manifold of bounded geometry is quasi-isometric to a leaf on a compact* C^1*-foliated Riemannian manifold, then its complexity growth is at most exponential.*

The following example is due to Zeghib [**Ze**]. Roughly speaking, his construction consists in distributing around the hyperbolic plane ("an ocean") groups ("archipelagos") of discs ("islands") containing positively curved domains ("mountains"). The size of archipelagos, islands and mountains grows polynomially in such a way that for each size we have super-exponentially many archipelagos with different distribution of mountains of large but different size ("height"), so no two sufficiently large archipelagos are ε-quasi-isometric for an arbitrary fixed ε. Therefore, such an ocean has the super-exponential type of complexity growth. More precisely, we have the following

Example 6.1.7. Take the hyperbolic plane H^2, a number $r > 1$ and a ball $B \subset H^2$ of radius r^5. Geometric properties of H^2 imply the existence of a constant $c > 1$ such that any ball of radius r^m ($m \geq 2$, $r \geq 1$) contains $k = c^r$ pairwise disjoint balls of radius $r^{m-1}/2$. Therefore, one can choose k points $x_1, \ldots, x_k \in B$ such that $d(x_i, x_j) \geq r^4$ for $i \neq j$. Modify B by attaching standard spheres $S^2(r)$ of radius r (with a disc of a given radius $\delta << 1$ removed) at the points $x_1, \ldots, x_j$ and standard spheres $S^2(r^2)$ at the points $x_{j+1}, \ldots, x_{k-1}$. Let $D_0(j, r)$ denote the obtained Riemannian manifold while $D_1(j, r)$ — our manifold $D_0(j, r)$ modified by attaching at x_k a sphere $S^2(r^3)$. With these manifolds $D_0(j, r)$ and $D_1(j, r)$ in hand, construct for any map $\alpha : \{1, \ldots, k\} \to \{0, 1\}$ a Riemannian manifold $M(\alpha, r)$ by taking a hyperbolic ball B' of radius r^7, k points $y_1, \ldots, y_k \in B'$ with $d(y_i, y_j) \geq r^6$ for $i \neq j$, and replacing any ball $B(y_j, r^5)$ by $D_{\alpha(j)}(j, r)$. Finally, choose a geodesic ray $\gamma : [0, +\infty) \to H^2$ and a sequence (t_n) increasing so fast that $t_{n+1} - t_n > 100 \cdot n^8$, and attach at the points $\gamma(t_n)$ all the possible manifolds $M(\alpha, r)$ with $r = 1, 2, \ldots$. That is, for each n you have to remove the ball $B(\gamma(t_n), r^7) \in H^2$ and to replace it with some $M(\alpha, r)$.

Denote by M the Riemannian manifold obtained by the above construction. Clearly, M is complete and has bounded geometry. We claim that it has super-exponential type of complexity growth. That is, the complexity growth of M is exactly that of the function

$$n \mapsto A^{C^n}$$

with some constants A and C greater than 1.

To prove this claim it is sufficient to show that if $\varepsilon > 1$ is fixed and r is large enough, then the manifolds $M(\alpha, r)$ and $M(\alpha', r)$ are ε-quasi-isometric if and only if $\alpha = \alpha'$. Since the number of all possible functions $\alpha : \{1, \ldots, k\} \to \{0, 1\}$ equals $2^k = 2^{c^r}$, this would show that $N_M(r^7, \varepsilon) \geq 2^{c^r}$, proving the claim.

So, fix ε, take a large positive r and two functions $\alpha, \alpha' : \{1, \ldots, k\} \to \{0, 1\}$. Assume that $\alpha(j) \neq \alpha'(j)$, say $\alpha(j) = 1$ and $\alpha'(j) = 0$, for some j. Then, there exists a point $x_0 \in D_1(j, r)$ (the "north pole" of the attached sphere of radius r^3 or — if you prefer this terminology — the summit of the highest mountain) such that $\operatorname{vol} B(x_0, r^3) \leq a \cdot r^6$ for some universal constant $a > 0$ ($a \approx 4\pi$). On the other hand, if $x \in D_0(j,r)$, then $B(x,r^3) \subset M$ contains a hyperbolic ball of radius r and, therefore, $\operatorname{vol} B(x,r^3) \geq b^r$ for another universal constant $b > 1$. An argument similar to that of the proof of Proposition 2.2.16 shows that any ε-quasi-isometry has to map x_0 onto a point of $D_1(i,r)$ for some $i \neq j$.

The same argument shows that the "north poles" of all the standard spheres $S^2(r^2)$ contained in $D_1(i,r)$ have to be mapped by any ε-quasi-isometry onto some points of the spheres $S^2(r^2)$ contained in $D_1(j,r)$. The same holds for north poles of spheres $S^2(r)$ in $D_1(i,r)$ and $D_1(j,r)$. Since $i \neq j$ and the distances between different north poles exceed r^4, this is impossible when r is large. □

By Corollary 6.1.6, the surface M in the example above is not quasi-isometric to a leaf of any C^1-foliation of a compact manifold. The first example of this sort is due to Attie and Hurder [**AH**]. Their manifold with super-exponential complexity growth is of dimension 6 and the estimate of its type of complexity growth is obtained by comparing Pontrjagin characteristic classes of pieces inserted into H^6, the 6-dimensional hyperbolic space. Also, Schweitzer [**Sch**] has shown that a Zeghib's type construction performed over any complete non-compact surface Σ provides a Riemannian structure on Σ with respect to which Σ is not quasi-isometric to a leaf of a codimension-one foliated 3-manifold. Schweitzer's argument is that Σ contains short loops which can be contracted to points only through very long curves, while any leaf L of any codimension-one foliation of a compact Riemannian manifold has the following property: For any $a > 0$ there exists $A > 0$ such that any homotopically trivial loop γ on L of length $l(\gamma) \leq a$ admits a homotopy (γ_t), $0 \leq t \leq 1$, such that $\gamma_0 = \gamma$, $\gamma_1 = \{\text{pt}\}$ and $l(\gamma_t) \leq A$. Recall that Cantwell and Conlon [**CC4**] proved that every surface is homeomorphic to a leaf of a foliation of a compact 3-dimensional manifold, while Ghys [**Gh1**] and Inaba, Nishimori, Takamura and Tsuchiya [**INTT**] provided examples of open 3-dimensional manifolds which are not homeomorphic (nor, homotopy equivalent) to leaves of codimension-one foliations of compact manifolds. It seems that examples of topological non-leaves in higher codimension do not exist and that it is pretty hard to construct some of them.

6.2 Expansive systems

In the classical theory of dynamical systems, expansive mappings have been known and studied for a long time. The list of relevant references is too long to be presented here, see [**Wa**] for some information.

Definition 6.2.1. A homeomorphism f of a compact metric space (X,d) is said to be *expansive* when there exists a positive constant δ (called a *constant of expansion* for f) such that for any x and y in X, $x \neq y$, there exists $n \in \mathbb{Z}$ for which $d(f^n(x), f^n(y)) \geq \delta$.

The reader can easily modify this definition to get the notion of an expansive map (not necessarily one-to-one). Also, it is possible to generalize this notion to the case of groups (or, pseudogroups) of homeomorphisms.

Definition 6.2.2. A (finitely generated) group G acting on a compact metric space (X, d) is called *expansive* when there exists $\delta > 0$ such that for any x and y in X, $x \neq y$, one can find $g \in G$ for which $d(g(x), g(y)) \geq \delta$.

Rewriting this definition by replacing formally groups with pseudogroups could be confusing. By property (iv) in Definition 1.1.1, if $x \neq y$, U and V are disjoint neighbourhoods of x and y, respectively, g and h are elements of a pseudogroup $\mathcal{G}$ for which $U \subset D_g$ and $V \subset D_h$, then $k = g|U \cup h|V$ is an element of $\mathcal{G}$ which very likely expands x and y to the distance δ. That is, with such a definition almost all pseudogroups would be expansive and the notions of expansiveness for groups and pseudogroups generated by them would be completely inconsistent. Therefore, it is better to accept the following

Definition 6.2.3. Let $\mathcal{G}$ be a good pseudogroup on a compact metric space (X, d) generated by a good finite symmetric set $\mathcal{G}_1$. $\mathcal{G}$ is *expansive* with a constant $\delta > 0$ when for any two points x and y of X there exist $n \in \mathbb{N}$ and $g \in \mathcal{G}_n^c$ for which $\{x, y\} \subset D_g$ and $d(g(x), g(y)) \geq \delta$.

Working as in the proof of Lemma 2.4.3 one can show that expansiveness is independent of the choice of a good generating set $\mathcal{G}_1$. However, a constant of expansion may change while varying a generating set $\mathcal{G}_1$. Similarly, expansiveness is independent of the choice of the metric d on X. If d' is another metric on X (providing the same topology as d), then

$$\delta' = \inf\{d'(x, y);\ x, y \in X \text{ and } d(x, y) \geq \delta\}$$

is positive and becomes a constant of expansion on (X, d') for any expansive system with a constant of expansion δ on (X, d).

Also, if two good Lipschitz pseudogroups $\mathcal{G}$ on (X, d_X) and $\mathcal{H}$ on (Y, d_Y) are Lipschitz isomorphic, then one of them is expansive if and only if the other one is such. In fact, suppose that $\mathcal{H}$ is expansive with a constant δ, $\Phi : \mathcal{G} \to \mathcal{H}$ is an isomorphism generated by a finite set Φ_0, $\mathcal{G}_1$ and $\mathcal{H}_1$ are good finite sets generating $\mathcal{G}$ and $\mathcal{H}$, respectively, $\lambda \geq 1$ is a Lipschitz constant for all the maps of Φ_0, $\mathcal{G}_1$ and $\mathcal{H}_1$, and η is a Lebesgue number for the covers of X by the domains of $\phi \in \Phi_0$ and of Y by the ranges of these maps. If x and x' are arbitrary points of X, $x \neq x'$, then either $d_X(x, x') \geq \eta$ or there exists $\phi \in \Phi_0$ for which x and x' lie in D_ϕ. In the first case, x and x' are certainly expanded to the distance η by $\mathrm{id}_X \in \mathcal{G}$. In the second case, $\phi(x) \neq \phi(x')$ in Y and there exists $h \in \mathcal{H}$ for which $\phi(x), \phi(x') \in D_h$ and $d_Y(h(\phi(x)), h(\phi(x'))) \geq \delta$. The element h of $\mathcal{H}$ can be chosen to be "minimal" in the sense that $h \in \mathcal{H}_n$ and $d(h'(\phi(x)), h'(\phi(x'))) < \delta$ for all $h' \in \mathcal{H}_{n-1}$. Then, $d(h(\phi(x)), h(\phi(x'))) < \lambda\delta \leq \eta$ provided that δ is small enough. If so, there exists $\psi \in \Phi_0$ such that the points $h(\phi(x))$ and $h(\phi(x'))$ lie in R_ψ, the range of ψ. With this ψ we get $d_X(g(x), g(x')) \geq \delta/\lambda$, where $g = \psi^{-1} \circ h \circ \phi$. This shows that $\mathcal{G}$ is expansive with a constant $\delta' = \min\{\delta/\lambda, \eta/\lambda^2\}$.

The above observation allows us to define expansive foliations.

Definition 6.2.4. A C^1-foliation $\mathcal{F}$ of a compact manifold M is called *expansive* if its holonomy pseudogroup $\mathcal{H}_T$ is expansive for some (equivalently, any) complete transversal T. As usual, we equip T with the distance function defined by the Riemannian structure on T induced from a Riemannian structure on M.

Expansive foliations were defined and studied by Inaba and Tsuchiya [**IT**] who defined expansivity in terms of holonomy maps along curves. That is, $\mathcal{F}$ is expansive with a constant δ when, for any two points x and y lying on a small disc transverse to $\mathcal{F}$, there exists a leaf curve $\gamma : [0,1] \to L_x$ such that $\gamma(0) = x$ and its holonomy lift $\tilde{\gamma}$ to the leaf L_y through y satisfies the inequality $d(\gamma(1), \tilde{\gamma}(1)) \geq \delta$. Evidently, these two approaches are equivalent.

It is known that expansive homeomorphisms of the unit circle S^1 do not exist ([**Wa**], Thm. 5.27). However, one can easily construct an expansive group of homeomorphisms of S^1 as well as an expansive foliation of codimension 1.

Example 6.2.5. First, take two orientation preserving Morse–Smale diffeomorphisms f_0 and f_1 of S^1, each of them with exactly two periodic points x_i and y_i, $i = 0, 1$. Suppose that x_i's are sources, y_i's are sinks and that the sets $U_i = \{x \in S^1; f_i'(x) > 1 + \eta\}$, $i = 0, 1$, become, for some $\eta > 0$, open arcs covering S^1. Let δ be a Lebesgue number of the cover $\{U_0, U_1\}$. We shall show that the free group Γ generated by f_0 and f_1 is expansive with the constant δ. If $x, y \in S^1$, $x \neq y$ and $0 < d_0 = d(x, y) < \delta$, then there exists an index $i_1 \in \{0, 1\}$ such that $\{x, y\} \subset U_{i_1}$. Then, $d_1 = d(f_{i_1}(x), f_{i_1}(y)) \geq (1 + \eta)d_0$. If $d_1 < \delta$, then $\{f_{i_1}(x), f_{i_1}(y)\} \subset U_{i_2}$ for some $i_2 \in \{0, 1\}$, and so on. After a finite number of steps we end up with $g = f_{i_k} \circ \cdots \circ f_{i_1} \in \Gamma$ such that $d(g(x), g(y)) \geq (1 + \eta)^k d_0 \geq \delta$.

Now, take any compact manifold M with the fundamental group $\pi_1(M)$ containing a free subgroup generated by two elements z_0 and z_1 of $\pi_1(M)$. Choose any homomorphism $h : \pi_1(M) \to \mathrm{Diff}(S^1)$ for which $h(z_0) = f_0$ and $h(z_1) = f_1$. The suspension of h (see Section 1.3) provides an expansive foliation (in fact, a foliated bundle) of codimension 1.

Calculating entropy of expansive systems is somewhat easier than for arbitrary systems. This is because of

Proposition 6.2.6. *Let $\mathcal{G}$ be a good pseudogroup on a compact metric space (X, d) and $\mathcal{G}_1$ a finite good symmetric set generating $\mathcal{G}$. If $\mathcal{G}$ is expansive with a constant δ, then*

$$h(\mathcal{G}, \mathcal{G}_1) = s(\varepsilon, \mathcal{G}_1) \tag{6.2.1}$$

for all $\varepsilon < \delta$.

Proof. Recall that $s(\varepsilon, \mathcal{G}_1) = \limsup_{n \to \infty} \frac{1}{n} \log s(n, \varepsilon, \mathcal{G}_1)$, where $s(n, \varepsilon, \mathcal{G}_1)$ is the maximal cardinality of an (n, ε)-separated set $A \subset X$ and that two points x and

y of X are (n, ε)-separated when there exists $g \in \mathcal{G}_n$ such that $\{x, y\} \subset D_g$ and $d(g(x), g(y)) \geq \varepsilon$.

Take any $\varepsilon < \delta$. Since $\mathcal{G}$ is expansive with a constant δ and the complement of the ε-neighbourhood of the diagonal $\Delta \subset X \times X$ is compact, there exists $m = m(\varepsilon)$ such that the inequality $d(x, y) \geq \varepsilon$ implies that $d(g(x), g(y)) \geq \delta$ for some $g \in \mathcal{G}_m$. This shows that any (n, ε)-separated set $A \subset X$ is also $(n + m, \delta)$-separated and implies the inequalities

$$s(n, \varepsilon, \mathcal{G}_1) \leq s(n + m, \delta, \mathcal{G}_1) \quad \text{and} \quad s(\varepsilon, \mathcal{G}_1) \leq s(\delta, \mathcal{G}_1)$$

for all $\varepsilon < \delta$. Clearly, the above yields (6.2.1). □

Evidently, a similar argument applies to foliations of compact Riemannian manifolds and therefore we may write

Proposition 6.2.7. *If a foliation $\mathcal{F}$ of a compact Riemannian manifold M is expansive with a constant δ, then*

$$h(\mathcal{F}, g) = s(\varepsilon, \mathcal{F}) \tag{6.2.2}$$

for all $\varepsilon < \delta$. □

Here, $s(\varepsilon, \mathcal{F})$ is defined by (3.3.3) in Definition 3.3.1.

There exist examples of expansive systems with zero entropy. The simplest one is that of an isometry of a finite metric space. However, good one-dimensional expansive systems have positive entropy.

Theorem 6.2.8. [IT] *Let $\mathcal{G}$ be a good pseudogroup of orientation preserving local homeomorphisms of a finite union X of non-degenerate compact intervals $I_j \subset \mathbb{R}$. Suppose that $\mathcal{G}$ is generated by a good set $\mathcal{G}_1$ consisting of maps with connected domains. If $\mathcal{G}$ is expansive, then it has a resilient orbit and, therefore, its entropy is positive.*

Proof. Denote by δ a constant of expansion of $\mathcal{G}$. Cover X with a finite number of closed non-degenerate segments $K_i = [a_i, b_i]$, $i = 1, \dots, m$, of lengths less than $\delta/3$.

First, observe that the union of orbits $Y = \bigcup_{i=1}^{m} (\mathcal{G}(a_i) \cup \mathcal{G}(b_i))$ is dense in X. If not, $X \setminus \overline{Y}$ would contain an open interval (α, β), $\alpha < \beta$. Shrinking it, if necessary, we would be able to find $g \in \mathcal{G}_n^c$ such that $(\alpha, \beta) \subset D_g$ and $d(g(\alpha), g(\beta)) \geq \delta$. The interval $(g(\alpha), g(\beta))$ would contain some segment K_i and $g^{-1}(a_i)$ (as well as $g^{-1}(b_i)$) would lie in (α, β) contradicting $(\alpha, \beta) \cap Y = \emptyset$.

Next, note that X contains a local minimal set, i.e., a $\mathcal{G}$-invariant non-empty open set U such that any orbit $\mathcal{G}(x)$ with $x \in U$ is dense in U. If not, there would exist points s_i and t_i arbitrarily close to a_i and b_i, respectively, and with orbits $\mathcal{G}(s_i)$ and $\mathcal{G}(t_i)$ not locally dense . But, the argument similar to the above one shows the density (in X) of the set $Y' = \bigcup_{i=1}^{m} (\mathcal{G}(s_i) \cup \mathcal{G}(t_i))$, a contradiction.

Also, one can see that there exist $g \in \bigcup_n \mathcal{G}_n^c$ and $i \in \{1, \dots, m\}$ for which $K_i \subset \operatorname{int} g(K_i) \subset U$. In fact, we can find an interval $I = (\alpha, \beta)$, $\alpha < \beta$ entirely contained in U and such that there exists $h \in \mathcal{G}_l^c$, $l \in \mathbb{N}$, for which $I \subset D_h$ and $d(h(\alpha), h(\beta)) > \delta$. Then $K_{i_1} \subset h(I) \subset U$ for some i_1. Again, we can find $h_1 \in \mathcal{G}_{l_1}^c$ such that $K_{i_1} \subset D_{h_1}$ and $d(h_1(a_{i_1}), h_1(b_{i_1})) > \delta$. Then $K_{i_2} \subset \operatorname{int} h_1(K_{i_1})$ for some i_2. By induction, we can find infinite sequences $(i_1, i_2, \dots)$ of numbers $i_n \in \{1, \dots, m\}$ and $(h_1, h_2, \dots)$ of maps in $\bigcup_n \mathcal{G}_n^c$ such that $K_{i_{n+1}} \subset \operatorname{int} h_n(K_{i_n})$ for all n. There exist integers m and n such that $1 \leq m < n$ and $i_m = i_n$. The interval K_{i_m} and the map $g = h_{i_n} \circ \dots \circ h_{i_{m+1}}$ satisfy our requirements.

For these K_i and g we have $g(b_i) < b_i$, $g^2(b_i) < g(b_i)$ etc., so the sequence $g^n(b_i)$ converges to some x. Then $g(x) = x$ and, since the orbit $\mathcal{G}(x)$ is dense in a neighbourhood of x, there exists y in $(x, b_i) \cap \mathcal{G}(x)$. Evidently, $g^n(y) \to x$ and the orbit through x is resilient.

The second part of the statement follows directly from the first one and Proposition 2.4.10. □

This theorem applied to foliations yields

Corollary 6.2.9. *Any expansive codimension-one foliation contains a resilient leaf and has positive entropy.* □

Inaba and Tsuchiya [**IT**] proved also that some expansive foliations of codimension greater than 1 have also positive entropy. Such foliations should be "strongly expansive" in the sense that any arbitrarily small transverse disc at a point x should be mapped by some holonomy map h onto a set containing the transverse disc at $h(x)$ of radius δ, $\delta > 0$ being a constant of expansion.

6.3 Pseudo-orbits and pseudoleaves

Pseudo-orbits for classical dynamical systems were introduced by Bowen [**Bow2**] and were used by several authors as a powerful conceptual tool in dynamics. In particular, Misiurewicz [**Mis2**], Barge and Swanson [**BS**] showed how to calculate topological entropy $h_{\text{top}}(f)$ of a continuous transformation of a compact metric space (X, d) in terms of pseudo-orbits. In this section, we discuss a similar question for foliations and finitely generated groups of transformations. We begin with the classical case of a single transformation.

Let α be a non-negative number and G_1 a finite set generating a group G of transformations of a compact metric space (X, d).

Definition 6.3.1. A map $z : G \to X$ is said to be an *α-pseudo-orbit* when

$$d(z(hg), h(z(g))) \leq \alpha \tag{6.3.1}$$

for all $g \in G$ and $h \in G_1$.

If $G = \mathbb{Z}$ is generated by a single transformation f, then (6.3.1) reduces to

$$d(z_{n+1}, f(z_n)) \leq \alpha \tag{6.3.2}$$

for all $n \in \mathbb{Z}$. Therefore, our definition is consistent with Bowen's.

Clearly, 0-pseudo-orbits coincide with genuine G-orbits of points of X.

Now, let us say that two pseudo-orbits z and $z' : G \to X$ are *(n, ε)-separated* when

$$d(z'(g), z(g)) \geq \varepsilon \tag{6.3.3}$$

for some $g \in G_n$. Let X_α be the the space of all α-pseudo-orbits (so, X_0 is just the space of orbits) and $N_\alpha(n, \varepsilon)$ the maximal cardinality of a subset A of X_α such that any two pseudo-orbits z and z' of A, $z \neq z'$, are (n, ε)-separated.

The following result is originally due to Misiurewicz [**Mis2**].

Theorem 6.3.2. *If $f : X \to X$ is a homeomorphism of a compact metric space (X, d), then*

$$h_{\text{top}}(f) = \lim_{\varepsilon \to 0} \lim_{\alpha \to 0} \limsup_{n \to \infty} \frac{1}{n} \log N_\alpha(n, \varepsilon). \tag{6.3.4}$$

We split the proof into some lemmas. First, equip the spaces X_α, $\alpha \geq 0$, with a distance function d_0,

$$d_0(z, z') = \sum_{k \in \mathbb{Z}} \frac{1}{2^{|k|}} d(z_k, z'_k), \tag{6.3.5}$$

and consider the shifts $\sigma_\alpha : X_\alpha \to X_\alpha$ given by

$$(\sigma_\alpha(z))_k = z_{k+1} \tag{6.3.6}$$

for all $k \in \mathbb{Z}$.

Lemma 6.3.3. $h_{\text{top}}(\sigma_0) = h_{\text{top}}(f)$.

Proof. Denote by $s(n, \varepsilon)$ and $s_\alpha(n, \varepsilon)$ the maximal cardinalities of subsets of X and X_α which are (n, ε)-separated with respect to to f and σ_α, respectively.

By definition, if points x and y of X are (n, ε)-separated by f, then their orbits z^x and z^y are (n, ε)-separated by σ_0 with the same n and ε. Therefore, $s_0(n, \varepsilon) \geq s(n, \varepsilon)$ for all n and ε, and this yields the inequality $h_{\text{top}}(f) \leq h_{\text{top}}(\sigma_0)$.

On the other hand, if the orbits z^x and z^y are (n, ε)-separated with respect to σ_0, then

$$\sum_{j \in \mathbb{Z}} \frac{1}{2^{|j|}} d(f^{k+j}(x), f^{k+j}(y)) = d_0(\sigma_0^k(z^x), \sigma_0^k(z^y)) \geq \varepsilon$$

for some $k \leq n$. There exists $m = m(\varepsilon) \in \mathbb{N}$ such that

$$\sum_{j > m} \frac{1}{2^j} < \frac{\varepsilon}{4 \operatorname{diam} X}.$$

For this m we obtain

$$\sum_{j=-m}^{m} \frac{1}{2^{|j|}} d(f^{k+j}(x), f^{k+j}(y)) \geq \frac{\varepsilon}{2}$$

and

$$d(f^{k+j}(x), f^{k+j}(y)) \geq \frac{\varepsilon}{8}$$

for some $k \leq n$ and $j \in \{-m, \dots, m\}$. This shows that the points x and y are $(n+m, \varepsilon/8)$-separated. In this way, we have obtained the inequality $s_0(n, \varepsilon) \leq s(n + m(\varepsilon), \varepsilon/8)$ which yields the relation $h_{\text{top}}(\sigma_0) \leq h_{\text{top}}(f)$. □

Lemma 6.3.4. $N_\alpha(n, \varepsilon) \leq s_\alpha(n, \varepsilon)$.

Proof. The statement follows directly from the inequality $d_0(z, z') \geq d(z_0, z_0')$ which holds evidently in view of (6.3.5) and implies $d_0(\sigma_\alpha^k(z), \sigma_\alpha^k(z')) \geq d(z_k, z_k')$ for all $k \in \mathbb{Z}$, z and z' in X_α. □

As usually when defining entropy (compare Section 3.1.1), put

$$s_\alpha(\varepsilon) = \limsup_{n\to\infty} \frac{1}{n} \log s_\alpha(n, \varepsilon).$$

The most important step in the proof of the theorem is the following

Lemma 6.3.5. $s_0(\varepsilon/2) \geq \inf_{\alpha>0} s_\alpha(\varepsilon)$.

Proof. Let $\mathcal{U}_\varepsilon$ be the cover of X by all the open balls of radius $\varepsilon/2$. By the argument of the proof of Theorem 3.1.3,

$$s_\alpha(n, \varepsilon) \leq N(\mathcal{U}_\varepsilon^{(n)}) \leq s_\alpha(n, \varepsilon/2)$$

and

$$s_\alpha(\varepsilon) \leq \lim_{n\to\infty} \frac{1}{n} \log N(\mathcal{U}_\varepsilon^{(n)}) = \inf_{n\in\mathbb{N}} \frac{1}{n} \log N(\mathcal{U}_\varepsilon^{(n)}),$$

where the last equality follows from elementary calculus and from the relation $N(\mathcal{U} \vee \mathcal{V}) \leq N(\mathcal{U}) \cdot N(\mathcal{V})$ which — as was mentioned in Section 3.1.1 — holds for arbitrary open covers $\mathcal{U}$ and $\mathcal{V}$ of X. It follows that

$$\inf_{\alpha>0} s_\alpha(\varepsilon) \leq s_\alpha(\varepsilon) \leq \frac{1}{n} \log s_\alpha(n, \varepsilon/2)$$

for all n. Since all the spaces X_α are compact and $X_0 = \bigcap_{\alpha>0} X_\alpha$, the distance between X_0 and the complement of X_α (in, say, X_1) decreases to 0 as $\alpha \to 0$. Therefore, if A is a maximal (n, ε)-separated subset of X_0 and α is small enough, then any ε-ball centred at a point of X_α intersects at least one of the ε-balls centred at points of A and this implies that $s_\alpha(n, \varepsilon) \to s_0(n, \varepsilon)$ when $\alpha \to 0$. Passing to suitable limits yields the statement. □

The inequality "$\leq$" in (6.3.4) is evident. To prove the reverse just compile Lemmas 6.3.3 through 6.3.5. □

A good part of the above argument can be adapted to the case of an arbitrary finitely generated group G of transformations of X. For instance, we can define a distance in X_α by

$$d_0(z, z') = \sum_{k=0}^{\infty} \frac{1}{2^k} \max_{g \in G_k} d(z(g), z'(g)) \tag{6.3.7}$$

and lift the action of G to X_α by

$$\sigma_g(z)(h) = z(gh) \tag{6.3.8}$$

for all g and h of G. Clearly, the maps σ_g are supposed to play the role of the shifts σ_α in the proof above. It is easy to rewrite the proofs of Lemmas 6.3.3 and 6.3.4 to get analogous results for G. In particular, the entropies of G with respect to any finite generating set G_1 on X and X_0 are equal. However, the argument in the proof of Lemma 6.3.5 is purely "one-dimensional", in the sense that the group there is just $\mathbb{Z}$ and has linear growth. Moreover, the example below shows that the equality analogous to (6.3.4) does not hold for general groups.

Example 6.3.6. Consider the group Γ described in Example 6.2.5 and use the notation from there. Take any point $x \in S^1$, $\alpha > 0$, $\varepsilon > 0$ and $n \in \mathbb{N}$. Choose any $\gamma = f_{i_n} \circ \cdots \circ f_{i_1}$ in Γ_n and let $y = \gamma(x)$. Then, $y \in U_{i_{n+1}}$ for some $i_{n+1} \in \{0, 1\}$. Put $y_1 = f_{i_{n+1}}(y) + \alpha$. Again, $y_1 \in U_{i_{n+2}}$ for some $i_{n+2} \in \{0, 1\}$ and we put $y_2 = f_{i_{n+2}}(y_1) + \alpha$. We continue defining consequent points $y_3, y_4, \dots$ in such a way that $y_{k+1} = f_{i_{k+1}}(y_k)$ when $y_k \in U_{i_{k+1}}$. The Mean Value Theorem shows that $d(y_m, f_{i_{n+m}} \circ \cdots \circ f_{i_{n+1}}(y)) \geq \varepsilon$ for some $m \in \mathbb{N}$ which depends on α, ε and η, a constant involved in Example 6.2.5, only.

Define an α-pseudo-orbit $z = z(\gamma, x)$ of Γ as follows:

(1) $z(g) = g(x)$ for all $g \in \Gamma_n$ and any $g \in \Gamma_k \setminus \Gamma_{k-1}$, $k > n$, of the form $g = g'\gamma'$, where $g' \in \Gamma_{k-n}$, $\gamma' \in \Gamma_n$ and $\gamma' \neq \gamma$.

(2) If $\gamma(x) \in U_0$, then put $h_1 = f_0$, otherwise $h_1 = f_1$ and, in both cases, $z(h_1\gamma) = h_1(\gamma(x)) + \alpha$.

(3) Assume that $h_1, \dots, h_k \in \Gamma_1$ and $z(h_k \dots h_1\gamma)$ are already defined. If $z(h_k \dots h_1\gamma)$ lies in U_0, then put $h_{k+1} = f_0$, otherwise $h_{k+1} = f_1$ and, in both cases,
$$z(h_{k+1}h_k \dots h_1\gamma) = h_{k+1}(z(h_k \dots h_1\gamma)) + \alpha.$$

(4) If $k \in \mathbb{N}$, $m \in \mathbb{N}$, $h = h'_m \dots h'_1 h_k \dots h_1\gamma$, where $h'_i \in \Gamma_1$, $h'_1 \neq h_{k+1}$ and $h'_1 \neq h_k^{-1}$, then $z(h) = h'_m \dots h'_1(z(h_k \dots h_1\gamma))$.

Roughly speaking, $z(\gamma, x)$ is obtained by a suitable modification of the orbit $\Gamma(x)$ along a single, originated at γ, branch of the Cayley graph of Γ.

The Mean Value Theorem shows that $d(z(h_k \dots h_1\gamma), h_k \dots h_1\gamma(x)) \geq \varepsilon$ whenever k satisfies the inequality $k > \log(\varepsilon/\alpha)/\log(1+\eta)$, η being the constant from Example 6.2.5 as before.

It follows that for any (n, ε)-separated set $A \subset S^1$ the set

$$A_\alpha = \{z(\gamma, x), \ \gamma \in \Gamma_n \quad \text{and} \quad x \in A\}$$

of all the pseudo-orbits built as above is $(n+m, \varepsilon)$-separated for some m independent on n. Consequently,

$$s_\alpha(n+m, \varepsilon) \geq 2^n s(n, \varepsilon)$$

and

$$\lim_{\varepsilon \to 0} \lim_{\alpha \to 0} \limsup_{n \to \infty} \frac{1}{n} \log s_\alpha(n, \varepsilon) \geq \log 2 + h(\Gamma, \Gamma_1).$$

The suitable result concerning calculation of $h(G, G_1)$ in terms of separated pseudo-orbits reads as follows.

Theorem 6.3.7. *For any group G of Lipschitz transformations of a compact metric space X and any finite symmetric set G_1 generating G we have the equality*

$$h(G, G_1) = \lim_{\varepsilon \to 0} \inf_\beta \mathfrak{s}_\beta(\varepsilon), \tag{6.3.9}$$

where

$$\mathfrak{s}_\beta(\varepsilon) = \limsup_{n \to \infty} \frac{1}{n} \log s_{\beta_n}(n, \varepsilon) \tag{6.3.10}$$

and $\beta = (\beta_n)$ ranges over the set of all the sequences of positive reals which decrease to 0 as $n \to \infty$.

Note that the "error" β_n decreases to 0 as $n \to \infty$ and this is the principal difference between this result and the classical one (Theorem 6.3.2).

Proof. The proof is quite analogous to that of Theorem 6.3.2, so we just sketch it here emphasizing some differences.

As we mentioned before, G acts on the space X_α, $\alpha \geq 0$, of α-pseudo-orbits and the entropies of G (with respect to G_1) acting on X and on X_0, the space of genuine orbits, are equal. Take any α-pseudo-orbit z, the genuine orbit z' of the point $z(e)$ and any $g \in G_n$. Since all the maps of G are Lipschitz, the inequality

$$d(z(g), z'(g)) \leq \alpha \cdot (1 + L + \cdots + L^{n-1}),$$

where L is a Lipschitz constant for all the maps $g \in G_1$, holds.

This shows that if $A = \{z_1, \ldots, z_k\}$ is an (n, ε)-separated subset of X_α and the product

$$\alpha \cdot (1 + L + \cdots + L^{n-1})$$

is small enough, then the set $A' = \{z'_1, \ldots z'_k\}$ of corresponding orbits z'_i, $z'_i(e) = z_i(e)$, is (n, ε')-separated with ε' sufficiently small, say $\varepsilon' = \varepsilon/10$. Clearly, $\#A' = \#A$, so

$$s_0(n, \varepsilon') \geq s_\alpha(n, \varepsilon)$$

if α is small enough. This implies inequality "$\geq$" in (6.3.9).

The converse inequality is obvious since any genuine orbit belongs to X_α for any $\alpha > 0$. □

Note that the same result holds for pseudogroups of locally Lipschitz transformations of X (see [**BW1**]). However, one has to define carefully pseudo-orbits of pseudogroups and to find a good distance function on X_α. It occurs that for our purposes one can define α-pseudo-orbits of a finitely generated pseudogroup $\mathcal{G}$ as maps $z : D_z \to X$, $D_z \subset \mathcal{G}_\infty = \bigcup_n \mathcal{G}_n$ being the domain, such that

(i) $e \in D_z$,

(ii) $D_z \subset \mathcal{G}(z(e)) = \{g \in \mathcal{G}_\infty;\ z(e) \in D_g\}$, and

(iii) if $g \in D_z$ and $h \in \mathcal{G}_1$, then $hg \in D_z$ if and only if $z(g) \in D_h$, and then $d(z(hg), h(z(g))) \leq \alpha$.

Since the domains of different pseudo-orbits can differ, the formula analogous to (6.3.7) does not define any distance function on X_α and one has to modify it, for instance in the following way:

$$d_0(z, z') = \sum_{n=0}^{\infty} \frac{1}{2^n} \max\{d(z(g), z'(g));\ g \in \mathcal{G}_n \cap D_z \cap D_{z'}\}$$

and

$$d_1(z, z') = \inf\Big\{\sum_{j=0}^{m-1} d_0(z_j, z_{j+1});\ z_j \in X_\alpha, z_0 = z \text{ and } z_m = z'\Big\}.$$

Note that the function d_0 is symmetric and non-negative but *a priori* does not satisfy the triangle inequality while d_1 satisfies all the axioms of distance. Therefore, (X_α, d_1), $\alpha \geq 0$, become metric spaces. One can check that they are compact and therefore we can repeat the proof of Theorem 6.3.7 to get the same result for pseudogroups.

Applying the above to holonomy pseudogroups of foliations one can consider their pseudo-orbits. Since orbits of holonomy pseudogroups correspond to leaves one can expect the analogous correspondence between holonomy pseudo-orbits and submanifolds called pseudoleaves in [**In2**].

Definition 6.3.8. An *α-pseudoleaf* ($\alpha \geq 0$) on a compact foliated Riemannian manifold $(M, \mathcal{F})$ is a complete immersed submanifold N of M such that $\dim N = \dim \mathcal{F}$ and the angle between tangent spaces T_xN and $T_x\mathcal{F}$ ($x \in N$) is at most α. This condition can be expressed by the following inequality:

$$\max\Big\{\min\{|v - w|;\ v \in T_xN, |v| = 1\};\ w \in T_x\mathcal{F}, |w| = 1\Big\} \leq \sqrt{2(1 - \cos\alpha)}. \tag{6.3.11}$$

It is easy to see that any α-pseudoleaf N equipped with a base point $x \in N$ gives rise to a pseudo-orbit z_N of the holonomy pseudogroup if only α is sufficiently small. To see this better let us take a nice covering $\mathcal{U}$ of $(M, \mathcal{F})$ such that any two overlapping plaques P and Q of $\mathcal{U}$ are contained in a geodesically convex ball on the corresponding leaf of $\mathcal{F}$ and consider the holonomy pseudogroup $\mathcal{H} = \mathcal{H}_{\mathcal{U}}$ acting on the transversal $T = T_{\mathcal{U}}$ determined by $\mathcal{U}$. First, put $z_N(e) = x$. Then, if $z_N(g)$, $g \in \mathcal{H}$, is already defined and $h = h_{UV}$ is an elementary holonomy map corresponding to suitable overlapping charts U and V of $\mathcal{U}$, then define $z_N(hg)$ as the end point of the lift $\tilde{\gamma}$ to N of a unique minimal leaf geodesic γ joining $z_N(g)$ to $h(z_N(g))$. By definition, $\tilde{\gamma}$ is a unique curve in N originated at $z_N(g)$ and projecting orthogonally onto γ. Such $\tilde{\gamma}$ exists if only α is small enough. It is clear that z_N becomes a β-pseudo-orbit of $\mathcal{H}$ for some $\beta = \beta(\alpha)$ such that $\beta \searrow 0$ as $\alpha \searrow 0$. The correspondence $\alpha \mapsto \beta(\alpha)$ depends on $\mathcal{U}$ and the geometry of $(M, \mathcal{F})$.

Constructing pseudoleaves from holonomy pseudo-orbits is not so obvious in general. We shall show how to perform such a construction in the case of foliated bundles. The construction consists in suitably "stretching" a submanifold (seen as a membrane) on a "frame" built from all the points of a holonomy pseudo-orbit. It provides non-trivial (that is different from the leaves themselves and manifolds obtained by tiny local deformations of leaves) examples of pseudo-leaves.

First, let us recall the notion of the centre of mass in Riemannian geometry (see [**BK**] for more details).

Let N be any Riemannian manifold and $x \in N$. Assume that $r > 0$ is so small that $r < r_x = \min\{r(x), \frac{1}{4}\pi/\kappa\}$, where $r(x)$ is the injectivity radius of N at x and $\kappa^2 > 0$ bounds from above the sectional curvature of N on $B(x,r)$. For any points $x_1, \dots, x_m$ of $B(x,r)$ and any non-negative reals $\lambda_1, \dots, \lambda_m$ (masses of the points $x_1, \dots, x_m$) with $\sum_i \lambda_i = 1$ the function

$$x \mapsto \sum_i \lambda_1 d(x, x_i)^2$$

is strictly convex and admits a unique minimum point $c = c(\{(x_i, \lambda_i)\})$. This point c is called the *centre of mass* of x_i's with respect to the mass distribution (λ_i). It depends smoothly on x_i's and λ_i's. That is, if N' is any other manifold, $(\lambda_1, \dots, \lambda_m)$ — a smooth partition of unity subordinated to an open cover $(U_1, \dots, U_m)$ of N' and $f_i : U_i \to B(x,r)$, $i = 1, \dots, m$ — differentiable maps, then the map

$$N' \ni y \mapsto c(\{(f_i(y), \lambda_i(y))\})$$

is smooth.

Now, let $\mathcal{F}$ be the suspension of a homomorphism $f : \Gamma \to \mathrm{Diff}(F)$, where $\Gamma = \pi_1(B)$, B and F are compact Riemannian manifolds (Example 1.3.8). Lift the Riemannian structure of B to $\tilde{B}$, the universal covering of B, equip $\tilde{B} \times F$ with the product Riemannian structure and project it to $M = (\tilde{B} \times F)/\Gamma$. Let Γ_1 be a finite symmetric set generating Γ and $z : \Gamma \to F$ be an α-pseudo-orbit, $\alpha > 0$. Cover $\tilde{B}$ by open sets U_γ, $\gamma \in \Gamma$, in such a way that $U_\gamma \cap U'_\gamma \neq \emptyset$ if and only if $\gamma'\gamma^{-1} \in \Gamma_1$. This can be achieved, for instance, by defining the sets U_γ as η-neighbourhoods of the sets

$$V_\gamma = \{\tilde{x} \in \tilde{B};\ d(\tilde{x}, \gamma(\tilde{x}_0)) \leq d(\tilde{x}, \gamma'(\tilde{x}_0)) \text{ for all } \gamma' \in \Gamma\},$$

where $\tilde{x}_0$ is a fixed base point of $\tilde{B}$ and $\eta > 0$ is sufficiently small. Next, choose any smooth partition of unity λ_γ, $\gamma \in \Gamma$, with $\operatorname{supp} \lambda_\gamma \subset U_\gamma$ and define the map $\iota : \tilde{B} \to M$ by the formula

$$\iota(\tilde{x}) = \pi(\tilde{x}, c(\{(\gamma^{-1}z(\gamma), \lambda_\gamma(\tilde{x}))\})), \tag{6.3.12}$$

where $\pi : \tilde{B} \times F \to M$ is the canonical projection. Note that if z is a genuine orbit, then $\gamma^{-1}z(\gamma)$ is the same for all $\gamma \in \Gamma$. Since for any $\tilde{x} \in \tilde{B}$ the number of strictly positive weights $\lambda_\gamma(\tilde{x})$ is finite, all the points $\gamma^{-1}z(\gamma)$ corresponding to positive weights are close and ι is well defined whenever α is sufficiently small. Clearly, ι maps $\tilde{B}$ onto a leaf of $\mathcal{F}_0$ when z is a genuine orbit of Γ. Therefore, again if α is small enough, ι is an immersion of $\tilde{B}$ into M which provides us with a β-pseudoleaf of $\mathcal{F}$, $\beta = \beta(\alpha)$ being a positive number which depends on α and the geometry of $(M, \mathcal{F})$. As before, $\beta(\alpha) \searrow 0$ as $\alpha \searrow 0$.

In spite of the fact that the above construction cannot be performed in the case of an arbitrary foliation, a result analogous to Theorem 6.3.7 holds for foliations and pseudoleaves. To formulate it we have to define the notion of separation of pseudoleaves.

Definition 6.3.9. Let $(M, \mathcal{F}, g)$ be a compact foliated Riemannian manifold. Two α-pseudoleaves N_1 and $N_2 \subset M$ equipped with base points $x_1 \in N_1 \cap T$ and $x_2 \in N_2 \cap T$, T being a complete transversal, are (R, ε)-*separated* when either $d_M(x_1, x_2) \geq \varepsilon_0$ or there exists a curve $\gamma : [0,1] \to N_1$ (resp, $\gamma : [0,1] \to N_2$) originated at x_1 (resp., at x_2), of length $l(\gamma) \leq R$ and such that its orthogonal projection $\tilde{\gamma}$ onto N_2 (resp., onto N_1) originated in the ball $B_{N_2}(x_2, \varepsilon_1)$ (resp., in $B_{N_1}(x_1, \varepsilon_1)$) exists and satisfies

$$d_M(\gamma(1), \tilde{\gamma}(1)) \geq \varepsilon. \tag{6.3.13}$$

Here, ε_0 and ε_1 are positive constants which depend on the geometry of $(M, \mathcal{F})$ and provide the existence and uniqueness of suitable orthogonal projections $N_1 \ni x \mapsto \tilde{x} \in B_{N_2}(y, \varepsilon_1)$ whenever N_1 and N_2 are α-pseudoleaves, $\alpha \leq \alpha_0$, $y \in N_2$ and $d_M(x, y) < \varepsilon_0$.

We shall denote by $s_\alpha(R, \varepsilon, \mathcal{F})$ the maximal cardinality of a set of pairwise (R, ε)-separated pairs (N, x), N being an α-pseudoleaf of $\mathcal{F}$ and x being a point of $N \cap T$. Also, for any function $\beta : (0, \infty) \to (0, \infty)$ such that $\beta(R) \searrow 0$ as $R \to \infty$ we put

$$\mathfrak{s}_\beta(\varepsilon, \mathcal{F}) = \limsup_{R \to \infty} \frac{1}{R} \log s_{\beta(R)}(R, \varepsilon, \mathcal{F}). \tag{6.3.14}$$

With this notation we have

Theorem 6.3.10. *The geometric entropy $h(\mathcal{F},g)$ a foliation $\mathcal{F}$ of a compact Riemannian manifold (M,g) satisfies the equality*

$$h(\mathcal{F},g) = \lim_{\varepsilon\to 0+} \inf_{\beta} \mathfrak{s}_{\beta}(\varepsilon,\mathcal{F}). \tag{6.3.15}$$

Note that here again the "error", this time seen as an angle, $\beta(R)$ decreases to 0 as $R \to \infty$.

Proof. Since genuine leaves of $\mathcal{F}$ become α-pseudoleaves for any $\alpha > 0$, the inequality "$\leq$" becomes obvious.

To prove the converse cover M by "coins" U_y, $y \in Y$, as in the proof of Theorem 3.4.1. Recall that the coins are defined by

$$U_y = \bigcup_{x\in T_y} B^{\mathcal{F}}(x,\Delta),$$

where

$$T_y = \exp(B^{\perp}(0_y,\rho)),$$

and Δ and ρ are small enough.

Fix a small number $\eta > 0$ and make $Y \subset M$ finite and dense in such a way that any point of M is at most $\eta/2$ apart in d_M-distance from one of the points y of Y and $\eta/2$ apart in the leaf distance from a point of $T = \bigcup_{y\in Y} T_y$.

Next, take $\alpha > 0$ and $\varepsilon > 0$ sufficiently small, two (R,ε)-separated α-pseudoleaves N_1, N_2 with base points x_1, x_2 and the corresponding $\beta(\alpha)$-pseudo-orbits z_1, z_2 of the holonomy pseudogroup $\mathcal{H} = \mathcal{H}_{\mathcal{U}}$, where $\mathcal{U} = \{U_y, y \in Y\}$. Here $z_i(e) = x_i$, so if $d_M(x_1,x_2) \geq \varepsilon_0$, then the pseudo-orbits z_1 and z_2 are (n,ε)-separated for any n. If $\gamma : [0,1] \to N_1$, $\gamma(0) = x_1$, $l(\gamma) \leq R$, is a curve separating N_1 from N_2, then shorten γ if necessary in such a way that $d_M(\gamma(1),\tilde{\gamma}(1)) = \varepsilon$ and $d(\gamma(t),\tilde{\gamma}(t)) < \varepsilon$ for all $t \in (0,1)$. Split γ into $n = [1 + R/(2\Delta - 2\eta)]$ pieces γ_j of length $l(\gamma_j) \leq 2\Delta - 2\eta$. Set $x'_0 = x_1$, $x'_{n+1} = \gamma(1)$ and denote by x'_j the mid-point of γ_j, $j = 1,\dots,n$. Find points $y_j \in Y$ for which $d_M(y_j,x'_j) \leq \eta/2$.

If α is small enough, each piece γ_j lies entirely in $U_j = U_{y_j}$. If $h_j = h_{U_jU_{j+1}}$ is the holonomy map and $g_j = h_j \circ \dots \circ h_0$, then, again if α is small enough, $d_{N_1}(x'_j, z_1(g_j)) \leq \eta$, $z_1(g_j) \in U_j$ for all j and $g \in D_{z_1}$. Similarly, $g_n \in D_{z_2}$ if ε is small enough. Moreover, since $d_{N_1}(z_1(g_n),\gamma(1)) \leq \eta$ and the similar inequality holds for $z_2(g_n)$ and $\tilde{\gamma}(1)$ on N_2, $d_M(z_1(g_n), z_2(g_n)) \geq C\varepsilon$ for some $C > 0$. Hence, the pseudo-orbits z_1 and z_2 are $(n+1, C\varepsilon)$-separated and, consequently,

$$s_{\alpha}(R,\varepsilon,\mathcal{F}) \leq s_{\beta(\alpha)}([2 + R/(2\delta - 2\eta)], C\varepsilon,\mathcal{H}) \tag{6.3.16}$$

for α and ε sufficiently small.

Passing to suitable limits we obtain from Theorem 6.3.7 the estimate of the right-hand side in (6.3.15) by $(2\Delta - 2\eta)^{-1}h(\mathcal{H},\mathcal{H}_1)$. Since $\operatorname{diam}\mathcal{U} = 2\Delta$ and $\eta > 0$ is arbitrarily small, the required inequality follows from Theorem 3.4.1. $\square$

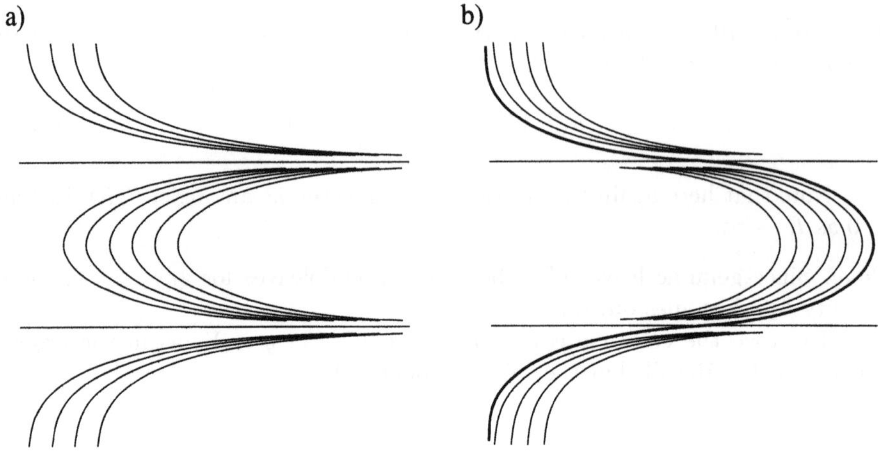

Figure 6.1: Orienting leaves of the Reeb component: a) unsuitable, b) suitable.

Remark 6.3.11. The construction described above of a pseudoleaf from a pseudo-orbit allows us to get an example of a "virtual leaf" which can be defined as a family of quasi-isometric α-pseudo-leaves ($\alpha > 0$) with super-exponential complexity growth.

To this end, choose disjoint open arcs $I_1, I_2 \subset S^1$ and two diffeomorphisms f_1, f_2 preserving the arcs and satisfying the following condition:

(*) If $\alpha > 0$, then there exists $m = m(\alpha) \in \mathbb{N}$ such that for any sequence $k_1, \ldots, k_m$ of elements of the set $\{1, 2\}$ and any points $x \in I_1$ and $y \in I_2$ there exists a suitable (connecting x to y) "piece" of an α-pseudo-orbit of the group generated by f_1 and f_2, that is a sequence $z_1, \ldots, z_m$ of points of S^1 for which $z_1 = x$, $z_m = y$ and $d(f_{k_{i+1}}(z_i), z_{i+1}) \leq \alpha$ for any $i = 1, \ldots, m-1$.

Define a homomorphism of the fundamental group $\Gamma = \pi_1(\Sigma_2)$ of the closed oriented surface Σ_2 of genus 2 into the group $\mathrm{Diff}(S^1)$ by

$$a_i \mapsto f_i, \quad b_i \mapsto \mathrm{id}_{S^1}, \quad i = 1, 2,$$

where a_i, b_i are the standard generators of Γ for which $a_1 b_1 a_1^{-1} b_1^{-1} a_2 b_2 a_2^{-1} b_2^{-1} = e$. Suspend this homomorphism to get a foliated S^1-bundle $(M, \mathcal{F}_0)$ over Σ_2. Then, construct α-pseudo-orbits z_α of Γ which stay in I_1 outside $m(\alpha)$-neighbourhoods of some elements $g_1, g_2, \ldots$ of Γ mapped by z_α onto y. Equip M with a Riemannian metric for which all the leaves of $\mathcal{F}_0$ become hyperbolic (curvature $= -1$). Extend each z_α to an α-pseudoleaf $\iota_\alpha : \tilde{\Sigma}_2 \to M$.

Next, choose a loop γ passing through $y \in I_2 \subset S^1 \subset M$ transverse to $\mathcal{F}_0$ and turbulize $\mathcal{F}_0$ in a neighbourhood $\mathcal{N}(\gamma)$ of γ. The obtained foliation $\mathcal{F}$ possesses a Reeb component $\mathcal{R}(\gamma)$ contained in $\mathcal{N}(\gamma)$. Modify the Riemannian structure of M inside $\mathcal{N}(\gamma)$ making the leaves of $\mathcal{F}|\mathcal{R}(\gamma)$ non-negatively curved. Finally, modify ι_α

by pasting to $\tilde{\Sigma}_2$ large pieces of the leaves of the Reeb component at all the points mapped onto y by ι_α. Make this carefully enough to get α-pseudoleaves of $\mathcal{F}$. In particular, choose properly (see Figure 6.1) orientations of leaves within and around $\mathcal{R}(\gamma)$ to make the construction performable. Also, choose properly sizes of our "non-negatively curved islands" and their distribution all over the $\tilde{\Sigma}_2$ to get pairwise quasi-isometric pseudoleaves of super-exponential complexity growth (compare Example 6.1.7). See [**Wal3**] for a more detailed description.

Remark 6.3.12. Hurley [**Hur**] defined several different entropies of inverses of maps. If a map is not one-to-one, its inverse is just a relation, not a map. In [**LW1**], the authors introduced a notion of entropy for some relations. When applied to inverses of maps, it coincides with one of Hurley's entropies. This entropy has been shown by Langevin and Przytycki [**LP**] to vanish for inverses of arbitrary maps of a circle as well as for inverses of rational maps of the Riemann sphere. Hurley has shown also some relations between his entropies expressed, for example, in terms of pseudo-orbits. More relations of this sort were obtained recently by Nitecki and Przytycki [**NP**] who provided also several striking examples which show how different is the behaviour of entropies of inverses in comparison to the classical topological entropy of maps. In [**LW2**] and [**LW3**], the authors introduced more dynamical invariants (like transverse entropy of a flow tangent to a foliation, pressure of cocycles for pseudogroups, strength of resiliency for leaves, etc.) for different systems and obtained some relations between them. We can imagine that some of them can be expressed also in terms of pseudo-orbits and pseudoleaves.

6.4 Generic leaves

As was mentioned at the very beginning of this chapter, we shall complete it by a presentation of an application of harmonic measures to a description of the topology of generic leaves. The main result presented here is the following

Theorem 6.4.1. [**Gh3**] *Let $\mathcal{F}$ be a foliation of a compact manifold M equipped with a Riemannian metric and with a harmonic Borel probability measure μ. The union of all the leaves of $\mathcal{F}$ with* 0, 1, 2 *or a Cantor set of ends is of μ-measure* 1.

First, let us recall that an end of a locally compact topological space Z is a map E which assigns to each compact subset K of Z an unbounded (that is such that its closure in noncompact) connected component $E(K)$ of $Z \setminus K$ in such a way that

$$K_1 \subset K_2 \quad \Longrightarrow \quad E(K_2) \subset E(K_1).$$

If Z admits a countable basis, then the space $\mathcal{E}(Z)$ of all the ends of Z can be equipped with a metric topology. It can be induced by the distance function d defined by the formula

$$d(E_1, E_2) = \exp(-\sup\{j; E_1(K_j) = E_2(K_j)\}),$$

where $K_1 \subset K_2 \subset \cdots$ is a countable family of increasing compact subsets of Z for which $Z = \bigcup_j K_j$. The topology induced by d is independent of the family (K_j). One can show that $\mathcal{E}(Z)$ equipped with this topology is compact and totally discontinuous (see [**Ep**], for instance). We say that Z has a Cantor set of ends, when $\mathcal{E}(Z)$ has no isolated points.

Next, let us consider a harmonic measure μ on $(M, \mathcal{F})$. Proposition 6.4.4 plays a fundamental role in the proof of the above theorem. In its proof we shall use a slightly delicate version of the classical Poincaré Recurrence Theorem.

Let $f : X \to X$ be a measure preserving transformation of a probability space (X, μ). The classical Poincaré Recurrence Theorem reads as follows.

Theorem 6.4.2. *If $A \subset X$ is a measurable set and $\mu(A) > 0$, then almost all the points of A return infinitely often to A under positive iterates of f, i.e., there exists a subset B of A such that $\mu(B) = \mu(A)$ and for any point x of B one can find an infinite sequence $n_1 < n_2 < n_3 < \cdots$ of natural numbers for which all the points $f^{n_i}(x)$, $i \in \mathbb{N}$, lie in A.* □

This theorem is pretty well known and it is easy either to manage to prove it or to find its proof in the literature ([**Wa**], p. 26, for instance).

Theorem 6.4.2 implies that the function $R_A : A \to \mathbb{N}$ assigning to each x of A the time of its first return to A can be defined almost everywhere on A:

$$R_A(x) = \inf\{n > 0; f^n(x) \in A\}.$$

The following property of R_A will be used later.

Theorem 6.4.3. [**Ka**] *The first return time function R_A is measurable and*

$$\int_A R_A d\mu = 1 - \mu(\cap_{n=0}^{\infty} B_n) \leq 1, \tag{6.4.1}$$

where $B_n = f^{-n}(X \setminus A)$.

This theorem seems to be not so popular and its proof is a bit delicate, hence we decided to give the proof here to make the book more self-contained.

Proof. The proof of measurability of R_A is rather standard. To establish (6.4.1) let $A_n = f^{-n}(A) = X \setminus B_n$, $a_0 = 1$ and $a_n = \mu(B_0 \cap \cdots \cap B_{n-1})$ when $n \geq 1$. Then

$$\mu(A \cap B_1 \cap \cdots \cap B_n \cap A_{n+1}) = a_n - 2a_{n+1} + a_{n+2}.$$

In fact, set theoretic manipulations and f-invariance of μ implies that

$$\begin{aligned}
\mu(A \cap B_1 \cap \cdots \cap B_n \cap A_{n+1}) &= \mu(A \cap B_1 \cap \cdots \cap B_n) - \mu(A \cap B_1 \cap \cdots \cap B_{n+1}) \\
&= (\mu(B_1 \cap \cdots \cap B_n) - \mu(B_0 \cap \cdots \cap B_n)) \\
&\quad - (\mu(B_1 \cap \cdots \cap B_{n+1}) - \mu(B_0 \cap \cdots \cap B_{n+1})) \\
&= (\mu(B_0 \cap \cdots \cap B_{n-1}) - \mu(B_0 \cap \cdots \cap B_n)) \\
&\quad - (\mu(B_0 \cap \cdots \cap B_n) - \mu(B_0 \cap \cdots \cap B_{n+1})) \\
&= a_n - 2a_{n+1} + a_{n+2}.
\end{aligned}$$

Since $R_A \equiv k+1$ on $A \cap B_1 \cap \cdots \cap B_k \cap A_{k+1}$, it follows that

$$\begin{aligned}\int_A R_A d\mu &= \sum_{k=0}^{\infty}(k+1)\mu(A \cap B_1 \cap \cdots \cap B_k \cap A_{k+1}) \\ &= \lim_{n\to\infty}\left(\sum_{k=0}^{n-1}(k+1)a_k - 2\sum_{k=1}^{n} ka_k + \sum_{k=2}^{n+1}(k-1)a_k\right) \\ &= \lim_{n\to\infty}(a_0 - (n+1)a_n + na_{n+1}) \\ &= 1 - \lim_{n\to\infty}(n(a_n - a_{n+1}) + a_n).\end{aligned} \tag{6.4.2}$$

It is clear that the above partial sums increase, hence the terms

$$n(a_n - a_{n+1}) + a_n, \quad n \in \mathbb{N},$$

decrease and, consequently, converge. The series

$$\sum_{n=0}^{\infty}(a_n - a_{n+1})$$

converges as well. In fact, the difference $a_n - a_{n+1}$ is equal to the measure of the set of all the points $x \in X \setminus A$ which enter A at time n for the first time and these sets are pairwise disjoint. Therefore,

$$\sum_{k=1}^{n}(a_k - a_{k+1}) \leq \mu(X) = 1$$

for all $n \in \mathbb{N}$. It follows that

$$\lim_{n\to\infty} n(a_n - a_{n+1}) = 0. \tag{6.4.3}$$

In fact, if $n(a_n - a_{n+1}) \to c > 0$ as $n \to \infty$, then $\frac{c}{2n} \leq a_n - a_{n+1}$ for n large enough and the series $\sum_n n^{-1}$ would converge.

Relations (6.4.2) and (6.4.3) imply (6.4.1). □

Proposition 6.4.4. *If $B \subset M$ is a Borel set, then, for μ-a.e. x of M, the intersection $L_x \cap B$ of B with the leaf L_x of $\mathcal{F}$ passing through x either is empty or approaches all the ends of L_x, i.e., any non-compact domain $V \subset L_x$ with compact boundary $\partial V = \bar{V} \setminus \operatorname{int} V$ intersects B.*

Proof. Let us consider the space $\Omega^{\mathcal{F}}$ of all the leaf curves γ defined on the interval $[0, +\infty)$ and its subspaces $\Omega^{\mathcal{F}}_A$, $A \subset M$, of all the members γ of $\Omega^{\mathcal{F}}$ with $\gamma(0) \in A$. Put $\Omega^{\mathcal{F}}_x = \Omega^{\mathcal{F}}_{\{x\}}$ for all x of M. Equip the spaces $\Omega^{\mathcal{F}}_x$ with the Wiener measures w_x and define the probability measure $\bar{\mu}$ on $\Omega^{\mathcal{F}}$ by

$$\bar{\mu} = \int_M w_x d\mu(x)$$

as in Section 4.7.

For any Borel set $B \subset M$ define a function $\mathcal{R}_B : \Omega^{\mathcal{F}} \to \mathbb{N} \cup \{\infty\}$ by

$$\mathcal{R}_B(\gamma) = \inf\{k \in \mathbb{N}; \gamma(k) \in B\} = \inf\{k \in \mathbb{N}; S_k(\gamma) \in \Omega_B^{\mathcal{F}}\},$$

where S_τ, $\tau \geq 0$, is an endomorphism of $\Omega^{\mathcal{F}}$ given by $S_\tau(\gamma)(t) = \gamma(t+\tau)$ (compare (4.7.31) in Chapter 4). Since $S_k = (S_1)^k$ and, by Proposition 4.7.13, $\bar{\mu}$ is S_τ invariant for all τ, Theorem 6.4.3 implies that

$$\int_{\Omega_B^{\mathcal{F}}} \mathcal{R}_B d\bar{\mu} < +\infty$$

and, by the Fubini Theorem,

$$\int_{\Omega_x^{\mathcal{F}}} \mathcal{R}_B dw_x < +\infty \tag{6.4.4}$$

for μ-a.e. x of M.

Also, for any compact subset K of a non-compact leaf L of $\mathcal{F}$ define a function $r_K : \Omega_L^{\mathcal{F}} \to [0, +\infty]$ by

$$r_K(\gamma) = \inf\{t \geq 0; \gamma(t) \in K\} = \inf\{t \geq 0; S_t\gamma \in \Omega_K^{\mathcal{F}}\}.$$

We intend to prove the following claim: The equality

$$\int_{\Omega_x^{\mathcal{F}}} r_K dw_x = +\infty \tag{6.4.5}$$

holds for all x of any unbounded connected component of $L \setminus K$.

For this purpose, denote by $f(x)$ the left-hand side of (6.4.5) and take any small ball $B \subset L \setminus K$. For any $\gamma \in \Omega_B^{\mathcal{F}}$ we have

$$r_K(\gamma) = r_{\partial B}(\gamma) + r_K(S_{r_{\partial B}(\gamma)}(\gamma))$$

and therefore $f(x)$, $x \in B$, can be decomposed as

$$f(x) = f_1(x) + f_2(x),$$

where

$$f_1(x) = \int_{\Omega_x^{\mathcal{F}}} r_{\partial B} dw_x \quad \text{and} \quad f_2(x) = \int_{\Omega_x^{\mathcal{F}}} r_K(S_{r_{\partial B}(\gamma)}(\gamma)) dw_x(\gamma).$$

From the theory of diffusion and Brownian motion ([**Fe**], [**IW**] etc.) it follows that f_2 is either infinite at all the points of B or is finite and harmonic in B. Similarly, f_1 is finite, C^2-differentiable in B and $\Delta f_1 = -1$, Δ being the Laplace operator on L. Therefore, in each connected component of $L \setminus K$, f is either identically infinite or finite, C^2-differentiable and $\Delta f = -1$ there.

Since L is a leaf on a compact foliated manifold M, its geometry is bounded, in particular its injectivity radius is bounded from below by a positive constant ε_0. If $\varepsilon < \varepsilon_0$ and $S(x, \varepsilon) \subset L$ is the sphere of centre x and radius ε, then

$$\int_{\Omega_x^{\mathcal{F}}} r_{S(x,\varepsilon)} dw_x \geq c,$$

where c is a positive constant which depends on ε and the geometry bounds of L only. It follows that $f(x) > mc$ if only $d_L(x, K) \geq 2m\varepsilon$, d_L being, as usual, the Riemannian distance on L. This means that

$$f(x) \to \infty \quad \text{when} \quad d_L(x, K) \to \infty.$$

Assume now that V is such a component of $L \setminus K$ that f is finite on V. If the set K is contained in a smooth manifold $K_0 \subset L$, $\dim K_0 = \dim L$, with the boundary ∂K_0, then — since $\Delta f = -1$ — the inequality $f \geq \eta$ along the boundary of $V \cap K_0$ implies the same inequality all over $V \setminus K_0$. Therefore, all the sets $V \cap f^{-1}(a)$, $a \geq \eta_0 = \inf f|V \cap K_0$, are compact and the sets $V \cap f^{-1}((0, a])$ are relatively compact in L.

Passing to a twofold covering, if necessary, we may assume that L is oriented in such a way that all the fluxes Φ_a, $a > \eta_0$, of the gradient ∇f over the submanifolds $V \cap f^{-1}(a)$ corresponding to all the regular values a of f are non-negative. If $a < b$ are such regular values, then by Stokes' Theorem

$$-\operatorname{vol}(V \cap f^{-1}([a, b])) = \int_{V \cap f^{-1}([a,b])} \Delta f = \Phi_b - \Phi_a$$

and consequently

$$\operatorname{vol}(V \cap f^{-1}([a, b]) \leq \Phi_a$$

for all such $b > a$. It follows that $\operatorname{vol}(V \cap f^{-1}(a, \infty)) \leq \Phi_a < \infty$. Since the set $V \cap f^{-1}(0, a]$ is relatively compact, its volume is finite, $\operatorname{vol} V < \infty$ and V is bounded. (If not, by recurrence of leaves in compact foliated manifolds, it would contain an infinite family of pairwise disjoint balls $B_1, B_2, \ldots$ of radius $\varepsilon > 0$, $\operatorname{vol} B_i \geq v(\varepsilon) > 0$ for all i and $\operatorname{vol} V = \infty$.) This proves the claim.

To complete the proof of the proposition take first a Borel set $B \subset M$, a point x of B and a connected non-compact set $V \subset L_x$ with compact boundary ∂V. Let

$$\Omega_{x,V}^{\mathcal{F}} = \{\gamma \in \Omega_x^{\mathcal{F}}; \gamma(1) \in V\}.$$

Clearly, $w_x(\Omega_{x,V}^{\mathcal{F}}) > 0$. Let ν_x be the measure on L_x induced by w_x and the map $\Omega_x^{\mathcal{F}} \ni \gamma \mapsto \gamma(1)$. Then

$$\nu_x(V) = w_x(\Omega_{x,V}^{\mathcal{F}}) > 0.$$

Assume that $B \cap V = \emptyset$. Then

$$\mathcal{R}_B(\gamma) \geq 1 + \inf\{t \in [1, \infty); \gamma(t) \in \partial V\}$$

for any $\gamma \in \Omega^{\mathcal{F}}_{x,V}$. Therefore,

$$\int_{\Omega^{\mathcal{F}}_x} \mathcal{R}_B dw_x \geq \int_{\Omega^{\mathcal{F}}_{x,V}} \mathcal{R}_B dw_x \geq \int_V \left(\int_{\Omega^{\mathcal{F}}_y} (1 + r_{\partial V}) dw_y \right) d\nu_x(y).$$

By (6.4.5), if $y \in \operatorname{int} V$, then

$$\int_{\Omega^{\mathcal{F}}_y} r_{\partial V} dw_y = \infty.$$

Consequently,

$$\int_{\Omega^{\mathcal{F}}_x} \mathcal{R}_B dw_x = \infty. \tag{6.4.6}$$

Relations (6.4.4) and (6.4.6) show that $B \cap V \neq \emptyset$ and, therefore, the intersection $B \cap L_x$ approaches all the ends of L_x for μ-a.e. x of B.

Finally, denote by $B(\eta)$, $\eta > 0$, the set of all the points $x \in M$ for which the L_x-distance from B does not exceed η. Considering collar neighbourhoods of ends one can see that if $B(\eta) \cap L_x$ approaches all the ends of L_x, then the same holds for $B \cap L_x$. Denote by $A(\eta)$ the set of all the points x of $B(\eta)$ for which $B(\eta) \cap L_x$ approaches all the ends of L_x and by A the set of all the points x of M for which $B \cap L_x$ approaches all the ends of L_x. Then $A = \bigcup_\eta A(\eta)$ and $\mu(A(\eta)) = \mu(B(\eta))$ by the previous argument. Therefore,

$$\mu(A) = \lim_{\eta \to \infty} \mu(A(\eta)) = \lim_{\eta \to \infty} \mu(B(\eta)) = \mu(\tilde{B}),$$

$\tilde{B}$ being the $\mathcal{F}$-saturation of B. □

To begin the proof of Theorem 6.4.1 let us choose a nice covering $\mathcal{U}$ of $(M, \mathcal{F})$. Denote by T the complete transversal determined by $\mathcal{U}$ and by $\mathcal{H}$ the holonomy pseudogroup of $\mathcal{F}$ acting on T. Recall (Section 4.7) that a harmonic measure on M decomposes in each chart U of $\mathcal{U}$ as

$$\mu | U = \int_{T \cap U} m_x d\sigma_U(x), \tag{6.4.7}$$

where $m_x = f_x \cdot \mathrm{vol}_{\mathcal{F}}$, f_x is a harmonic function on the plaque $P_x \subset U$ ($x \in P_x$), $\mathrm{vol}_{\mathcal{F}}$ is the Riemannian volume element on the leaves and σ_U is a Borel measure on the transversal $T_U \subset U$. The measures σ_U, $U \in \mathcal{U}$, constitute a measure σ on T.

The following fact is evident from (6.4.7).

Lemma 6.4.5. *If $A \subset T$ is $\mathcal{H}$-invariant and $\sigma(A) = 0$, then $\mu(\tilde{A}) = 0$, where $\tilde{A}$ is the $\mathcal{F}$-saturation of A, that is $\tilde{A}$ is the union of all the leaves of $\mathcal{F}$ passing through the points of A.* □

Therefore, we can reduce the problem to a discrete case, that of the orbits of $\mathcal{H}$.

Let $\mathcal{H}_1$ be the generating set consisting of all the elementary holonomy maps corresponding to overlapping charts of $\mathcal{U}$. Given $x \in T$, define a distance function $d_{\mathcal{H}}$ in $\mathcal{H}(x)$, the $\mathcal{H}$-orbit of x, as follows:

$$d_{\mathcal{H}}(y,z) = \inf\{m;\ y = h(z) \text{ and } h \in \mathcal{H}_m\}, \quad y,z \in \mathcal{H}(x),$$

where, as in Chapter 1, $\mathcal{H}_n$ denotes the set of all the superpositions of n generating elements. Then, consider the graph $\tilde{\mathcal{H}}(x)$ which has vertices at all the points of the orbit $\mathcal{H}(x)$ and edges of length 1 connecting any vertices y and z for which $d_{\mathcal{H}}(y,z) = 1$. As we have seen in previous chapters, the leaf L_x and the graph $\tilde{\mathcal{H}}(x)$ are quasi-isometric. This observation implies directly the following

Lemma 6.4.6. *The spaces of ends $\mathcal{E}(L_x)$ and $\mathcal{E}(\tilde{\mathcal{H}}(x))$ are homeomorphic for any point x of M.* □

Therefore, Theorem 6.4.1 is equivalent to

Theorem 6.4.7. *The set of all the points x of T for which $\tilde{\mathcal{H}}(x)$ has $0,1,2$ or a Cantor set of ends is of full σ-measure in T.*

Observe also that Proposition 6.4.4 implies directly its discrete version which can be formulated as

Proposition 6.4.8. *If $B \subset T$ is a Borel set, then, for σ-a.e. x of T, the intersection $B \cap \tilde{\mathcal{H}}(x)$ either is empty or approaches all the ends of $\tilde{\mathcal{H}}(x)$.* □

Let us consider for a moment an arbitrary locally finite graph Γ. Fix $k \in \mathbb{N}$. For any finite connected graph $\Delta \subset \Gamma$ containing at most k vertices, define $\omega(\Gamma, \Delta)$ as the union of all the connected components Γ_0 of $\Gamma \setminus \Delta$ such that the closure $\overline{\Gamma_0}$ has exactly one end. Moreover, let $\omega_k(\Gamma, \Delta)$ be the set of all the points ξ of $\omega(\Gamma, \Delta)$ for which $d_{\mathcal{H}}(\xi, \Delta) > k$ and let $\omega_k(\Gamma)$ denote the union of all such sets $\omega_k(\Gamma, \Delta)$. Finally, let $B_k(\Gamma) = \Gamma \setminus \omega_k(\Gamma)$.

Lemma 6.4.9. *If $\mathcal{E}(\Gamma)$ contains an isolated element and k is large enough, then $B_k(\Gamma) \neq \Gamma$ and $B_k(\Gamma)$ does not approach some of the ends of Γ.*

Proof. If $\mathcal{E}(\Gamma)$ contains an isolated end E_0, then there exist a connected finite graph $\Delta \subset \Gamma$ and a connected component Γ_0 of $\Gamma \setminus \Delta$ such that $\overline{\Gamma_0}$ has just one end. Therefore, $\omega(\Gamma, \Delta) \neq \emptyset$ and $\omega_k(\Gamma) \neq \emptyset$ whenever k is bigger than the number of vertices of Δ. For such k, $B_k(\Gamma) \neq \Gamma$. Obviously, $B_k(\Gamma)$ does not approach E_0. □

Lemma 6.4.10. *If $\#\mathcal{E}(\Gamma) \geq 3$, then $B_k(\Gamma) \neq \emptyset$ for k large enough.*

Proof. Let $K \subset \Gamma$ be a connected compact set for which $\Gamma \setminus K$ consists of pairwise disjoint connected components $V_1, \ldots, V_m$, where $m \geq 3$. Take any $k \geq \operatorname{diam}(K) + 1$. We shall show that $K \subset B_k(\Gamma)$.

Suppose that this is not true. Then there exists a connected finite graph $\Delta \subset \Gamma$ containing at most k vertices and such that $\omega_k(\Gamma, \Delta) \cap K \neq \emptyset$. There exist also a connected component Γ_0 of $\Gamma \setminus \Delta$ and a point x of K such that $\overline{\Gamma_0}$ has exactly one end, $x \in \Gamma_0$ and $d_{\mathcal{H}}(x, \Delta) > k$. Since k is bigger than the diameter of K, $K \cap \Delta = \emptyset$ and $K \subset \Gamma_0$. Also, Δ is contained in a connected component, say V_1, of $\Gamma \setminus K$. Since K is connected, the union $K \cup V_2 \cup \cdots \cup V_m$ is connected as well. Therefore, $K \cup V_2 \cup \cdots \cup V_m \subset \Gamma_0$ and $\overline{\Gamma_0}$ has at least two ends (contained in V_2 and V_3), a contradiction. □

Let us come back to our foliated manifold $(M, \mathcal{F})$ and let

$$B_k = \bigcup_{x \in T} B_k(\tilde{\mathcal{H}}(x)).$$

It is clear from Lemmas 6.4.9 and 6.4.10 that applying Proposition 6.4.8 to one of the sets B_k, $k = 1, 2, \ldots$, yields Theorem 6.4.7 (and Theorem 6.4.1 at the same time). Therefore, it remains to show that we are allowed to do this, i.e., that B_k is a Borel set for any $k \in \mathbb{N}$.

Lemma 6.4.11. *B_k is a Borel subset of T for any $k = 1, 2, \ldots$.*

Proof. Let us make first some observations which lead towards the proof of our lemma.

(i) The collection W^m of all the sequences $(x_1, \ldots, x_m)$, $x_j \in T$, such that $x_1, \ldots, x_m$ belong to the same orbit of $\mathcal{H}$ is a Borel set. In fact, if $\mathcal{H}_\infty = \bigcup_{n \in \mathbb{N}} \mathcal{H}_n$, then

$$W^m = \bigcup_{g_2, \ldots, g_m \in \mathcal{H}_\infty} \{(x, g_2(x), \ldots, g_m(x)); x \in D_{g_2} \cap \cdots \cap D_{g_m}\},$$

occurs to be a countable union of Borel subsets of T^m.

(ii) The distance function $d_{\mathcal{H}} : W^2 \to \mathbb{N}$ is Borel measurable.

(iii) The collection of all the points x of T for which $\tilde{\mathcal{H}}(x)$ has exactly one end is a Borel set. In fact, $\tilde{\mathcal{H}}(x)$ has one end when $\mathcal{H}(x)$ is infinite and for any $k_1 \in \mathbb{N}$ there exists $k_2 \in \mathbb{N}$ such that arbitrary points y and z of $\tilde{\mathcal{H}}(x)$ satisfying the relations

$$d_{\mathcal{H}}(x, y),\ d_{\mathcal{H}}(x, z) \geq k_2$$

can be joined by a path contained entirely in the complement of the ball $B(x, k_1) \subset \tilde{\mathcal{H}}(x)$. These properties can be described by countable operations of union and intersection of Borel sets. For instance, $\mathcal{H}(x)$ is infinite whenever

$$x \in \bigcap_{n \in \mathbb{N}} \bigcup_{g \in \mathcal{H}_\infty \setminus \mathcal{H}_n} \left(D_g \setminus g^{-1}(\mathcal{H}_n(x))\right).$$

(iv) Given a finite connected set $\Delta \subset \mathcal{H}_\infty$, the collection of all the triples $(x, y, z) \in W^3$ such that $x \in \bigcap_{g \in \Delta} D_g$ while y and z belong to the same connected component of $\tilde{\mathcal{H}}(x) \setminus \Delta(x)$ is a Borel set. In fact, points y and z belong to the same component of $\tilde{\mathcal{H}}(x) \setminus \Delta(x)$ if and only if there exists a path embedded in $\tilde{\mathcal{H}}(x) \setminus \Delta(x)$ and connecting x to y; and the set of all such paths is countable.

Combining observations (i)–(iv) we get

(v) Given $k \in \mathbb{N}$ and Δ as above, the collection of all the pairs $(x, y) \in W^2$ for which $x \in \bigcap_{g \in \Delta} D_g$, $d_{\mathcal{H}}(y, \Delta(x)) \geq k$ and the connected component of $\mathcal{H}(x) \setminus \Delta(x)$ containing y has exactly one end is a Borel set.

Now, the statement of our lemma follows since B_k can be seen as the union of all the sets considered in (v) for all connected finite subgraphs Δ and the family of all such subgraphs is countable. □

This completes definitely the proof of Theorem 6.4.1.

Now, let us restrict our attention to the case of an orientable foliation $\mathcal{F}$ of dimension 2. As before, consider a nice covering $\mathcal{U}$ of $(M, \mathcal{F})$ and the corresponding holonomy pseudogroup $\mathcal{H}$. For any chain $\gamma = (h_1, \ldots, h_k)$ of holonomy maps $h_j \in \mathcal{H}_1$ define V_γ as the set of all the points x of $T \cap D_{h_k \circ \cdots \circ h_1}$ which are fixed by $h_k \circ \cdots \circ h_1$. For any two such chains γ and γ' let $V_{\gamma,\gamma'}$ be the set of all $x \in V_\gamma \cap V_{\gamma'}$ for which the corresponding free homotopy classes of loops on L_x have intersection number ± 1. The set $V_{\gamma,\gamma'}$ is Borel and the number of possible pairs (γ, γ') is countable. Therefore, Proposition 6.4.8 implies that, for σ-a.e. x of T and any pair (γ, γ'), either $\mathcal{H}(x) \cap V_{\gamma,\gamma'} = \emptyset$ or $\tilde{\mathcal{H}}(x) \cap V_{\gamma,\gamma'}$ approaches all the ends of $\tilde{\mathcal{H}}(x)$.

If a leaf L_x has non-zero genus, then there exist on L_x loops with intersection number ± 1 and then $x \in V_{\gamma,\gamma'}$ for some chains γ and γ'. Therefore, L_x has genus 0 if only $\mathcal{H}(x) \cap V_{\gamma,\gamma'} = \emptyset$ for any γ and γ'; in this case, all the ends of L_x are planar.

If $\mathcal{H}(x) \cap V_{\gamma,\gamma'}$ approaches all the ends of $\tilde{\mathcal{H}}(x)$, then there exist on L_x pairs of loops which have intersection number ± 1 and which approach all the ends of L_x. In this case, L_x has no planar ends.

These observations, Theorem 6.4.1 and the well-known classification of non-compact orientable surfaces (see [**Ri**]) yield the following

Theorem 6.4.12. [**Gh3**] *If μ is a harmonic measure on a compact foliated manifold $(M, \mathcal{F})$, $\mathcal{F}$ is orientable and $\dim \mathcal{F} = 2$, then μ-almost all points x of M belong to the union of all the leaves diffeomorphic to one of the following surfaces (see Figure* 6.2*):*

(i) *a closed surface,*

(ii) *the plane $\mathbb{R}^2$,*

(iii) *the cylinder $\mathbb{R} \times S^1$,*

(iv) *the "Cantor tree", i.e., the sphere S^2 with a Cantor set removed,*

(v) *the "Loch Ness monster", i.e., the plane with infinitely many handles attached,*

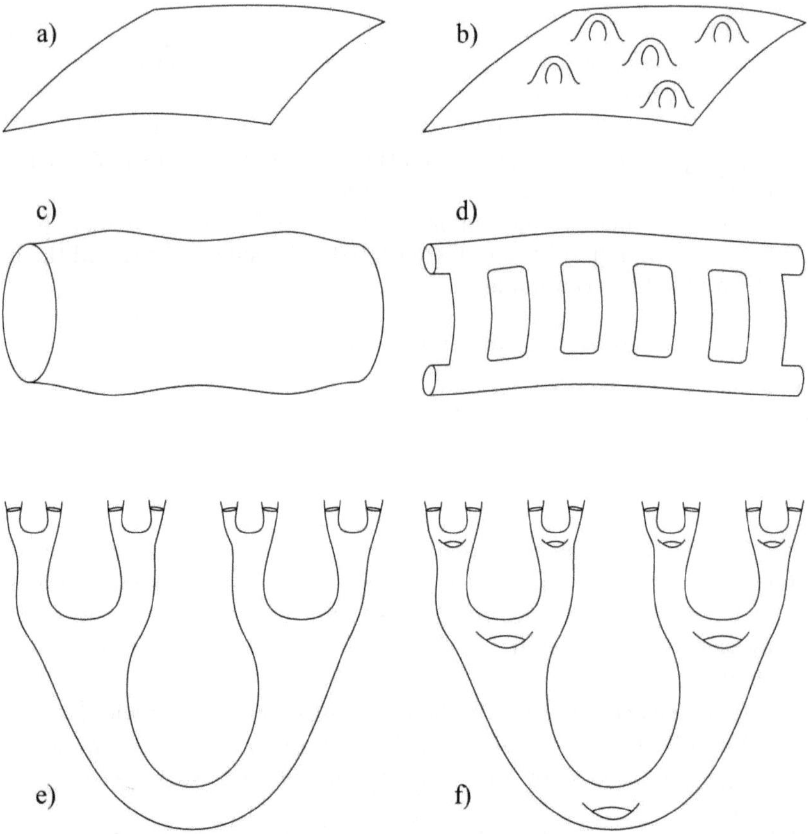

Figure 6.2: Generic leaves: a) plane, b) Loch Ness monster, c) cylinder, d) Jacob's ladder, e) Cantor tree, f) blooming Cantor tree

(vi) *the "Jacob's ladder", i.e., the cylinder with infinitely many handles attached in the direction of both ends,*

(vii) *the "blooming Cantor tree", i.e., the Cantor tree with infinitely many handles attached in the direction of all the ends.* □

Roughly speaking, Theorem 6.4.12 says that μ-almost all the leaves of $\mathcal{F}$ are of one of the types (i)–(vii).

Finally, let us note that Theorems 6.4.1 and 6.4.12 could be formulated for laminations rather than foliations. To get such generalization one should rewrite a good part of the theory of harmonic measures (Section 4.7) in terms of laminations. Since this is not done here, we decided to formulate the results of this section in terms of foliations and to leave generalizing them to the reader.

Bibliography

[**AKM**] R. L. Adler, A. G. Konheim and M. H. McAndrew, *Topological entropy*, Trans. Amer. Math. Soc. **114** (1965), 309–319.

[**Al**] A. Alexiewicz, Functional Analysis, PWN, Warsaw 1969.

[**An**] D. Anosov, *Geodesic flows on closed Riemannian manifolds with negative curvature*, Trudy Mat. Inst. Steklov **90** (1967), 1–235 (in Russian, English transl. in Proc. Steklov Inst. Mat. **90**, Amer. Math. Soc., 1969).

[**Ar**] V. I. Arnold, *Small denominators. I. Mappings of the circumference onto itself*, Izv. Akad. Nauk SSSR, Ser. Mat. **25** (1961), 21–86 (in Russian, English transl.: Amer. Math. Soc. Transl. **46** (1965), 213–284).

[**AH**] O. Attie and S. Hurder, *Manifolds which cannot be leaves of foliations*, Topology **35** (1996), 335–353.

[**Bad**] M. Badura, *Prescribing growth type of complete Riemannian manifolds of bounded geometry*, Ann. Polon. Math. **75** (2000), 167–175.

[**BGS**] W. Ballmann, M. Gromov and V. Schröder, Manifolds of Nonpositive Curvature, Birkhäuser, Boston 1985.

[**Bar**] K. Barański, *Hausdorff dimension and measures on Julia sets of some meromorphic functions*, Fund. Math. **147** (1995), 239–260.

[**BS**] M. Barge and R. Swanson, *Pseudo-orbits and toplogical entropy*, Proc. Amer. Math. Soc. **109** (1990), 559–566.

[**Be1**] A. F. Beardon, The Geometry of Discrete Groups, Springer Verlag, Berlin – Heidelberg – New York, 1983.

[**Be2**] A. F. Beardon, Iteration of Rational Functions, Springer Verlag, New York – Berlin – Heidelberg etc., 1991.

[**Ben1**] M. Benson, *Growth series of finite extensions of $\mathbb{Z}^n$ are rational*, Invent. Math. **73** (1983), 251–269.

[**Ben2**] M. Benson, *On the rational growth of virtually nilpotent groups*, in Combinatorial Group Theory and Topology, Ann. Math. Studies **111**, Princeton Univ. Press, Princeton 1987, 185–196.

[BC] R. Bishop, R. Crittenden, Geometry of Manifolds, Academic Press, New York – London, 1964.

[BJ] C. J. Bishop and P. W. Jones, *Hausdorff dimension and Kleinian groups*, Acta Math. **179** (1997), 1–39.

[Bi] A. Biś, *Entropy of topological directions*, Ann. Fac. Sci. Toulouse **6** (1997), 59–75.

[BW1] A. Biś and P. Walczak, *Pseudo-orbits, pseudo-leaves and geometric entropy of foliations*, Erg. Th. and Dynam. Sys. **18** (1998), 1335–1348.

[BW2] A. Biś and P. Walczak, *Entropies of hyperbolic groups and some foliated spaces*, in Foliations – Geometry and Dynamics, World Sci. Publ., Singapore 2002, 197–211.

[BHM] F. Blanchard, B. Host and A. Maas, *Topological complexity*, Erg. Th. & Dynam. Sys. **20** (2000), 641–662.

[BZ] O. Bodart and M. Zinsmeister, *Quelques résultats sur la dimension de Hausdorff des ensembles de Julia des polynômes quadratiques*, Fund. Math. **151** (1996), 121–137.

[Bo] R. Bott, *Lectures on characteristic classes and foliations*, in Lectures on Algebraic and Differential Topology, Lecture Notes in Math. vol. **279**, Springer Verlag, Berlin – Heidelberg – New York, 1972, pp. 1–94.

[Bou] N. Bourbaki, Groupes et algebres de Lie, Chapt. 4, 5 et 6, Hermann, Paris 1968.

[Bow1] R. Bowen, *Entropy for group endomorphisms and homogeneous spaces*, Trans. Amer. Math. Soc. **153** (1971), 401–414.

[Bow2] R. Bowen, Equilibrium states and the ergodic theory of Axiom A diffemorphisms, Lecture Notes in Math., vol. **470**, Springer Verlag, New York, 1975.

[BK] P. Buser and H. Karcher, *Gromov's almost flat manifolds*, Astérisque **81** (1981), 1–148.

[CL] C. Camacho and A. Lins Neto, Geometric Theory of Foliations, Birkhäuser, Boston – Basel – Stuttgart, 1985.

[Ca1] A. Candel, *Uniformization of surface laminations*, Ann. Sci. École Norm. Sup. **26** (1993), 489–516.

[Ca2] A. Candel, *The harmonic measures of Lucy Garnett*, preprint, 2000.

[CaC1] A. Candel and L. Conlon, Foliations I, Amer. Math. Soc., Providence 2000.

[CaC2] A. Candel and L. Conlon, Foliations II, Amer. Math. Soc., Providence 2003.

[**CW**] J. Cannon, Ph. Wagereigh, *Growth functions of surface groups*, Math. Ann. **293** (1992), 239–257.

[**CC1**] J. Cantwell and L. Conlon, *Poincaré-Bendixon theory for leaves of codimension one*, Trans. Amer. Math. Soc. **265** (1981), 181–209.

[**CC2**] J. Cantwell and L. Conlon, *The dynamics of open foliated manifolds and a vanishing theorem for the Godbillon-Vey class*, Adv. in Math. **53** (1984), 1–27.

[**CC3**] J. Cantwell and L. Conlon, *Nonexponential leaves at finite level*, Trans. Amer. Math. Soc. **269** (1982), 637–651.

[**CC4**] J, Cantwell and L. Conlon, *Every surface is a leaf*, Topology **26** (1987), 265–285.

[**CC5**] J. Cantwell and L. Conlon, *Foliations and subshifts*, Tohoku Math. J. **40** (1988), 165–187.

[**CC6**] J. Cantwell and L. Conlon, *Generic leaves*, Comment. Math. Helv. **73** (1998), 306–336.

[**CC7**] J. Cantwell and L. Conlon, *Endsets of exceptional leaves; a theorem of G. Dumimy*, in Foliations: Geometry and Dynamics, Proc. of the Conf., Warsaw 2000, World Sci. Publ. Singapore 2002, 225–262.

[**CE**] J. Cheeger, D. Ebin, Comparison Theorems in Riemannian Geometry, North-Holland Publ. Co, Amsterdam – Oxford, 1975.

[**Co**] L. Conlon, Differentiable manifolds. Birkhäuser Boston Inc., Boston 2001.

[**Cn**] J. Conway, A Course in Functional Analysis, Springer Verlag, 1990.

[**Con**] J. P. Conze, *Entropie d'un groupe abélien de transformations*, Z. Wahrsch. **25** (1972), 11–30.

[**Coo**] M. Coornaert, *Mesures de Patterson- Sullivan sur le bord d'une espace hyperbolique au sens de Gromov*, Pacific J. Math. **159** (1993), 241–270.

[**De**] A. Denjoy, *Sur les courbes definies par leséquations différentielles á la surface du tore*, J. Math. Pures Appl. **11** (1932), 333–375.

[**DS**] M. Denker and C. Seck, *A short proof of a theorem of Ruelle*, Monatsh. Math. **4** (1989), 295–299.

[**DU1**] M. Denker and M. Urbański, *Hausdorff measures on Julia sets of subexpanding rational maps*, Israel. J. Math. **76** (1991), 193–214.

[**DU2**] M. Denker and M. Urbański, *Hausdorff and conformal measures on Julia sets with a rationally indifferent periodic points*, J. London Math. Soc. **43**, (1991), 107–118.

[Di] E. I. Dinaburg, *The realtion between topological entropy and metric entropy*, Dokl. Akad. Nauk SSSR **190** (1970), 19–22 (in Russian, English translation in Soviet Math. Dokl. **11** (1970), 13–16).

[Do] J. L. Doob, Measure theory. Graduate Texts in Mathematics, 143. Springer-Verlag, New York 1994.

[Du1] G. Duminy, *Sur les cycles feuilletes de codimension un*, a manuscript.

[Du2] G. Duminy, *L'invariant de Godbillon–Vey se localise sur les feuilles resort*, Preprint, Lille 1982.

[Eb] P. Eberlein, *Visibility manifolds*, Pacific J. Math. **46** (1973), 45–109.

[EHS] P. Eberlein, U. Hamenstädt and V. Schröder, *Manifolds of nonpositive curvature*, in Differential geometry: Riemannian geometry (Los Angeles 1990), Proc. Symp. Pure Math. **54**, Amer. Math. Soc., Providence 1993, 179–227.

[Ed] G. A. Edgar, Measure, Topology and Fractal Geometry, Springer Verlag, New York – Berlin – Heidelberg etc., 1990.

[Eg1] S. Egashira, *Expansion growth of foliations*, Ann. Fac. Sci. Toulouse **2** (1993), 15–52.

[Eg2] S. Egashira, *Expansion growth for horospherical foliations*, J. Fac. Sci. Univ. Tokyo, Sect. Math., **40** (1993), 663–682.

[En] R. Engelking, Outline of general topology. Interscience Publishers Division John Wiley & Sons, Inc., New York 1968.

[Ep] D. B. A. Epstein, *Ends*, in Topology of 3-manifolds and Related Topics, Prentice Hall, 1961, pp. 110–117.

[EL] A. E. Eremenko and M. Yu. Lyubich, *The dynamics of analytic transformations*, Algebra i Analiz, **1** (1989), 1–70 (in Russian), an English transl. in Leningrad Math. J. **1** (1990), 563–634.

[Es] A. Eskin, *Counting problems and semisimple groups*, in Proc. Int. Congress Math., Berlin 1998, vol. II, Geronimo GmbH, Rosenheim, 1998, pp. 539–552.

[Fa] K. J. Falconer, The Geometry of Fractal Sets, Cambridge Univ. Press, Cambridge, 1985.

[Fe] W. Feller, An Introduction to Probability Theory and its Applications, vol. 2, Wiley and Sons Inc., New York, 1966.

[FP] W. Floyd and S. Plotnik, *Growth functions on Fuchsian groups and the Euler characteristic*, Invent. Math. **88** (1987), 1–29.

[Fr] S. Friedland, *Entropy of graphs, semigroups and groups*, in Ergodic Theory of $\mathbb{Z}^d$ actions, ed. M. Pollicott and K. Schmidt, London Math. Soc., Lect. Notes Ser., vol. **228**, London, 1996, pp. 319–343.

[Fri] H. Frings, *Generalized entropy for foliations*, Thesis, Dsseldorf, 1991.

[Gan] F. R. Gantmacher, The Theory of Matrices, vol. 2, Chelsea Publ. Co., New York, 1959.

[Ga] V. L. Garber, *On the iteration of rational functions*, Math. Proc. Cambridge Philos. Soc. **84** (1978), 497–505.

[Gar1] L. Garnett, *Foliations, the ergodic theorem and Brownian motion*, J. Func. Anal. **51** (1983), 285–311.

[Gar2] L. Garnett, *A computer algorithm for determining the Hausdorff dimension of certain fractals*, Math. Comp. **51** (1988), 291–300.

[Gh1] E. Ghys, *Une variété qui n'est pas une feuille*, Topology **24** (1985), 67–73.

[Gh2] E. Ghys, *Gauss-Bonnet theorem for 2-dimensional foliations*, J. Funct. Anal. **77** (1988), 51–59.

[Gh3] E. Ghys, *Topologie des feuilles génériques*, Ann. of Math. **141** (1995), 387–422.

[GH] E. Ghys, P. de la Harpe, Sur les Groupes Hyperboliques d'apres Mikhael Gromov, Birkhäuser, Boston – Basel – Berlin, 1990.

[GLW1] E. Ghys, R. Langevin and P. Walczak, *Entropie et partitions de l'unité*, C. R. Acad. Sci. Paris **303** (1986), 251–254.

[GLW2] E. Ghys, R. Langevin and P. Walczak, *Entropie géométrique des feuilletages*, Acta Math. **160** (1988), 105–142.

[Go1] C. Godbillon, Dynamical Systems on Surfaces, Springer Verlag, Berlin – Heidelberg – New York, 1983.

[Go2] C. Godbillon, Feuilletages, Birkhäuser, Basel – Boston – Berlin, 1991.

[GV] C. Godbillon and J. Vey, *Un invariant des feuilletages de codimension 1*, C. R. Acad. Sci. Paris **273** (1971), 92–95.

[Gb] S. Gołąb, *Über den Begriff der Pseudogruppe von Transformationen*, Math. Ann. **116** (1939), 768–780.

[Gd] T. N. T. Goodman, *Relating topological entropy and measure entropy*, Bull. London Math. Soc. **3** (1971), 176–180.

[GHa] M. J. Greenberg, J. R. Harper, Algebraic topology. A first course. Benjamin/Cummings Publishing Co., Inc., Reading 1981.

[GrH] R. Grigorchuk and P. de la Harpe, *On problems related to growth, entropy and spectrum in group theory*, J. Dyn. and Control Sys. **3** (1997), 55–89.

[GrN] R. Grigorchuk and T. Nagnibeda, *Complete growth functions of hyperbolic groups*, Invent. Math. **130** (1997), 159–187.

[GKM] D. Gromoll, W. Klingenberg, W. Meyer, Riemannsche Geometrie im Grossen, Springer Verlag, Berlin – Heidelberg – New York, 1968.

[Gro1] M. Gromov, *Groups of polynomial growth and expanding maps*, Publ. Math. IHES **53** (1981), 53–78.

[Gro2] M. Gromov, *Hyperbolic groups*, in "Essays in group theory", S. M. Gersten (Editor), MSRI Publ., vol. **8**, Springer Verlag, Berlin – Heidelberg – New York, 1987, pp. 75–263.

[Gro3] M. Gromov, *Assymptotic invariants of infinite groups*, in Geometric group theory, vol. 2, London Math. Soc. Lecture Notes **182**, Cambridge Univ. Press, Cambridge, 1993.

[Gro4] M. Gromov, Metric Structures for Riemannian and non-Riemannian Spaces, Birkhäuser, Boston – Basel – Berlin, 1999.

[GLP] M. Gromov, Structures métriques pour les variétés riemanniennnes, edited by J. Lafontaine and P. Pansu, Cedic/Fernand Nathan, Paris, 1981.

[Hae1] A. Haefliger, *Variétés feuilletées*, Ann. Scuola Norm. Pisa **16** (1962), 367–397.

[Hae2] A. Haefliger, *Some remarks on foliations with minimal leaves*, J. Diff. Geom. **15** (1980), 269–284.

[Hae3] A. Haefliger, *Groupoides d'holonomie et classifants*, Astérisque **116** (1984), 70–97.

[Hae4] A. Haefliger, *Foliations and compactly generated pseudogroups*, Foliations: Geometry and Dynamics, Proc. of the Conf., Warsaw 2000, World Sci. Publ., Singapore 2002, 275–296.

[Ham1] U. Hamenstädt, *Entropy-rigidity of locally symmetric spaces of negative curvature*, Ann. of Math. **131** (1990), 35–51.

[Ham2] U. Hamenstädt, *A geometric characterization of negatively curved locally symmetric spaces*, J. Diff. Geom. **34** (1991), 193–221.

[Ham3] U. Hamenstädt, *Time preservig conjugacies of geodesic flows*, Erg. Th. and Dynam. Sys. **12** (1992), 67–74.

[Ham4] U. Hamenstädt, *Regularity at infinity of compact negatively curved manifolds*, Erg. Th. and Dynam. Sys. **14** (1994), 493–514.

[Ham5] U. Hamenstädt, *Harmonic measures, Hausdorff measures and positive eigenfunctions*, J. Diff. Geom. **44** (1996), 1–31.

[Har] P. de la Harpe, *Free groups in linear groups*, L'Enseignement math. **29** (1983), 129–144.

[Hau] F. Hausdorff, *Dimension und äusseres Mass*, Math. Ann. **79** (1918), 157–179.

[He] G. Hector, *Architecture des feuilletages de classe C^2*, Astérisque **107–108** (1983), 243–258.

[HH] G. Hector and U. Hirsch, Introduction to the Geometry of Foliations, Part A and B, Vieweg & Sohn, Braunschweig – Wiesbaden, 1981 and 1983.

[Her] M. Hermann, *The Godbillon–Vey invariant of foliations by planes of T^2*, in Lecture Notes in Math. **597**, Springer Verlag, New York, 1977, pp. 294–307.

[Hig] N. Higson, C*-*algebra extension theory and duality*, J. Funct. Anal. **129** (1995), 349–363.

[HR] N. Higson and J. Roe, Lectures on K-homology, Oxford Univ. Press., 1994.

[Hu1] S. Hurder, *Ergodic theory of foliations and a theorem of Sacksteder*, in Dynamical Systems: Proceedings, Univ. of Maryland 1986–1987, Lecture Notes in Math. **1342**, Springer Verlag, Berlin – Heidelberg – New York, 1988, pp. 291–328.

[Hu2] S. Hurder, *Exceptional minimal sets of $C^{1+\alpha}$ groups on the circle*, Erg. Th. and Dynam. Sys. **11** (1991), 455–467.

[Hu3] S. Hurder, *Coarse geometry of foliations* in Geometric Study of Foliations, Proc. Tokyo 1993, World Sci., Singapore, 1994, pp. 35–96.

[Hu4] S. Hurder, *Entropy and dynamics of C^1-foliations*, preprint 2000, available from http://www.math.uic.edu/~hurder/publications.

[Hur] M. Hurley, *On topological entropy of maps*, Erg. Th. and Dynam. Sys. **15** (1995), 557–568.

[IW] N. Ikeda and S. Watanabe, Stochastic Differential Equations and Diffusion Process, North-Holland, Amsterdam, 1981.

[In1] T. Inaba, *Examples of exceptional minimal sets*, A Fete of Topology, ed. Y. Matusmoto, T. Mizutani and S. Morita, Academic Press 1988, pp. 95–100.

[In2] T. Inaba, *Expansivity, pseudoleaf tracing property and semistability of foliations*, Tokyo J. Math. **23** (2000), 311–323.

[INTT] T. Inaba, T. Nishimori, M. Takamura and N. Tsuchiya, *Open manifolds which are not realizable as leaves*, Kodai Math. J. **8** (1985), 112–119.

[IT] T. Inaba and N. Tsuchiya, *Expansive foliations*, Hokkaido Math. J., 21 (1992), 39–49.

[IWa] T. Inaba and P. Walczak, *Transverse Hausdorff dimension of codim-1 C^2-foliations*, Fund. Math. **149** (1996), 239–244.

[Ja] K. Jacobs, Lecture Notes on Ergodic Theory, Aarhus Lecture Notes Ser., no. **1**, Part I, Aarhus, 1992/93.

[JM] P. W. Jones and N. G. Makarov, *Density properties of harmonic measures*, Ann. of Math. **142** (1995), 427–455.

[Ka] M. Kac, *On the notion of recurrence in discrete stochastic processes*, Bull. Amer. Math. Soc. **53** (1947), 1002–1010.

[KT] F. W. Kamber and Ph. Tondeur, Foliated Bundles and Characteristic Classes, Lecture Notes in Math. vol. **493**, Springer Verlag, Berlin – Heidelberg – New York, 1975.

[Ke] J. L. Kelley, General topology. Graduate Texts in Mathematics, No. 27. Springer-Verlag, New York-Berlin 1975.

[Kl] W. Klingenberg, Riemannian Geometry, Walter de Gruyter, Berlin – New York, 1982.

[Kn] H. Kneser, *Reguläre Kurvenscharen auf den Ringflächen*, Math. Ann. **91** (1923), 135–154.

[KN] S. Kobayashi and K. Nomizu, Foundations of Differential Geometry, vol. I and II, Interscience Publ., New York – London, 1963 and 1969.

[Ko1] J. Kotus, *On the Hausdorff dimension of Julia sets of meromorphic functions, I*, Bull. Soc. Math. France **122** (1994), 305–331.

[Ko2] J. Kotus, *On the Hausdorff dimension of Julia sets of meromorphic functions, II*, Bull. Soc. Math. France **123** (1995), 33–46.

[LSU] O. A. Ladyzenskaja, V. A. Solonnikov and N. N. Uralceva, Linear and Quasi-linear Equations of Parabolic Type, Transl. of Math. Monographs **23**, Amer. Math. Soc., Providence, 1967.

[La] R. Langevin (ed.), *A list of questions about foliations*, in Differential Topology, Foliations and Group Actions, Contemp. Math. **152**, Amer. Math. Soc., Providence, 1994.

[LP] R. Langevin and F. Przytycki, *Entropie de l'image inverse d'une application*, Bull. Soc. Math. France **120** (1992), 237–250.

[LW1] R. Langevin and P. Walczak, *Entropie d'une dynamique*, C. R. Acad. Sci. Paris, Ser. I **312** (1991), 141–144.

[LW2] R. Langevin and P. Walczak, *Entropy, transverse entropy and partitions of unity*, Erg. Th. and Dynam. Sys. **14** (1994), 551–563.

[LW3] R. Langevin and P. Walczak, *Some invariants measuring dynamics of codimension-one foliations*, in Geometric Study of Foliations, Proc. Tokyo 1993, World Sci., Singapore, 1994, pp. 345–358.

[Le] G. Levitt, *La dynamique des pseudogroupes de rotations*, Invent. math. **113** (1993), 633–670.

[McK] H. MacKean, Stochastic Integrals, Academic Press, New York, 1969.

[MKS] W. Magnus, A. Karrass and D. Solitar, Combinatorial Group Theory, Intersci. Publ., New York – London – Sydney, 1966.

[Ma] R. Mañe, Ergodic Theory and Differentiable Dynamics, Springer Verlag, Berlin – Heidelberg – New York, 1987.

[Mat1] S. Matsumoto, *Measure of exceptional minimal sets of codimension-one foliations*, in A Fete of Topology, Academic Press, Boston, 81–94.

[Mat2] S. Matsumoto, *Numerical invariants for semiconjugacy of homeomorphisms of the circle*, Proc. Amer. Math. Soc. **98** (1986), 163–168.

[Mat3] S. Matsumoto, *Some remarks on foliated* S^1 *bundles*, Invent.Math. **90** (1987), 343–358.

[Mi1] J. Milnor, *A note on curvature and volume*, J. Diff. Geom. **2** (1968), 1–7.

[Mi2] J. Milnor, *Growth of finitely generated solvable groups*, J. Diff. Geom. **2** (1968), 447–449.

[Mis1] M. Misiurewicz, *A short proof of the variational principle for a* $\mathbb{Z}_+^N$ *action on a compact space*, Asterisque **40** (1976), 147–187.

[Mis2] M. Misiurewicz, *Remark on the definition of topological entropy*, in Dynamical Systems and Partial Differential Equations (Caracas 1984), Caracas 1986, 65–67.

[MMT] T. Mizutani, S. Morita and T. Tsuboi, *The Godbillon–Vey classes of codimension one foliations which are almost without holonomy*, Ann. of Math. **113** (1981), 515–527.

[MT] S. Morita and T. Tsuboi, *The Godbillon–Vey class of codimension one foliations without holonomy*, Topology **19** (1980), 43–49.

[Mo] J. Moser, *A Harnack inequality for parabolic differential equations*, Comm. Pure Appl. Math. **17** (1964), 101–134.

[Ni] P. J. Nicholls, The Ergodic Theory of Discrete Groups, Cambridge Univ. Press, Cambridge, 1989.

[NP] Z. Nitecki and F. Przytycki, *Preimage entropy for mappings*, Int. J. Bifur. Chaos **9** (1999), 1815–1843.

[Pa] J. Palis, Smooth Dynamical Systems, Springer Verlag, Berlin – Heidelberg – New York, 1981.

[Pan] P. Pansu, *Métriques de Carnot–Carathéodory et quasiisométries des espaces symétriques de rang un*, Ann. of Math. **129** (1989), 1–60.

[Par] L. Paris, *Growth series of Coxeter groups*, in Group Theory from Geometrical Point of View, E. Ghys, A. Haefliger and A. Verjovsky, eds., World.Sci., Singapore 1991, 302–310.

[Pry] W. Parry, *Intrinsic Markov chains*, Trans. Amer. Math. Soc. **112** (1964), 55–65.

[Pat1] S. J. Patterson, *The limit set of a Fuchsian group*, Acta Math. **136** (1976), 241–273.

[Pat2] S. J. Patterson, *Lectures on measures on limit sets of Kleinian groups*, in Analytical and geometric aspects of hyperbolic space, London Math. Sos. Lecture Notes **111**, Cambridge Univ. Press, Cambridge 1987, 281–323.

[Pl] J. Plante, *Foliations with measure preserving holonomy*, Ann. of Math. **102** (1975), 327–361.

[Ri] I. Richards, *On the classification of noncompact surfaces*, Trans. Amer. Math. Soc. **106** (1963), 259–269.

[Ro1] J. Roe, *Finite propagation speed and Conne's foliation algebra*, Math. Proc. Camb. Phil. Soc. **102** (1987), 459–466.

[Ro2] J. Roe, *Remark on a paper of E. Ghys: "Gauss–Bonnet theorem for 2-dimensional foliations"*, J. Funct. Anal. **89** (1990), 150–153.

[Rud] W. Rudin, Real and Complex Analysis, McGraw-Hill, New York – St. Louis etc., 1966.

[Ru1] D. Ruelle, *Statistical mechanics on a compact set with $\mathbb{Z}^\nu$ action satisfying expansiveness and specification*, Trans. Amer. Math. Soc. **185** (1973), 237–251.

[Ru2] D. Ruelle, *Repellers for real analytic maps*, Erg. Th. and Dynam. Sys., **2** (1982), 99–107.

[Sa] R. Sacksteder, *Foliations and pseudogroups*, Amer. J. Math. **87** (1965), 79–102.

[Sc] A. J. Schwartz, *A generalization of a Poincaré–Bendixon theorem to closed two dimensional manifolds*, Amer. J. Math. **85** (1963), 453–458.

[Sch] P. Schweizer, *Surfaces not quasi-isometric to leaves of foliations of compact 3-manifolds*, in Analysis and Geometry in Foliated Manifolds (E. Macias-Virgos and J. A. Alvarez Lopez, eds.), World Sci., Singapore – New Jersey – London – Hong Kong, 1995, 223–238.

[Se] C. Series, *Geometrical Markov coding of geodesics on surfaces of constant negative curvature*, Ergod. Th. and Dynam. Sys. **6** (1986), 601–625.

[Sh] M. Shapiro, *A geometric approach to growth and almost convexity in some nilpotent groups*, Math. Ann. **285** (1989), 601–624.

[St] S. Sternberg, Lectures on Differential Geometry, Prentice Hall Inc., Englewood Cliffs, N.J., 1964.

[Su1] D. Sullivan, *Cycles for the dynamical study of foliated manifolds and complex manifolds*, Invent. Math. **36** (1976), 225–256.

[Su2] D. Sullivan, *The density at infinity of a discrete group of hyperbolic isometries*, Publ. Math. IHES **50** (1979), 171–209.

[Su3] D. Sullivan, *Conformal dynamical systems*, in Geometric Dynamics, Rio de Janeiro 1981, Lect. Notes in Math., **1007**, Springer Verlag, Berlin – New York, 1983, pp. 725–752.

[Su4] D. Sullivan, *Entropy, Hausdorff measures old and new, and limit sets of geometrically finite Kleinian groups*, Acta Math. **153** (1984), 259–277.

[Su5] D. Sullivan, *Quasiconformal homeomorphisms and dynamics I: Solution of the Fatou–Julia problem on wandering domains*, Ann. of Math. **122** (1985), 401–418.

[Su6] D. Sullivan, *Bounds, quadratic differentials and renormalization conjectures*, in Mathematics into 21st Century, vol.2, Amer. Math. Soc., Providence 1991, 417–466.

[Sv] A. S. Svarc, *A volume invariant of coverings*, Dokl. Akad. Nauk SSSR **105** (1955), 32–34 (in Russian).

[Ta] M. Takamura, *Reeb stability for leaves with uncountable endset* (in Japnese), Master Thesis, Hokkaido Univ. 1984.

[Tam] I. Tamura, Topology of Foliations: An Introduction, Amer. Math. Soc., Providence, 1992.

[Th] W. Thurston, *Existence of codimension one foliations*, Ann. of Math. **104** (1976), 249–268.

[To] Ph. Tondeur, Riemannian Geometry of Foliations, Birkhäuser, 1997.

[Tu] P. Tukia, *The Hausdorff dimension of the limit set of a geometrically finite Kleinian group*, Acta Math. **152** (1984), 127–140.

[Ur1] M. Urbański, *On the Hausdorff dimension of a Julia sets with a rationally indifferent periodic point*, Studia Math. **97** (1991), 167–188.

[Ur2] M. Urbański, *Measures and dimensions in conformal dynamics*, Bull. Amer. Math. Soc. **40** (2003), 281–321.

[VW] O. Veblen and J. H. C. Whitehead, The Foundations of Differential Geometry, Cambridge Tracts **29**, Cambridge, 1932.

[Wal1] P. Walczak, *Hausdorff dimension of Markov invariant sets*, J. Math. Soc. Japan **48** (1996), 125–133.

[Wal2] P. Walczak *Losing Hausdorff dimension while generating pseudogroups*, Fund. Math. **149** (1996), 211–237.

[Wal3] P. Walczak, *A virtual leaf*, Int. J. Bifur. and Chaos **9** (1999), 1845–1852.

[Wls] W. Waliszewski, *Categories, groupoids, pseudogroups and analytic structures*, Rozprawy Mat. (Dissert. Math.) **45** (1965), 1–40.

[Wlt] G. Wallet, *Nullité de l'invariant de Godbillon–Vey d'un tore*, C. R. Acad. Sci. Paris, Ser. A, **283** (1976), 821–823.

[Wa] P. Walters, An Introduction to Ergodic Theory, Springer Verlag, New York – Heidelberg – Berlin, 1982.

[War] F. Warner, Foundations of Differentiable Manifolds and Lie Groups, Scott, Foresman and Co., Glenview (Illinois) – London, 1971.

[Wat] N. Watanabe, *On topological entropy of group actions on* S^1, Proc. Amer. Math. Soc. **106** (1989), 245–249.

[Wi] B. Wirtz, *Entropies*, Thése, Université de Bourgogne, Dijon 1993.

[Wo1] J. A. Wolf, Spaces of Constant Curvature, Publish or Perish, Berkeley, 1977.

[Wo2] J. A. Wolf, *Growth of finitely generated solvable groups and curvature of Riemannian manifolds*, J. Diff. Geom. **2** (1968), 421–446.

[Yo] K. Yosida, Functional Analysis, Springer Verlag, Berlin – Heidelberg – New York, 1974.

[Zd] A. Zdunik, *Parabolic orbifolds and the dimension of the maximal measure for rational maps*, Invent. Math. **99** (1990), 627–649.

[Ze] A. Zeghib, *An example of a 2-dimensional no leaf*, in Geometric Study of Foliations, Proc. Tokyo 1993, World Sci., Singapore, 1994, pp. 475–477.

Index

Monografie Matematyczne

[1] S. Banach, *Théorie des opérations linéaires*, 1932
[2] S. Saks, *Théorie de l'integrale*, 1933
[3] C. Kuratowski, *Topologie* I, 1933
[4] W. Sierpiński, *Hypothèse de continu*, 1934
[5] A. Zygmund, *Trigonometrical Series*, 1935
[6] S. Kaczmarz, H. Steinhaus, *Theorie der Orthogonalreihen*, 1935
[7] S. Saks, *Theory of the integral*, 1937
[8] S. Banach, *Mechanika*, T. I, 1947
[9] S. Banach, *Mechanika*, T. II, 1947
[10] S. Saks, A. Zygmund, *Funkcje analityczne*, 1948
[11] W. Sierpiński, *Zasady algebry wyższej*, 1946
[12] K. Borsuk, *Geometria analityczna w n wymiarach*, 1950
[13] W. Sierpiński, *Działania nieskończone*, 1948
[14] W. Sierpiński, *Rachunek różniczkowy poprzedzony badaniem funkc elementarnych*, 1947
[15] K. Kuratowski, *Wykłady rachunku różniczkowego i całkoweg* T. I, 1948
[16] E. Otto, *Geometria wykreślna*, 1950
[17] S. Banach, *Wstęp do teorii funkcji rzeczywistych*, 1951
[18] A. Mostowski, *Logika matematyczna*, 1948
[19] W. Sierpiński, *Teoria liczb*, 1950
[20] C. Kuratowski, *Topologie* I, 1948
[21] C. Kuratowski, *Topologie* II, 1950
[22] W. Rubinowicz, *Wektory i tensory*, 1950
[23] W. Sierpiński, *Algèbre des ensembles*, 1951
[24] S. Banach, *Mechanics*, 1951
[25] W. Nikliborc, *Równania różniczkowe*, Cz. I, 1951
[26] M. Stark, *Geometria analityczna*, 1951
[27] K. Kuratowski, A. Mostowski, *Teoria mnogości*, 1952
[28] S. Saks, A. Zygmund, *Analytic functions*, 1952
[29] F. Leja, *Funkcje analityczne i harmoniczne*, Cz. I, 1952
[30] J. Mikusiński, *Rachunek operatorów*, 1953
* [31] W. Ślebodziński, *Formes extérieures et leurs applications*, 1954
[32] S. Mazurkiewicz, *Podstawy rachunku prawdopodobieństwa*, 1956
[33] A. Walfisz, *Gitterpunkte in mehrdimensionalen Kugeln*, 1957
[34] W. Sierpiński, *Cardinal and ordinal numbers*, 1965
[35] R. Sikorski, *Funkcje rzeczywiste*, 1958
[36] K. Maurin, *Metody przestrzeni Hilberta*, 1959
[37] R. Sikorski, *Funkcje rzeczywiste*, T. II, 1959
[38] W. Sierpiński, *Teoria liczb* II, 1959

* [39] J. Aczél, S. Gołąb, *Funktionalgleichungen der Theorie der geometrischen Objekte*, 1960
[40] W. Ślebodziński, *Formes extérieures et leurs applications*, II, 1963
[41] H. Rasiowa, R. Sikorski, *The mathematics of metamathematics*, 1963
[42] W. Sierpiński, *Elementary theory of numbers*, 1964
* [43] J. Szarski, *Differential inequalities*, 1965
[44] K. Borsuk, *Theory of retracts*, 1967
[45] K. Maurin, *Methods of Hilbert spaces*, 1967
[46] M. Kuczma, *Functional equations in a single variable*, 1967
[47] D. Przeworska-Rolewicz, S. Rolewicz, *Equations in linear spaces*, 1968
[48] K. Maurin, *General eigenfunction expansions and unitary representations of topological groups*, 1968
[49] A. Alexiewicz, *Analiza funkcjonalna*, 1969
* [50] K. Borsuk, *Multidimensional analytic geometry*, 1969
* [51] R. Sikorski, *Advanced calculus. Functions of several variables*, 1969
[52] W. Ślebodziński, *Exterior forms and their applications*, 1971
[53] M. Krzyżański, *Partial differential equations of second order*, vol. I, 1971
[54] M. Krzyżański, *Partial differential equations of second order*, vol. II, 1971
[55] Z. Semadeni, *Banach spaces of continuous functions*, 1971
[56] S. Rolewicz, *Metric linear spaces*, 1972
[57] W. Narkiewicz, *Elementary and analytic theory of algebraic numbers*, 1974
[58] Cz. Bessaga, A. Pełczyński, *Selected topics in infinite dimensional topology*, 1975
* [59] K. Borsuk, *Theory of shape*, 1975
[60] R. Engelking, *General topology*, 1977
[61] J. Dugundji, A. Granas, *Fixed point theory*, 1982
* [62] W. Narkiewicz, *Classical problems in number theory*, 1986

The volumes marked with * are available at the exchange department of the library of the Institute of Mathematics, Polish Academy of Sciences.

MIX
Papier aus verantwortungsvollen Quellen
Paper from responsible sources
FSC® C105338

If you have any concerns about our products,
you can contact us on
ProductSafety@springernature.com

In case Publisher is established outside the EU,
the EU authorized representative is:
Springer Nature Customer Service Center GmbH
Europaplatz 3, 69115 Heidelberg, Germany

Printed by Libri Plureos GmbH
in Hamburg, Germany